Optimal Planning and Operation in RES-Rich Power Systems under Electricity and Carbon Emission Market Environment

Optimal Planning and Operation in RES-Rich Power Systems under Electricity and Carbon Emission Market Environment

Editors

Xiuli Wang
Fushuan Wen

Basel • Beijing • Wuhan • Barcelona • Belgrade • Novi Sad • Cluj • Manchester

Editors
Xiuli Wang
School of Electric Power, Civil
Engineering and Architecture
Shanxi University
Taiyuan
China

Fushuan Wen
College of Electrical
Engineering
Zhejiang University
(Yuquan campus)
Hangzhou
China

Editorial Office
MDPI AG
Grosspeteranlage 5
4052 Basel, Switzerland

This is a reprint of articles from the Special Issue published online in the open access journal *Energies* (ISSN 1996-1073) (available at: https://www.mdpi.com/journal/energies/special_issues/6FU3GX778Y).

For citation purposes, cite each article independently as indicated on the article page online and as indicated below:

Lastname, A.A.; Lastname, B.B. Article Title. *Journal Name* **Year**, *Volume Number*, Page Range.

ISBN 978-3-7258-2339-0 (Hbk)
ISBN 978-3-7258-2340-6 (PDF)
doi.org/10.3390/books978-3-7258-2340-6

Contents

Preface

With the ever-increasing penetration of renewable energy sources (RESs), electric vehicles, and energy storage devices into modern power systems, power system planning and operation are facing new problems and even challenges. The establishment of electricity and carbon emission markets makes planning and operation issues more complicated and more challenging. It is believed that the present power system will gradually evolve to one with extensive RES generation integration, and the role that thermal generation units are currently playing will change accordingly. Given this background, this reprint will be devoted to research topics regarding optimal planning and operation in RES-rich power systems within the electricity and carbon emission market environments.

In this reprint, nine original articles and one review article are included. The articles involve the optimal planning and operation of power systems with RESs. Specifically, the following research works are addressed: (1) The shared energy storage among multiple islanded microgrid systems was investigated; (2) a decarbonization strategy was presented in relatively isolated power systems, while the evolution of the power system was analyzed, and plausible long-term trajectories were elaborated on within the energy systems of two European islands; (3) a frequency response estimation method for virtual power plants was presented; (4) an optimal scheduling strategy for distribution networks with mobile energy storage systems and offline control PVs was proposed; (5) the complementarity between offshore wind and other power generation sources in the Brazilian power system was explored; (6) a two-stage modeling framework for peer-to-peer (P2P) energy trading with voltage regulation service provision considered for the prevention of network congestion and nodal voltage violations was presented; (7) for power systems with high-penetration renewable energy generation, a data-driven approach combining experimental economics and machine learning was proposed, with mixed empirical learning employed to simulate the unit bidding strategies in electricity market competition; (8) a big data-based grid status analysis method for power systems with a high penetration rate of wind power was proposed; and (9) a gas–electricity price linkage analysis method considering the cost–benefit and gas–electricity price of gas units was proposed based on the attention mechanism. In addition, the review article mainly presents the characteristics and challenges of flexibility markets in electricity distribution networks.

We hope that the collection of papers in this reprint will help researchers and engineers in the power system community better understand the recent research progress on planning and operation issues in renewable energy-rich power systems.

Xiuli Wang and Fushuan Wen

Editors

Article

Natural Gas–Electricity Price Linkage Analysis Method Based on Benefit–Cost and Attention–VECM Model

Sheng Zhou [1], Wen Gan [2], Ang Liu [2], Xinyue Jiang [1], Chengliang Shen [2], Yunchu Wang [1], Li Yang [1,*] and Zhenzhi Lin [1]

1   School of Electrical Engineering, Zhejiang University, Hangzhou 310027, China
2   Huzhou Power Supply Company of State Grid Zhejiang Electric Power Co., Ltd., Huzhou 313000, China
*   Correspondence: eeyangli@zju.edu.cn

Abstract: In recent years, frequent linkage events of natural gas and electricity prices have seriously affected the stability of the energy supply market. In view of the increasingly close linkage relationship between gas and electricity prices, a gas–electricity price linkage analysis method considering the benefit–cost and price transmission of gas units is proposed. Firstly, a benefit–cost model of gas units based on risk premium theory is constructed to reflect the willingness of gas units to offer prices in different market environments. Secondly, a vector error correction model (VECM)-based gas–electricity price conduction analysis method was proposed to study the long-term co-integration relationship between gas–electricity prices and the short-term fluctuation conduction process. Then, aiming at the problem of delay in price transmission, a time-delay analysis method of gas–electricity price based on the attention mechanism is proposed. Finally, the gas price index and average gas power price data of China from August 2020 to November 2022 are used to verify the validity of the proposed model; the numerical results show that the proposed method can determine the long-term stable fitting relationship between gas and electricity prices, and can also analyze and judge the conduction direction of gas and electricity price fluctuations in different time periods.

Keywords: gas–electricity price linkage; gas unit; benefit–cost model; vector error correction model; co-integration expression; attention mechanism

Citation: Zhou, S.; Gan, W.; Liu, A.; Jiang, X.; Shen, C.; Wang, Y.; Yang, L.; Lin, Z. Natural Gas–Electricity Price Linkage Analysis Method Based on Benefit–Cost and Attention–VECM Model. *Energies* **2023**, *16*, 4155. https://doi.org/10.3390/en16104155

Academic Editor: Luigi Aldieri

Received: 2 April 2023
Revised: 14 May 2023
Accepted: 16 May 2023
Published: 17 May 2023

1. Introduction

In the context of the increasingly serious global energy crisis and the increasing importance of energy security, natural gas, as an important energy in the process of power system transformation, is of great significance in promoting the development of the power market [1]. According to the statistics of the International Energy Agency, the global demand for natural gas in 2021 has grown rapidly, the supply of the natural gas market has increased by 134 billion cubic meters compared with 2020, the demand ratio has increased by 140 billion cubic meters, and the gap between supply and demand has reached 31 billion cubic meters. The spot price of liquefied natural gas in Northeast Asia, JKM (Japan Korea Marker, JKM), and European Dutch natural gas prices hit record highs above $40/mmBtu; in 2022, the supply and demand of the global natural gas market is still tight, with the gap widening to 36 billion cubic meters [2]. Since the outbreak of the Ukraine crisis, the global natural gas trade pattern has been reshaped, and the natural gas price has soared. The European benchmark liquefied natural gas price once rose to 100 dollars per million British thermal units. The price of JKM soared to more than $80 per million British thermal units, causing a huge shock to the energy market, especially the electricity market [3]. In addition, recent crises such as soaring electricity prices in California and Texas and soaring natural gas prices in Europe are all related to gas–electricity price transmission. The price fluctuation is transmitted through the substitution effect of electricity and natural gas [4], which intensifies the volatility and uncertainty of gas price and electricity price [5,6].

Therefore, it is of great practical significance to study the linkage relationship between gas and electricity prices (gas–electricity price for short) [7].

At present, there are few analyses on the linkage between gas price and electricity price around the globe, and the research on price transmission mainly focuses on the price of coal and electricity. Most U.S. states and Japan adjust fuel prices and end-sale electricity prices directly on a monthly or quarterly basis, so fluctuations in coal prices are transmitted to end-users, avoiding the problem of blocked transmission of coal and electricity prices. Compared with the United States and Japan, China adjusts electricity coal price, online electricity price, and sales electricity price on an annual basis, which can ensure the stability of residential electricity price, but is not conducive to the transmission of coal price fluctuations [8,9]. In our country, there are three problems, such as excessively long interaction cycle and hysteresis, blocked price transmission, and electric coal price index and unmatching benchmark online power price [10]. Existing studies on gas–electricity price linkage mainly analyze the income of natural gas units and the pricing mechanism of natural gas power generation [11,12]. The input–output price model, break-even model, and equal-calorific-value direct cost model are constructed to analyze the income of units under different market environments and calculate the prices of various natural gas users and the corresponding affordability of gas prices when the break-even point is reached, providing a theoretical basis for gas price adjustment and reform [13–15]. Among them, reference [16] focuses on the price mechanism of domestic gas consumption. Considering the upstream and downstream price block of the natural gas industry in the linkage of gas–electricity prices [17], it proposes the basic ideas and specific pricing schemes to optimize and perfect the pricing mechanism of natural gas power generation [18], and studies the equilibrium of the gas–electricity market [19]. References [20–23] mainly analyze the operating mechanism and price conduction relationship of the foreign natural gas market and electric power market, and provides a reference for our natural gas and electric power marketization reform. It can be seen from the above literature that the existing research on gas–electricity price linkage as made a relatively comprehensive analysis of the income situation of gas units and the pricing mechanism of natural gas power generation, but has not considered the pricing behavior of gas units in the market, and lacks a quantitative analysis on the time lag of gas–electricity price linkage and the consideration of the change in the strength of the correlation degree of gas–electricity prices.

VAR and VECM models are relatively mature economic models to study the linkage relationship between prices, which are widely used in important fields such as energy, economy, transportation, and medicine [24]. Reference [25] studies how to use the VAR model to conduct a co-integration test among variables and further analysis to obtain the lag order. Reference [26] constructs the VAR model to prove that there is a long-term stable co-integration relationship between natural gas price and electricity price in the UK. Considering that most time series are not stationary, relevant scholars further proposed the VECM model based on the VAR model [27]. Reference [28] studies the short-term and long-term conduction relationships between variables based on the VECM model and machine-learning algorithm. Reference [29] constructs the VECM model to predict fuel prices, which proves that the long-term equilibrium relationship of fuel futures prices is helpful in improving the forecasting accuracy. Reference [30] adopts the co-integration test and VECM model to study the relationship between investment, international trade, and economic growth, and the causality test is further adopted. In reality, the coupling relationship between price variables is complex and the price transmission has a time lag. The VAR and VECM models can obtain the lag order, but the results are not accurate enough and cannot reflect more information about the time delay. The attention mechanism can analyze the correlation of historical price data information and establish dynamic weight parameters. The time delay of price transmission is measured according to the weight of time information between price variables [31,32].

This paper aims to quantitatively analyze the linkage relationship between natural gas price and electricity price in a certain province of China, and study the time lag of gas–

electricity price transmission. Firstly, the benefit–cost model of the gas unit is established to analyze the price changes of the natural gas market and electricity market, respectively. Secondly, based on the VECM model in econometrics, the gas–electricity price linkage relationship is studied from the static co-integration function expression and the dynamic price fluctuation response process. Then, the attention mechanism is used to calculate the delay weeks of the historical gas price to the current electricity price. Finally, the natural gas and electricity market in China is taken as an example to verify the rationality and validity of the proposed analysis method.

The paper is structured as follows: Section 2 studies the benefits and costs of natural gas units, and then it provides the basis for studying the tendency of the natural gas unit quotation. Section 3 builds the gas–electricity price linkage model and makes a comprehensive calculation and analysis of the linkage between gas and electricity prices. Section 4 calculates the delay of price conduction based on the attention mechanism. Section 5 makes an empirical analysis. Section 6 analyzes and summarizes this study.

2. Gas–Electricity Price Linkage Analysis Method Based on Benefit–Cost Model of Gas Unit

2.1. Revenue Model of Gas Unit

For the gas unit, when its own income is lower than its own comprehensive cost, the power generation willingness of the unit is low, and it may adopt a negative quotation strategy, which is embodied in the high price or even the ceiling price at the time of clearing. By constructing the benefit–cost model of the gas unit, the quotation behavior of the gas unit in different time scales can be analyzed, and the effective market information can be provided for each entity of the gas–electricity market.

The income R^i of the gas-generating unit in the electricity energy market can be divided into two parts: income R_1^i from electricity sales and income R_2^i from generation compensation. The income R_1^i from the sale of gas unit i in the electricity market is:

$$R_1^i = \sum_{t=1}^{S} \left(E_{i,t} P_{i,t} \right) \tag{1}$$

where S is the total time period; $E_{i,t}$ is the bid-winning electric quantity of unit i in time period t; and $P_{i,t}$ is the node marginal electricity price at time period t.

The winning unit will receive additional generation compensation income in addition to electricity sales income, which can be expressed as:

$$R_2^i = \sum_{t=1}^{S} \left(E_{i,t} M_{i,t} \right) \tag{2}$$

where R_2^i represents the total compensation income obtained by unit i due to power generation; $M_{i,t}$ represents the compensation income coefficient of each kilowatt-hour corresponding to the time period t of unit i.

To sum up, the total revenue R^i of the gas unit is:

$$R^i = R_1^i + R_2^i \tag{3}$$

2.2. Cost Model of Gas Market for Gas-Generating Units

The cost C^i of the gas unit in the natural gas market can be divided into spot market electricity purchase cost C_1^i and annual bilateral negotiation contract cost C_2^i.

Suppose that the gas purchased by the unit in the spot market at the t time period is $Q_{i,t}^{sp}$, and the kilowatt-hour fuel cost is $f_{i,t}$; then, the cost C_1^i of the unit in the spot market of natural gas is:

$$C_1^i = \sum_{t=1}^{S} Q_{i,t}^{SP} f_{i,t} \tag{4}$$

The kilowatt-hour fuel cost $f_{i,t}$ of the unit at time t is expressed as:

$$f_{i,t} = \frac{P_{Gas,t}}{(1 - \xi_i)b_{Gas,i}}\delta \tag{5}$$

where $P_{Gas,t}$ is the gas price of the gas unit at time period t; $b_{Gas,i}$ is the electrical generation consumption rate of unit i; ξ_i is the power consumption rate of unit i; and δ is the conversion coefficient.

Given the sharp volatility of electricity and natural gas prices, gas units hope to realize hedging strategies through medium- and long-term contracts. At present, there are two theories about the relationship between spot price and futures price: the warehousing theory and risk premium theory. The former considers the cost and convenience of goods holding inventory, while the latter deduces the relationship between short-term and long-term prices based on the spillover effect of price fluctuations and other methods. Combined with the actual situation of the natural gas commodity and market, this paper analyzes the corresponding relationship between futures price and spot price according to the risk premium theory, and its expression is as follows:

$$P^F_{t-T,t} = E_{t-T}(P^S_t)e^{(r_T - i_T)} = E_{t-T}(P^S_t)e^{-p_T} \tag{6}$$

where T is the time difference between the signing of the natural gas futures contract and the actual delivery; $P^F_{t-T,t}$ is the delivery price of natural gas futures signed under time period $t - T$ at time period t; $E_{t-T}(P^S_t)$ is the expected value of natural gas spot price P^S_t at time period $t - T$; i_T is the appropriate discount rate for investors to invest in futures contracts; and r_T is the risk-free rate.

If too many producers in the futures market want to avoid risks and hedge their products, the futures price may be lower than the expected future spot price, that is, $p > 0$. When the demand side is unwilling to take risks, the opposite situation will occur, that is, $p < 0$.

The futures cost of the unit needs to consider time value, interest, risk, and other factors; thus, it can be expressed as:

$$C^i_2 = \sum_{t=1}^{S} Q^{FP}_{i,t} E_{t-T}(P^S_t) \tag{7}$$

where $Q^{FP}_{i,t}$ is the gas consumption of futures contract at time period t.

To sum up, the total cost C^i of the gas unit is:

$$C^i = C^i_1 + C^i_2 + C^i_3 \tag{8}$$

where C^i_3 is the reduced fixed cost of the unit. For unit i, if the income it can obtain in the total time period S is lower than its own comprehensive cost $(R_i - C_i < 0)$, the willingness of unit i to generate electricity decreases, and the probability of the market behavior of offering a high price or even the top price increases.

3. VECM-Model-Based Gas–Electricity Price Linkage Analysis

The VAR model is an econometric model that uses the vector autoregressive method to analyze a system composed of various interrelated factors. It has been widely used in important fields such as economy, transportation, medicine, and the chemical industry [24]. The specific expression of the VAR model in the analysis of gas–electricity price linkage is as follows:

$$\begin{bmatrix} Y_t \\ X_t \end{bmatrix} = \varepsilon_t + \phi_0 + \sum_{i=1}^{p} \phi_i \begin{bmatrix} Y_{t-i} \\ X_{t-i} \end{bmatrix} \tag{9}$$

where Y_t is the time series of electricity price; X_t is the gas price time series; ε_t is the perturbation column vector of the model; ϕ_0 is the column vector composed of constant terms in the regression equation; p is the lag order of the gas–electricity price VAR model; ϕ_i is the coefficient matrix between gas price and electricity price and lag term data; and $t = 1, 2, \ldots, T$, where T is the total number of days for collecting price data.

The premise of using the VAR model is that the time series is stationary. Most time series in reality do not meet the requirement of stationarity, so the VECM model is usually used in the study of price transmission. The VECM model is a VAR model with co-integration constraints, which is mostly used for non-stationary time-series models with co-integration relations. The expression of the VECM model is as follows:

$$\begin{bmatrix} \Delta Y_t \\ \Delta X_t \end{bmatrix} = \varepsilon_t + \theta_0 + \sum_{i=1}^{p-1} \theta_i \begin{bmatrix} \Delta Y_{t-i} \\ \Delta X_{t-i} \end{bmatrix} + \Gamma \begin{bmatrix} Y_{t-1} \\ X_{t-1} \end{bmatrix} \tag{10}$$

where ΔY_t is the first-order difference term of the electricity price time series; ΔX_t is the first-order difference term of the gas price time series; θ_0 is a column vector of order 2; θ_i is a 2×2 matrix; and Γ is compression matrix.

VECM is essentially a constrained VAR model with a co-integration constraint in explanatory variables. When a large range of short-term fluctuations occurs, VECM will make endogenous variables converge to their long-term co-integration relationship [33]. Compared with other methods, the VECM model treats both gas price and electricity price as endogenous variables, thus reducing the uncertainty of simultaneous equations caused by subjective judgment errors. In addition, the expression of a static and stable co-integration relationship between gas and electricity prices and the transmission mechanism of dynamic price fluctuations can be obtained through the VECM model. The specific research methods can be divided into the stationarity test, co-integration test, Granger causality test, and dynamic characteristic analysis.

When constructing the VECM model and conducting research, it is necessary to conduct the stationarity test first to ensure that there will not be two invalid results of a negative model coefficient and pseudo regression in the research.

Most of the time series actually obtained cannot meet the requirements of stationarity in a strict sense. Therefore, the co-integration test should be conducted again after the stationarity test to ensure the existence of a stable co-integration relationship between gas price and electricity price, which meets the requirements of further research. In addition, in order to make the model more accurate, it is necessary to carry out the unit root test on the residual sequence in the co-integration test.

The co-integration test can obtain the long-term stable functional relationship between electricity price and gas price, while the Granger causality test can further judge the sequential relationship between gas price and electricity price in time. It should be noted that the result obtained by the Granger causality test is not a causal relationship of practical significance.

Each research method is described in detail below.

3.1. Gas–Electricity Price Stability Test

The analysis of the linkage relationship between gas and electricity prices requires the stationarity test of the data to ensure that the data meet the requirements of the VAR model test. Considering that there may be collinearity and heteroscedasticity in the gas–electricity price time series, it is necessary to clean the original data first, and let the processed natural gas and electricity price time series be X_t and Y_t, respectively.

Taking the stationarity test of power price time series Y_t as an example, the given regression equation of electricity price is:

$$Y_t = kY_{t-1} + \varepsilon_t \tag{11}$$

where k is the regression coefficient and ε_t is the error term.

The electricity price at time t can be expressed by historical data:

$$Y_t = k^n Y_{t-n} + \sum_{i=0}^{n-1} \varepsilon_{t-i} k^i \tag{12}$$

If $k = 1$, the variance of the electricity price time series in (11) will continue to increase, and the influence of residual ε_{t-i} cannot be eliminated, which will lead to the instability of the series. Therefore, $k = 1$ is called the unit root of the electricity price time series. The augmented Dickey–Fuller (ADF) test method can be used to determine whether there is a unit root in (11), so as to determine whether the time series of electricity price is stable.

Similarly, the same stationarity test should be conducted for gas price time series X_t.

3.2. Gas–Electricity Price Co-Integration Test

After the stationarity test, the linkage relationship of the gas–electricity price time series can be further analyzed. When there is a stable internal mechanism between gas price and electricity price, the gas price or electricity price can still maintain a stable equilibrium state even if there are short-term fluctuations. In order to directly reflect this equilibrium state, the co-integration test of the gas–electricity price can be carried out to reflect the linkage relationship of gas–electricity price in the form of a co-integration expression. The co-integration test mainly involves the following two steps:

First, the ordinary least squares (OLS) method is used to estimate the gas–electricity price fitting equation and calculate the corresponding residual value:

$$Y_t^* = a^* X_t + b^* \tag{13}$$

$$e_t = Y_t - Y_t^* \tag{14}$$

where a^* and b^* are the optimal parameters obtained by fitting; Y_t^* is the estimated electricity price; and e_t is the residual difference between the real value and the predicted value of electricity price.

Second, the stationarity of e_t is tested. If e_t is a stationary series, it is considered that there is a co-integration relationship between gas price and electricity price. In this case, there is no pseudo-regression problem in the equation obtained by regression.

3.3. Granger Causality Test of Gas–Electricity Price Based on VECM Model

In the VAR model, the Granger causality test is typically used to assess the temporal relationship between gas price and electricity price after performing a co-integration test. However, Granger causality tests require stable time series data, and direct testing of unstable data can lead to false regressions. The processed variables in the VECM model have stable characteristics and meet the necessary conditions for Granger causality tests, ensuring the accuracy of the results. Therefore, a VECM-model-based Granger causality test can be used to analyze variables $\Delta \ln Y_t$ and $\Delta \ln X_t$. It should be noted that the results obtained from the Granger causality test do not necessarily establish causality in a practical sense, but they can provide valuable insights for further causal analysis. For instance, considering the test of the first-order lag model "whether gas price $\Delta \ln X_t$ causes changes in electricity price $\Delta \ln Y_t$", the initial assumption is that "gas price $\Delta \ln X_t$ does not cause changes in electricity price $\Delta \ln Y_t$", and the specific steps are as follows:

(1) Estimate the unconstrained regression model (u) and the constrained regression model (r) of electricity price Y_t, respectively:

$$u: \Delta \ln Y_t = \phi_{11}(1)\Delta \ln Y_{t-1} + \phi_{12}(1)\Delta \ln X_{t-1} + \varepsilon_{1t} \tag{15}$$

$$r: \Delta \ln Y_t = k\Delta \ln Y_{t-1} + \varepsilon_{1t} \tag{16}$$

The estimation coefficients $\hat{\phi}_{11}(1)$, $\hat{\phi}_{12}(1)$, and $\hat{k}$ of Equations (14) and (15) are calculated, respectively. Then, the residual sum of squares of the model is obtained according to Equations (16) and (17) and the F statistic is constructed:

$$RSS_u = \sum_{i=1}^{T} \left(\Delta \ln Y_t - \hat{\phi}_{11}(1) \Delta \ln Y_{t-1} - \hat{\phi}_{12}(1) \ln X_{t-1} - \varepsilon_{1t} \right)^2 \tag{17}$$

$$RSS_r = \sum_{i=1}^{T} \left(\Delta \ln Y_t - \hat{k} \Delta \ln Y_{t-1} - \varepsilon_{1t} \right)^2 \tag{18}$$

$$F = \frac{(RSS_r - RSS_u)/p}{RSS_u/(n-k)} \sim F(p, n-k) \tag{19}$$

where $k = 2p$. In this paper, the lag order is order 1, so $p = 1$ and $k = 2$.

According to the statistical results, the original hypothesis "gas price $\Delta \ln X_t$ is not the reason for the change of electricity price $\Delta \ln Y_t$" can be judged. If $F < F_\alpha(p, n-k)$, it is believed that "gas price $\Delta \ln X_t$ is not the reason for the change of electricity price $\Delta \ln Y_t$" on the premise that the significance level is α; otherwise, reject the null hypothesis.

(2)   Change the sequence of causality between gas price and electricity price, and use the same method in (1) to test.

(3)   If the test results both reject "gas price is not the reason for the change of electricity price" and accept "electricity price is not the reason for the change of gas price", it can be concluded that "gas price is the Granger cause of electricity price".

The results of the Granger causality test can be compared with the actual supply and demand of the gas and electricity market to verify the accuracy of the results.

4. Time-Lag Analysis of Gas–Electricity Price Based on Soft Attention Mechanism

The linkage between gas and electricity prices is complex, so electricity price reflects gas price fluctuations with a time lag, and the attention mechanism can assign a weight value of influence on current electricity price according to each historical gas price, namely, attention weight value. The delay time of gas price conduction to electricity price can be judged according to the value of the attention weight. Attention mechanisms can be divided into hard attention and soft attention [34]. The hard attention mechanism assigns a weight value of 0 or 1 to the historical gas price, while the soft attention mechanism assigns a weight value of [0, 1], which is more flexible and more in line with the actual situation. Therefore, the soft attention mechanism is used in this paper to calculate the influence weight value of the historical gas price on current electricity price.

The attention mechanism can calculate the attention weight value of the gas price at each time node to current electricity price y_t according to electricity price Y_t and gas price X_t; the specific expression is:

$$\alpha_i = \frac{e^{-x_i}}{\sum\limits_{i=t-k+1}^{i=t} e^{-x_i}} \tag{20}$$

where α_i is the weight value of the attention of gas price x_i to current electricity price y_i, and $\sum\limits_{i=t-k+1}^{i=t} e^{-x_i} = 1$; considering that the gas price with a relatively long time has a small influence on the current electricity price, and this paper determines the transmission delay of gas–electricity price through the relative size of the attention weight value, it is necessary to estimate and determine an appropriate time period k and analyze some historical data.

a_i obtained from the analysis of Equation (21) is the weight value of the gas price in one day, and the weight value of weekly scale A_j can be further obtained through summation:

$$A_j = \sum_{m=t-7\mathrm{x}(j-1)}^{t-7j+1} a_m \tag{21}$$

where A_j is the weight value of the week j, and the delay time of gas–electricity price conduction can be judged according to the weight value.

5. Case Studies

This paper focuses on the linkage relationship between gas and electricity prices in China. Empirical data are derived from the Northeast Asia JKM natural gas price index of some months from August 2020 to November 2022 and the average price data of gas power generation in the spot market published by Power Trading Center (the spot market of electricity is not open in some months).

5.1. Analysis of Benefit–Cost Model of Gas Unit

As shown in Figure 1, from March to July 2022, when the spot market of electricity is open, the JKM price index affects the declaration of gas engine price by linking it with gas price, and the fluctuation trend of gas price and electricity price is the same. Since September, cities in northern China have been gradually taking heating measures. In order to ensure the supply of natural gas to the north, the supply of natural gas engines has been strained. Therefore, the JKM price index and the unit price fluctuate in an inconsistent trend, and the correlation is weakened.

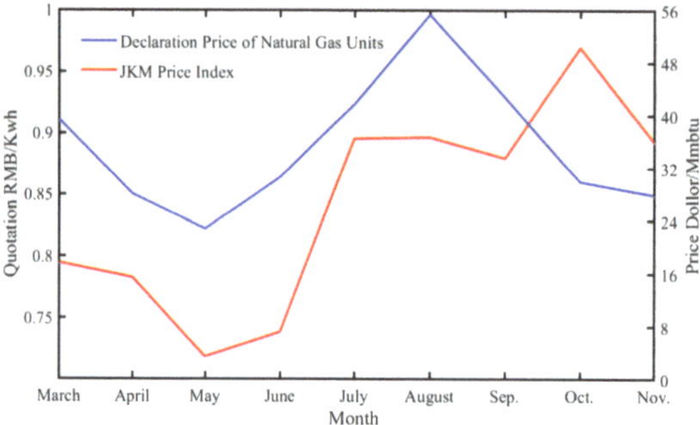

Figure 1. JKM price index with gas unit quotes.

Figure 2 shows the variation of kilowatt hour cost with the JKM price index and the average kilowatt-hour revenue of gas units in 3 months according to the benefit–cost model of gas units from March to May 2022. On the whole, the kilowatt-hour cost fluctuates with the increase of the JKM price index. According to the JKM price index of three months in Figure 1, the average kilowatt-hour cost in March was 0.794 yuan/kWh, significantly higher than the average kilowatt-hour income of 0.706 yuan/kWh. The average kilowatt-hour cost in April is 0.772 yuan/ kWh, which is slightly different from the average kilowatt-hour income of 0.759 yuan/ kWh. The average kilowatt-hour cost in May is 0.712 yuan/kWh, lower than the average kilowatt-hour income of 0.727 yuan/kWh, so the proportion of high capacity of gas units in March should be significantly higher than that in April and May. Figure 3 shows the actual quoted capacity of gas units from January to May 2022. It can be seen that the average daily high price capacity of gas units accounted for 42% in March, 26% in April, and 28% in May, which is in line with the expected results.

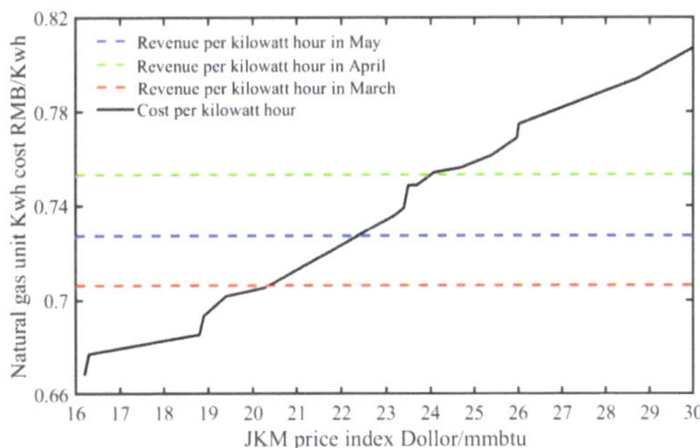

Figure 2. The relationship between average kilowatt-hour income and kilowatt-hour cost with JKM price index in March, April, and May 2022.

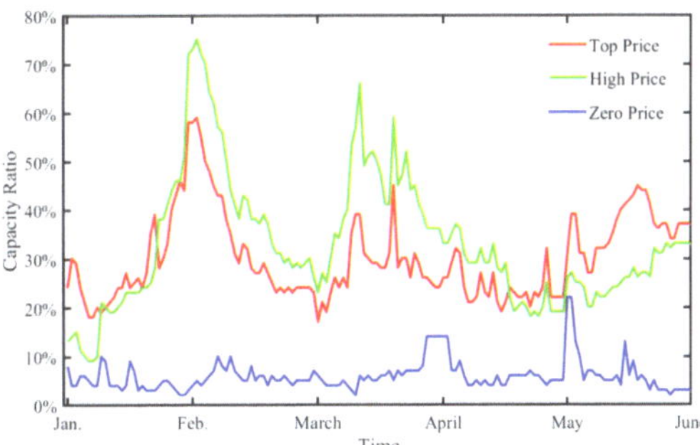

Figure 3. The quoted capacity ratio of gas units from January to May 2022.

5.2. Analysis of VECM Model of Gas–Electricity Price

As shown in Figure 1, gas–electricity prices from March to July 2022 have a strong correlation, so the VAR model focuses on analyzing the gas–electricity price linkage in the above time periods. It should be noted that the gas price X and electricity price Y studied are the data obtained from the original data after cleaning, normalization, logarithm, and other preprocessing.

First, gas price $\ln X_t$ and electricity price $\ln Y_t$ and their difference sequences are tested for stationarity, and the results are shown in Table 1. It can be seen from Table 1 that both the statistical probabilities of the original time series $\ln X_t$ and $\ln Y_t$ of gas price and electricity price at the significance level of 5% are greater than 0.05, indicating that the time series is not stable, and it is necessary to determine whether the VECM model can be analyzed according to the stationariness test results of the first-order difference series. The first-order difference sequences $d \ln X_t$ and $d \ln Y_t$ are tested as stationary sequences, so they can be further studied.

Table 1. Gas–electricity price time series stability test results.

Sequence of Variables	ADF Test Value	Statistical Probability at the Significance Level of 5%	Inspection Result
$\ln X_t$	0.686	0.8626	Unstable
$\ln Y_t$	−2.519	0.3185	Unstable
$d \ln X_t$	−10.489	0.0000	Stable
$d \ln Y_t$	−12.815	0.0000	Stable

Furthermore, the results of the co-integration test can be obtained, as shown in Equation (22) and Table 2.

$$\ln Y_t = 0.9404 \ln X_t + 2.3983 + ecm \qquad (22)$$

Table 2. Test results of gas–electricity price time series co-integration test.

Equation Coefficient	Optimal Parameter Value	*t*-Test Value	Statistical Probability
a	0.9404	17.2245	0.0000
b	2.3983	43.1492	0.0000

According to the numerical results, the co-integration relationship between gas prices and electricity prices can be expressed as: the unit root test shows that the error term ecm satisfies the stationarity assumption, implying a stable co-integration relationship. The goodness of fit $R^2 = 0.7036$ of the co-integration expression indicates a strong correlation between the two variables, and the fitted values are of practical significance. Moreover, the statistical significance level of the residual errors is less than 0.05, suggesting that the series is stable and there exists a stable equilibrium relationship between gas price and electricity price. Specifically, a 1% increase or decrease in gas price will lead to a 0.9404% increase or decrease in electricity price.

Once the co-integration relationship between variables is determined, a vector error correction model can be developed to verify their mutual adjustment rate and short-term interaction effects, and examine the causal relationship between them. The error correction term Coint Equation (1) represented in Table 3 reflects the short-term adjustment to long-term equilibrium. The outcomes demonstrate that Coint Equation (1) is negative, indicating that there is an error correction mechanism between gas and electricity prices, which stabilizes the short-term fluctuations and ultimately maintains the equilibrium relationship. It can be seen from the first column of Table 3 that the price of natural gas has a significant positive impact on the electricity price in the short term because the energy price directly impacts the spot market of electricity. Consequently, the rise in the price of natural gas leads to an increase in the power generation cost of gas-burning units, resulting in a corresponding increase in the electricity price. From the second column of Table 3, it is evident that the electricity price has a positive impact on the gas price at the 10% confidence level. This is because the increase of electricity price will increase the price of energy commodities to some extent. Despite the presence of a long-term stable co-integration relationship between gas and electricity prices, short-term price fluctuations still affect their respective markets. Therefore, market players need to pay close attention to the linkage between gas and electricity prices.

Table 3. Vector error correction model estimate results.

Error Correction Term	$d\ln Y_t$	$d\ln X_t$
Coint Equation (1)	−0.4218 (0.1941)	−0.1082 (0.1136)
$d \ln Y_{t-1}$	−0.0782 (0.2151)	1.3195 (0.0835)
$d \ln X_{t-1}$	2.1982 (0.0029)	−0.5194 (0.1837)

The co-integration expression and VECM model can reflect the specific linkage relationship between gas and electricity prices. However, they do not provide information on the direction of price transmission at the time level. Therefore, it is necessary to conduct a Granger causality test. Considering the differences in gas–electricity market environments across different months, the causality of gas–electricity prices in different time periods is analyzed, and the results are shown in Table 4.

Table 4. Causal test analysis of gas and electricity prices.

Time	Actual Situation of Gas and Electricity Market	Null Hypothesis	Statistical Probability	Conclusion
August 2020	Gas is cheap, and electricity supply environment is easy	Gas price $d\ln X_t$ is not the cause of electricity price $d\ln Y_t$ changes	0.79	Electricity price causes gas price change
		Electricity price $d\ln Y_t$ is not the cause of gas price $d\ln X_t$ changes	0.08	
May 2021	Gas price is high, and electricity supply environment is tight	Gas price $d\ln X_t$ is not the cause of electricity price $d\ln Y_t$ changes	0.09	There is a two-way causal relationship between gas price and electricity price
		Electricity price $d\ln Y_t$ is not the cause of gas price $d\ln X_t$ changes	0.06	
From March to July 2022	Gas price is high, and electricity supply environment is tight	Gas price $d\ln X_t$ is not the cause of electricity price $d\ln Y_t$ changes	0.08	Gas price causes electricity price change
		Electricity price $d\ln Y_t$ is not the cause of gas price $d\ln X_t$ changes	0.98	
From August to November 2022	Gas price is high, and electricity supply environment is easy	Gas price $d\ln X_t$ is not the cause of electricity price $d\ln Y_t$ changes	0.85	There is no obvious causal relationship between gas price and electricity price
		Electricity price $d\ln Y_t$ is not the cause of gas price $d\ln X_t$ changes	0.39	

When there is a co-integration relationship between the time series, it can be determined that there is a long-term equilibrium relationship between variables. Therefore, the VECM model can be further used to study the short-term nature of the time series. When there is no co-integration relationship in the time series, the Granger causality test can be used to establish the causal relationship between variables. In this paper, the Granger causality test is carried out on the processed time variables after the co-integration relation test. The processed gas price $d\ln X_t$ and electricity price $d\ln Y_t$ meet the stationarity requirements of the causality test, and the validity of the test results is guaranteed. Therefore, the Granger causality test based on the vector error correction model can be used to test the short-term relationship between gas and electricity prices [35,36]. The actual supply and demand of the gas and electricity market in different periods in Table 4 provides a reference for the results of the Granger causality test.

As can be seen from Table 4, the actual operation situation of the gas–electricity market is that the gas price is cheap and the electricity supply is loose in August 2020, and the electricity price mainly affects the change of the gas price. The gas price in May 2021 is expensive and electricity supply is tight, so the gas price interacts with electricity price during this period. From March to July 2022, the gas price continues to rise, becoming an important factor affecting electricity price. Although the gas price remained high from August to November, the correlation between gas and electricity price was low and there was no obvious causal relationship due to the policy of guaranteeing the supply of gas for winter in northern China. The result of Granger's causality test is in good agreement with the supply and demand situation reflected in the actual gas and electricity market.

5.3. Gas–Electricity Price Time Delay Analysis Results

Table 4 shows that, from March to July 2022, there exists a one-way Granger causality relationship affecting gas price and electricity price. Therefore, the time-lag relationship of the influence of gas price fluctuation on electricity price during this period can be analyzed. Table 5 shows the top five period gas prices with the greatest influence according to the weight value A_j, among which the influence weight in the second period is the highest, reaching 0.24. The influence weights of period 1 and period 3 are 0.21 and 0.19, respectively, while the weights of the remaining periods are relatively small. The sum of cumulative influence weights in the first three periods reaches 0.64, indicating that the time delay of gas–electricity price conduction is mainly in the first three periods. However, A_2 does not play an absolute role in the results of the example, which shows from the side that the price interaction between gas and electricity is weak, and the key problem lies in the high gas price and high volatility.

Table 5. The weight of the influence of weekly-scale gas price on the current electricity price.

Number of Cycles j	Cycle Scale Weight Value A_j	Cumulative Influence Weight
2	0.24	0.24
1	0.21	0.45
3	0.19	0.64
4	0.14	0.78
6	0.08	0.86

Table 6 shows the lag order determined according to the VECM model. The AIC of lag period 2 is the smallest, so the lag time of gas and electricity price is two periods. Although both methods can determine the optimal delay time, the proposed method can also reflect the weight relationship between different delay periods. Price transmission is an extremely complex process. The time delay of gas and electricity price transmission is affected by many factors, such as market supply and demand, and market price fluctuation. Therefore, the weight of the lag period can provide more price transmission information for market decision makers. Although the standardized coefficient method can analyze the relative importance of price variables, it lacks the ability to analyze the contribution degree in specific time periods. The variance decomposition method can disintegrate and analyze the total variance of the electricity price time series, so it can reflect the contribution trend of electricity price and gas price more accurately.

Table 6. Determination of the lag order of VAR model.

Order of Lag	AIC
1	−2.41
2 *	−5.13
3	−4.88
4	−4.64
5	−3.08
6	−3.32

* represents the optimal lag order determined according to AIC criterion.

6. Conclusions

The coupling degree of the natural gas market and electricity market is deepening, leading to a stronger linkage relationship between gas and electricity prices. In this context, it is crucial to accurately analyze the linkage mechanism between gas and electricity prices to enhance the ability of market participants to manage price risk events and gain a better understanding of the market environment. Therefore, this paper proposes a gas–electricity price linkage analysis method based on the benefit–cost and Attention–VECM models. An empirical analysis of natural gas and electricity markets in China confirms the effectiveness of this method in accurately capturing the linkage mechanism between gas and electricity prices.

(1) The gas unit benefit–cost model proposed in this paper captures the quotation behavior of gas units in the market by assessing the profitability of the units. This model provides valuable gas and electricity price information for the co-ordinated operation of the gas and electricity market, thereby helping market participants to manage market risks, such as an insufficient natural gas supply.

(2) Calculation reveals the co-integration relationship of long-term gas and electricity price stability. The results demonstrate a goodness of fit value greater than 0.7, indicating a strong correlation between the two. Additionally, this paper employs the VECM model to study the short-term volatility of gas and electricity prices. The results reveal the existence of an error correction mechanism, suggesting that short-term volatility will eventually tend to balance. This finding further confirms the presence of a long-term equilibrium relationship between gas and electricity prices. To analyze the causality relationship between prices in different periods, the processed gas and electricity price time series are subjected to a Granger causality test. The comparison of the results with actual market conditions reveals that the supply and demand situation in the gas and electricity market is continually changing.

(3) In this paper, an attention mechanism is used to quantitatively describe the conduction delay time of gas–electricity price fluctuations. Compared with determining the lag order by the VAR model, the proposed method comprehensively reflects the influence weight of the delay in each time period, and the results are more consistent with the actual price transmission process. An analysis of the weight value of each cycle and the cumulative influence weight reveals that the time delay is mainly within the first three weeks, and there is no time delay accounting for absolute proportion in the results. Therefore, it can be qualitatively concluded that the existing gas–electricity price linkage is not strong.

The proposed natural gas–electricity price linkage analysis method has practicability in the following aspects. Firstly, the benefit–cost model can reflect the quotation tendency of gas units in the market and the supply situation of the natural gas market. Secondly, the VECM model and the co-integration relationship of gas and electricity prices provide a reference for evaluating the long-term trend and short-term fluctuation range of prices. Thirdly, the Granger causality test results directly reflect the coupling relationship and supply and demand of the gas and electricity market. Lastly, the results of price conduction delay calculated by the attention mechanism can help market players develop reasonable trading strategies for natural gas and electricity.

In practice, the linkage between gas price and electricity price will also be affected by other energy resource prices, the upstream and downstream transmission relationship of the industrial chain, and the impact of macro policies. Therefore, the research method proposed in this paper is not applicable to the electricity market environment where the proportion of natural gas generation is relatively low. Moreover, price fluctuations themselves have relatively strong uncertain factors that were not considered in this study. In the future, the linkage relationship between gas price and electricity price could be further investigated through alternative methods.

Author Contributions: Conceptualization, S.Z. and W.G.; methodology, S.Z. and A.L.; software, S.Z. and Z.L.; validation, C.S. and L.Y.; formal analysis, X.J. and Z.L.; investigation, X.J., Y.W. and L.Y.; resources, S.Z. and W.G.; data curation, A.L.; writing—original draft preparation, C.S.; writing—review and editing, S.Z. and Y.W.; visualization, X.J.; supervision, W.G. and A.L.; project administration, W.G. and C.S.; funding acquisition, W.G. and A.L. All authors have read and agreed to the published version of the manuscript.

Funding: This work was funded by the Science and Technology Project of State Grid Zhejiang Electric Power Co., Ltd. (No. B311UZ220002).

Data Availability Statement: Not applicable.

Conflicts of Interest: The authors declare no conflict of interest.

References

1. Wu, C.; Wu, Z.; Yi, Z.; Yang, L. Strategic bidding of gas-fired units with tanks in coupled electricity and natural gas markets. *Sustain. Energy Grids Netw.* **2023**, *34*, 101021. [CrossRef]
2. Jing, Z. Analysis and prospect of global natural gas market in 2022. *Int. Pet. Econ.* **2021**, *29*, 34–43, 106.
3. Han, Y. Supply-demand situation of European natural gas market in winter 2022. *Int. Pet. Econ.* **2022**, *30*, 80–85.
4. Ruijin, Z.; Weilin, G.; Xuejiao, G. Short-term electricity load forecasting considering the correlation between natural gas and electricity load. *J. Electr. Power Syst. Autom.* **2019**, *31*, 27–32.
5. Chen, C.; Liang, H.; Zhai, X.; Zhang, J.; Liu, S.; Lin, Z.; Yang, L. Review of restoration technology for renewable-dominated electric power systems. *Energy Convers. Econ.* **2022**, *3*, 287–303. [CrossRef]
6. Poyrazoglu, P.G. Impact of gas price on electricity price forecasting via supervised learning and random walk. In Proceedings of the 2019 16th International Conference on the European Energy Market (EEM), Ljubljana, Slovenia, 18–20 September 2019; pp. 1–5.
7. Wang, T.; Qu, W.; Zhang, D.; Ji, Q.; Wu, F. Time-varying determinants of China's liquefied natural gas import price: A dynamic model averaging approach. *Energy* **2022**, *259*, 125013. [CrossRef]
8. Zhang, R.; Jiang, T.; Li, F.; Li, G.; Li, X.; Chen, H. Conic optimal energy flow of integrated electricity and natural gas systems. *J. Mod. Power Syst. Clean Energy* **2021**, *9*, 963–967. [CrossRef]
9. Liu, S.; You, S.; Lin, Z.; Zeng, C.; Li, H.; Wang, W.; Hu, X.; Liu, Y. Data-driven event identification in the U.S. power systems based on 2D-OLPP and RUSBoosted trees. *IEEE Trans. Power Syst.* **2022**, *37*, 94–105. [CrossRef]
10. Xiaolong, S.; Renan, J. Simulation and evaluation on implementation effect of coal and electricity linkage related control policies in Transition period. *Autom. Electr. Power Syst.* **2012**, *36*, 55–61.
11. Shan, L.; Qing, D.; Guojian, Y. Comprehensive benefits and evaluation of natural gas power plants. *J. Electr. Power Syst. Autom.* **2015**, *27*, 7–12, 60.
12. Yang, J.; Dong, Z.Y.; Wen, F.; Chen, Q.; Liang, B. Spot electricity market design for a power system characterized by high penetration of renewable energy generation. *Energy Convers. Econ.* **2021**, *2*, 67–78. [CrossRef]
13. Zhang, T.; Liu, S.; Qiu, W.; Lin, Z.; Zhu, L.; Zhao, D.; Qian, M.; Yang, L. KPI-based Real-time Situational Awareness for Power Systems with a High Proportion of Renewable Energy Sources. *CSEE J. Power Energy Syst.* **2022**, *8*, 1060–1073.
14. Wang, T.; Zhang, D.; Ji, Q.; Shi, X. Market reforms and determinants of import natural gas prices in China. *Energy* **2020**, *196*, 117105. [CrossRef]
15. Yin, J.; Wang, Y. Research on affordability of natural gas for domestic use. *Price Theory Pract.* **2012**, *336*, 47–48.
16. Dongyu, L.; Jun, Z.; Yan, Z. Study on Rationality of Chinese Residential Gas Ladder Pricing—An Analysis of Urumqi Residential Gas Ladder Price based on Utility function and resource allocation. *Price Theory Pract.* **2017**, *392*, 100–103.
17. Liu, G. Research on price block in upstream and downstream natural gas industries. *Price Theory Pract.* **2018**, *406*, 60–63.
18. Chen, C.; Li, Y.; Qiu, W.; Liu, C.; Zhang, Q.; Li, Z.; Lin, Z.; Yang, L. Cooperative-Game-Based Day-Ahead Scheduling of Local Integrated Energy Systems with Shared Energy Storage. *IEEE Trans. Sustain. Energy* **2022**, *13*, 1994–2011. [CrossRef]
19. Wang, C.; Wei, W.; Wang, J.; Wu, L.; Liang, Y. Equilibrium of interdependent gas and electricity markets with marginal price based bilateral energy trading. *IEEE Trans. Power Syst.* **2018**, *33*, 4854–4867. [CrossRef]
20. Nandakumar, N.; Annaswam, A.M. Impact of increased renewables on natural gas markets in eastern United States. *J. Mod. Power Syst. Clean Energy* **2017**, *5*, 424–438. [CrossRef]
21. Khani, H.; Farag, H.E.Z. Optimal Day-Ahead scheduling of power-to-gas energy storage and gas load management in wholesale electricity and gas markets. *IEEE Trans. Sustain. Energy* **2010**, *9*, 940–951. [CrossRef]
22. Shiri, A.; Afshar, M.; Rahimi-Kian, A.; Maham, B. Electricity price forecasting using support vector machines by considering oil and natural gas price impacts. In Proceedings of the 2015 IEEE International Conference on Smart Energy Grid Engineering (SEGE), Oshawa, ON, Canada, 17–19 August 2015; pp. 1–5.
23. Xiaoping, Z.; Jianing, L.; Hao, F. Reform and challenge of electricity retail market in UK. *Autom. Electr. Power Syst.* **2016**, *40*, 10–16.
24. Danistya, D.R.; Qaulifa, F.; Ramadani, Y.A.; Yulita, I.N.; Ardisasmita, M.N.; Agustian, D. Prediction New Cases of COVID-19 in Indonesia Using Vector Autoregression (VAR) and Long-Short Term Memory (LSTM) Methods. In Proceedings of the 2021 International Conference on Artificial Intelligence and Big Data Analytics, Bandung, Indonesia, 27–19 October 2021; pp. 1–4.
25. Toda, H.Y.; Taku, Y. Statistical inference in vector autoregressions with possibly integrated processes. *J. Econom.* **1995**, *66*, 225–250. [CrossRef]
26. Peng, W.; Mingyu, Z.; Bian, F. Study on the linkage relationship between gas and electricity prices in Britain. *Price Issue* **2020**, *522*, 23–29.
27. Zhang, J.; Hu, W.; Zhang, X. The Relative Performance of VAR and VECM Model. In Proceedings of the 2010 3rd International Conference on Information Management, Innovation Management and Industrial Engineering, Kunming, China, 26–28 November 2010; pp. 132–135.

28. Parvin, R.; Ferdows, R. The Nexus between Financial Development and Economic Growth: An Application of the VECM Model and Machine Learning Algorithms. In Proceedings of the 2022 25th International Conference on Computer and Information Technology (ICCIT), Cox's Bazar, Bangladesh, 17–19 December 2022; pp. 360–365.
29. Chen, Y.; Lu, J.; Ma, M. How Does Oil Future Price Imply Bunker Price—Cointegration and Prediction Analysis. *Energies* **2022**, *15*, 3630. [CrossRef]
30. Kang, Z.; Zhang, B. Analysis on Cointegration of FDI, International Trade and Chinese Economic Growth and VECM Model. *J. Int. Trade* **2006**, *2*, 73–78.
31. Zhao, B.; Wang, Z.P.; Ji, W.J.; Gao, X.; Li, X.B. CNN-GRU short-term Power load forecasting Method based on Attention mechanism. *Power Grid Technol.* **2019**, *43*, 4370–4376.
32. Huan, R.; Wang, X. Review on Attention Mechanism. *J. Comput. Appl.* **2021**, *41*, 1–6.
33. Zhou, Y.H.; Zou, L.G. Study on the Price relationship between Chinese soybean futures Market and international soybean futures Market—An Empirical Analysis based on VAR Model. *J. Agrotech. Econ.* **2007**, *1*, 55–62.
34. Luo, Z.; Li, J.; Zhu, Y. A Deep Feature Fusion Network Based on Multiple Attention Mechanisms for Joint Iris-Periocular Biometric Recognition. *IEEE Signal Process. Lett.* **2021**, *28*, 1060–1064. [CrossRef]
35. Li, K.J.; Qu, R.X. The impact of technological progress on China's carbon emission: An empirical study based on vector error correction model. *China Soft Sci.* **2012**, *258*, 51–58.
36. Asari, F.F.A.H.; Baharuddin, N.S.; Jusoh, N.; Mohamad, Z.; Shamsudin, N.; Jusoff, K. A vector error correction model (VECM) approach in explaining the relationship between interest rate and inflation towards exchange rate volatility in Malaysia. *World Appl. Sci. J.* **2011**, *12*, 49–56.

Article

A Grid Status Analysis Method with Large-Scale Wind Power Access Using Big Data

Dan Liu [1], Yiqun Kang [1], Heng Luo [1], Xiaotong Ji [1], Kan Cao [1] and Hengrui Ma [2,*]

[1] State Grid Hubei Electric Power Research Institute, Wuhan 430077, China
[2] School of Electrical and Automation, Wuhan University, Wuhan 430077, China
*   Correspondence: henry3764@foxmail.com

Abstract: Targeting the problem of the power grid facing greater risks with the connection of large-scale wind power, a method for power grid state analysis using big data is proposed. First, based on the big data, the wind power matrix and the branch power matrix are each constructed. Second, for the wind energy matrix, the eigenvalue index in the complex domain and the spectral density index in the real domain are constructed based on the circular law and the M-P law, respectively, to describe the variation of wind energy. Then, based on the concept of entropy and the M-P law, the index for describing the variation of the branch power is constructed. Finally, in order to analyze the real-time status of the grid connected to large-scale wind power, the proposed index is combined with the sliding time window. The simulation results based on the enhanced IEEE-33 bus system show that the proposed method can perform real-time analysis on the grid state of large-scale wind power connection from different perspectives, and its sensitivity is good.

Keywords: large-scale wind power access; power grid status analysis; wind power matrix; branch power matrix; random matrix; entropy

1. Introduction

At present, the world's petroleum energy is becoming increasingly scarce, and the problem of carbon emissions is becoming more and more serious. Under this background, the large-scale construction of the wind power base is of great significance to the adjustment of the energy structure, energy conservation and emission reduction [1]. However, the large-scale centralized access to wind power also increases the possibility of the power grid going black [2]. Therefore, a comprehensive and accurate grid condition analysis is conducive to controlling the operation of the grid under large-scale access.

Now, for a single power system, the literature [3–5] proposes a power grid state evaluation method based on entropy theory and points out that the load rate and network structure of the line are important factors affecting the power grid state evolution. The literature [6] combines the above two factors and considers the distribution balance of load rate and proposes a comprehensive evaluation method of a self-organized critical state of the power grid based on joint weighted entropy. However, the load rate zoning is greatly affected by human factors and the evaluation results are uncertain. For the power grid with large-scale wind power centralized access, the literature [7] constructs network topology entropy and power flow entropy based on entropy theory and analyzes the power grid state from the perspective of the influence of wind power fluctuation on power flow entropy. Additionally, for the above-mentioned situations, the literature [8] takes weighted power flow entropy, network topology entropy, wind wave dynamic entropy and other entropy physical indexes as inputs and constructs the mapping relationship between key physical quantity indexes and the power grid state through neural networks. However, it requires a large number of training samples, and the generalization of machine learning methods is poor, so its application in practice is limited.

Citation: Liu, D.; Kang, Y.; Luo, H.; Ji, X.; Cao, K.; Ma, H. A Grid Status Analysis Method with Large-Scale Wind Power Access Using Big Data. *Energies* **2023**, *16*, 4802. https://doi.org/10.3390/en16124802

Academic Editors: Akhtar Kalam, Fushuan Wen and Xiuli Wang

Received: 4 March 2023
Revised: 17 May 2023
Accepted: 8 June 2023
Published: 19 June 2023

Random matrix theory (RMT) is a mathematical statistical method based on big data. It analyzes the big data matrix from the perspective of feature distribution to avoid information loss due to dimension reduction or feature extraction. It has achieved good application results in data-driven power grid situation awareness [9], transient stability analysis [10], power grid weak point identification [11], power consumption behavior analysis [12] and power grid state analysis, which verifies the effectiveness of random matrix theory in power grid state analysis.

Therefore, based on the random matrix theory and making full use of the big data of power grid measurement, a power grid state evaluation method including the power grid state analysis method of large-scale wind power centralized access is proposed in this paper. The contributions of this paper are as follows:

(1) Based on the ring law and M-P law, the evaluation indexes of wind power change level in the complex domain and real domain are constructed, respectively. The comprehensive index of wind power change rate is constructed by combining the two.

(2) Based on the concept of entropy and M-P law, the state evaluation index of the power network is constructed to represent the power change level of the branch.

(3) Based on the synthesis of the wind power change index and branch power change index, the power grid state analysis index under large-scale wind power access is constructed. Combining the proposed index with the sliding time window, the real-time state analysis of the power grid connected with large-scale wind power is carried out from different angles.

2. Random Matrix Theory

Random matrix theory is a high-dimensional mathematical statistical method. It analyzes the correlation degree of the whole system from the perspective of probability and can reflect the operating state of the system from the global perspective.

2.1. Sample Covariance Matrix and Empirical Spectrum Distribution

For a matrix Y with dimension N and sample number T, the sample covariance matrix M of the matrix is:

$$M = \frac{1}{T}YY^*\tag{1}$$

where Y^* is the complex conjugate transposition of Y. In the analysis of large-dimensional data, the analysis quantity in many multivariate statistics can be expressed as the function of the empirical spectrum distribution of the sample covariance, and so the sample covariance is widely used in the random matrix.

The empirical spectrum distribution function is as follows:

$$F^{\mathbf{M}}(x) = \frac{1}{p}\sum_{i=1}^{p} I\left\{\lambda_i^{\mathbf{M}} \leq x\right\}, x \in R\tag{2}$$

where $F^M(x)$ is the empirical spectrum distribution function; p is the number of eigenvalues; λ_i^M is the i-th eigenvalue of matrix M; $I\{\cdot\}$ is an indicative function.

When the elements of a high-dimensional random matrix are all independent identically distributed (IID) random variables, we call the limit of the empirical spectral distribution as the limit spectral distribution. Furthermore, when the mean value is 0 and the variance is σ^2, the limit spectrum distribution of the sample covariance matrix of the random matrix converges to the following formula with a probability of 1:

$$f_c(\lambda) = \begin{cases} \frac{1}{2\pi c\lambda\sigma^2}\sqrt{(b-\lambda)(\lambda-a)}, & a\leq\lambda\leq b \\ 0, & others \end{cases}\tag{3}$$

where $a = \sigma^2\sqrt{1-c^2}$, $b = \sigma^2\sqrt{1+c^2}$, and where $c = p/n$ and σ^2 is the variance. This is the M-P law with parameters c and σ^2, which is a statistical property of high-dimensional matrices.

The M-P law analyzes the correlation within the data from the perspective of the limit spectrum distribution. When the data inside the high-dimensional matrix conforms to the independent identity distribution condition and there is no correlation between the elements in the matrix, the limit spectrum distribution of the matrix will be consistent with the M-P law. On the contrary, when the data in the matrix do not conform to the condition of independent identity distribution and there is correlation within the data, the limit spectral distribution will deviate from the M-P law. Figure 1 is an M-P law diagram corresponding to a random matrix of 80×200, where red is the spectral distribution result estimated with the Gaussian kernel, and blue is the M-P law.

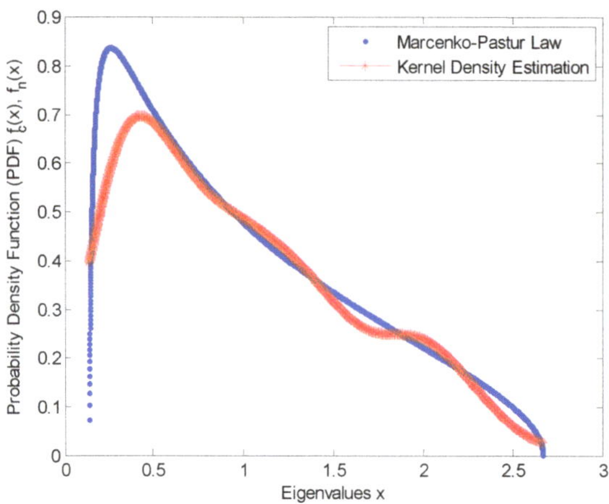

Figure 1. The M-P law diagram.

2.2. The Ring Law

The torus law describes how much the data deviates from the randomness in the random matrix.

Let matrix $Y = \{x_{ij}\}$ be non-Hermitian random matrix of order $N \times T$. It satisfies $N, T \to \infty$, $N/T \in (0, 1)$. In addition, each element in the matrix satisfies the independent identity distribution (IID) with a mean of 0 and a variance of 1. Let the singular value equivalent matrix of Y be $\hat{Y}$, then the empirical spectral distribution of the covariance matrix of $\hat{Y}$ almost always converges to the one ring theorem, and its probability density is:

$$f(\lambda_i) = \begin{cases} \frac{1}{\pi c}|\lambda_i|^{-2}, & (1-c)^{1/2} \le |\lambda_i| \le 1 \\ 0, & others \end{cases} \tag{4}$$

where λ_i is the characteristic value of $\hat{Y}$. The ring theorem states that if the elements of Y satisfy the independent and identical distribution, the eigenvalues of the covariance of Y will be distributed in the ring with the outer ring radius of 1 and the inner ring radius of $(1-c)^{1/2}$. Otherwise, if there is correlation within the system, some eigenvalues will converge to the center of the ring. The stronger the correlation, the stronger the degree of convergence to the center.

2.3. Average Spectral Radius

The randomness of a single eigenvalue of a matrix cannot reflect the randomness of the matrix. The mean spectral radius (MSR) is a linear statistical measure that can be used to reflect the statistical characteristics of the eigenvalues. It is the average of the distances between all eigenvalues on the complex plane and the center point.

$$\text{MSR} = \frac{1}{N}\sum_{i=1}^{N}|\lambda_i| \tag{5}$$

where N is the number of eigenvalues, and the number of eigenvalues is the same as the dimension of Y.

3. Data Model Construction and Data Processing

3.1. State Analysis Matrix Construction and Preprocessing Method

In this paper, the power of each wind turbine and the power of each branch are used to construct the state analysis matrix, then the data characteristics are extracted from it. Lastly, we analyze the state change in the power grid with large-scale wind power access.

3.1.1. Wind Power Matrix

It is assumed that there are p nodes in the power grid connected with wind turbines, and each measuring point has a power value at the sampling time t_i. Then, the power values of all measuring points can form a time vector $x(t_i)$, as shown in Equation (6):

$$x(t_i) = [x_{1t_i}, x_{2t_i} \dots x_{pt_i}] \tag{6}$$

For the N sampling results, all time vectors are combined to form a high-dimensional data matrix X with a size of $p \times n$:

$$X = \begin{bmatrix} x_{11} & x_{12} & \cdots & x_{1n} \\ x_{21} & x_{22} & \cdots & x_{2n} \\ \vdots & \cdots & \cdots & \cdots \\ x_{p1} & x_{p2} & \cdots & x_{pn} \end{bmatrix} \tag{7}$$

From Equation (7), each line of X is the same measurement point, and each column is the same sampling time.

3.1.2. Wind Power Matrix

Suppose there are q branches in the power grid and each branch has a power value at the sampling time t_i. Then, the power values of all branches can form a time vector $y(t_i)$. For n sampling results, the branch power matrix Z is the same as the structure of matrix X, as shown in Formula (8):

$$Z_{q \times n} = \begin{bmatrix} z_{11} & z_{12} & \cdots & z_{1n} \\ z_{21} & z_{22} & \cdots & z_{2n} \\ \vdots & \cdots & \cdots & \cdots \\ z_{q1} & z_{q2} & \cdots & z_{qn} \end{bmatrix} \tag{8}$$

3.2. State Matrix Preprocessing Method

Normalize X according to Formula (9) to obtain a normalization matrix $\tilde{X}$ with a mean value of 0 and a variance of 1:

$$\tilde{X}_i = (X_i - \mu(X_i))/\sigma(X_i), \\ i = 1, 2, \dots p \tag{9}$$

where $\mu(X_i)$ and $\sigma(X_i)$, respectively, represent the mean and variance of X_i, and X_i is the i-th line of X.

In the same way, the normalized matrix $\tilde{Z}$ can be obtained.

In large-dimensional data analysis, many multivariate statistical analysis quantities can be expressed as empirical spectral distribution functions of sample covariance. Therefore, the covariance matrices of $\tilde{X}$ and $\tilde{Z}$ can be obtained, respectively, and on this basis, the state of the power grid with large-scale wind power centralized access can be analyzed by applying the random matrix theory.

4. Power Grid State Analysis under Large-Scale Wind Power Connection

4.1. Power Grid State Evaluation Index Based on Wind Power Matrix

4.1.1. Wind Power Change Level Index Based on Ring Law

Under large-scale wind power connection, the output change of the wind turbine will affect the state of the power grid. Specifically, the power change in the wind turbine will cause the change in the power flow distribution of the power grid, and then cause the change in the state of the power grid.

The ring law points out that when there are only white noise, small disturbances and measurement errors in the system, the data distribution of the system will show a statistical random characteristic. When the data satisfy the independent identity distribution, the singular eigenvalues of the state matrix of the system will be distributed in the ring with the inner ring radius of $(1 - c)^{1/2}$ and the outer ring radius of 1 on the complex plane. On the contrary, when there are signal sources (events) in the system, this randomness will be broken, and the data will no longer meet the condition of independent and identical distribution, and the singular eigenvalues will fall into the inner loop. The higher the degree of deviation of the elements in the matrix from the independent uniform distribution, the higher the degree of convergence of the eigenvalues to the center of the circle. The singular eigenvalue analysis of the wind power matrix $\tilde{X}$ is carried out. When the power of the large-scale connected wind turbines in the power grid changes little, the singular value of the covariance matrix corresponding to x will be distributed between the inner and outer rings, and the average spectral radius (MSR) is between the inner ring radius and 1. On the contrary, when the overall power of the wind turbine changes, whether there is an overall increase or decrease, the singular value of the covariance matrix corresponding to x will fall into the inner loop. Moreover, the greater the power change, the greater the degree of convergence of singular values to the center of the circle, and the smaller the value of MSR. Therefore, MSR can be used to construct an index to characterize the change level in wind power.

4.1.2. Wind Power Change Level Index Based on M-P Law

The M-P law states that when the elements in the matrix conform to the independent identical distribution condition, the spectral distribution corresponding to the covariance matrix will converge to the M-P law. When the elements in the matrix deviate from the independent uniform distribution condition, the spectral distribution corresponding to the covariance matrix will also deviate from the M-P law. As described in Section 4.1.1, if the wind power matrix X is subjected to spectral analysis, the degree to which the spectral distribution deviates from the M-P law indicates the degree to which the wind power changes.

In this paper, the probability density similarity index of spectral distribution is constructed with reference to document [13,14]. Let the spectral distribution function of the standard M-P law be $f_1(x), x \in [c_1, d_1]$, and the spectral distribution function corresponding to the wind power matrix be $f_2(x), x \in [c_2, d_2]$. Define the difference degree V between the two functions as:

$$v = \int_c^d |f_1(x) - f_2(x)|^n \, \mathrm{d}x \tag{10}$$

where $c = \min(c_1, c_2)$, $d = \max(d_1, d_2)$ and n is the order. For the finite dimensional wind power matrix, there is a discrete form of Formula (10), as shown in Formula (11).

$$V = \frac{1}{N}\sum_{i=1}^{N}|f_1(ki) - f_2(ki)|^n \tag{11}$$

where $k = (d - c)/N$.

The smaller v and V, the closer $f_1(x)$ and $f_2(x)$ are, the smaller the change in wind power. On the contrary, the greater the difference between $f_1(x)$ and $f_2(x)$, the greater the change in wind power.

4.1.3. Comprehensive Index of Wind Power Change Rate

Based on the indicators in Sections 3.1.1 and 3.1.2, this section synthesizes the two indicators. Thus, the change in wind power in the grid connected with large-scale wind power is comprehensively described from a more comprehensive perspective. The constructed indicators are shown in Formula (12):

$$W = MSR + c/V \tag{12}$$

where C is a constant. Then, when the wind power in the power grid does not change significantly, the singular value corresponding to the wind power matrix is distributed between the inner and outer rings, and the MSR is also between the radius of the inner and outer rings, and the value is relatively large. At the same time, the spectral distribution obtained based on wind power is similar to the M-P law. If V is small, the comprehensive index w of wind power change rate is large. On the contrary, a large wind power change rate will cause the odd-5-odd value corresponding to the wind power matrix to drop to the center of the circle, and the MSR is small. The deviation between the spectral distribution obtained based on wind power and M-P law is large. If V is large, W is small.

4.2. Power Grid Status Evaluation Index Based on Branch Power Matrix

The change in wind power or the abnormal state of the power grid (such as grounding fault, load abnormality, etc.) will cause the redistribution of power flow in the power grid. Therefore, in this section, based on the branch power matrix, the M-P law and entropy theory are used to construct the power grid state evaluation index that represents the branch power change.

The definition of entropy [15] is:

$$H = -C\sum_{i=1}^{q} p(\omega_i)\ln p(\omega_i) \tag{13}$$

where C is a constant, l is the number of states and $p(\omega_i)(i = 1,\ldots,l)$ is the state occurrence probability. Entropy represents the disorder degree of the system. The larger its value, the greater the disorder degree of the system.

Among them, the calculation method of $p(\omega_i)$ is referred to [16,17]. That is, calculate the covariance matrix of the branch power matrix Z, and then calculate the contribution rate of each branch power to the anomaly based on the M-P law. The details are as follows:

$$p(\omega_i) = \frac{\sum\limits_{\lambda_k > b} \lambda_k W_{ik}^2}{\sum\limits_{\lambda_k > b} \lambda_k} \tag{14}$$

where λ_k is the k-th eigenvalue of the covariance matrix of the branch power matrix, W_{ik} represents the i-th element of the eigenvector W_k corresponding to the k-th eigenvalue and b is the value calculated according to the size of the branch power matrix in Equation (3).

Then, $p(\omega_i)$ represents the contribution rate of the i-th branch to the abnormal state of the power grid, and H represents the abnormal level of the whole power grid from the perspective of entropy. Additionally, the smaller and more uniform the change in branch power, the smaller the corresponding entropy index H. On the contrary, the larger the side path rate of each branch and the more uneven the change in each branch, the larger the corresponding entropy index H.

4.3. Comprehensive Indicators of Power Grid Status under Large-Scale Wind Power Access

Under the large-scale access of wind power, the power change in wind power and the power change in each branch in the power grid reflect the state of the whole power grid. Therefore, this section synthesizes the indicators mentioned in Sections 4.1 and 4.2 to construct the power grid status analysis indicators under large-scale wind power access, as shown in Formula (15):

$$S = W/H \tag{15}$$

As mentioned above, the more stable the grid state of large-scale wind power access, the greater the s value. On the contrary, the greater the change in wind power, the greater the degree of abnormality and the smaller the s value.

4.4. Description of Power Grid State Trajectory Based on Sliding Time Window

Combining the state evaluation index of the power grid with the sliding time window can evaluate the state of the power grid and depict the operation track of the power grid in real time. When the power grid structure is determined, the number of fans P and the number of branches Q of the power grid are also determined. The value of n is determined by $c_1 = p/n \in (0,1)$ and $c_2 = q/n \in (0,1)$ where n is the width of the time window.

Among the N columns of sampling values, there is one column of current time data and $n-1$ column of historical data. Let the step size of the sliding time window be 1, that is, one column of historical data is moved out and one column of data at the current time is moved in at each sampling time.

With the movement of time, in each sliding time window, the power grid state evaluation index based on the wind power matrix, the power grid state evaluation index based on the branch power matrix and the comprehensive power grid state index under the large-scale wind power access are calculated, respectively, and the response is analyzed.

5. Example Analysis

In this paper, based on reference [18], the 33 node distribution system was improved to verify the effectiveness of the method proposed in this paper, as shown in Figure 2. Among them, the line parameters of the original system remain unchanged, so that the upper limit of the active output of the wind turbine was 0.6 MW, and the range of the reactive output was $-0.02{\sim}0.06$ Mvar. We installed 10 groups of fans at node 10 and 10 groups of fans at node 24. In the process of simulation verification, this paper built the model based on PSASP. When the wind speed increases and single-phase grounding occurs at node 6, the proposed indexes were analyzed. The simulation sampling frequency was 100 Hz and the simulation duration was 5S.

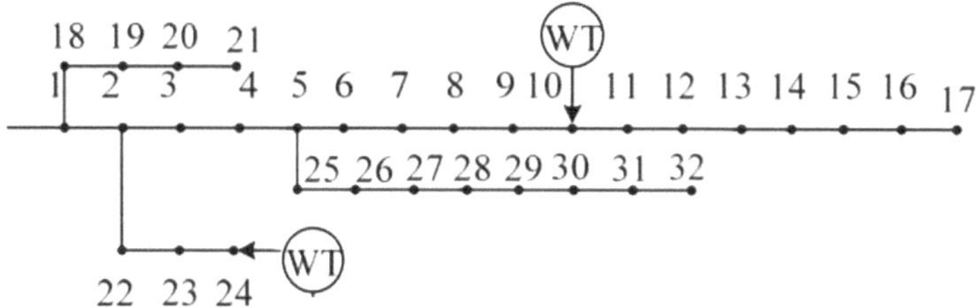

Figure 2. Improved IEEE-33 node power distribution system.

The change in wind speed will cause the change in wind turbine output, and the change in wind turbine output will affect the change in power flow distribution in the power grid, thus causing the change in the power grid state. This section assumes that the wind speed slowly increases from 6 to 10 m/s. The change in wind speed during simulation is shown in Figure 3. That is, at 0–1 s, the wind speed is maintained at about 6 m/s; at 1–3 s, the wind speed increases linearly to 10 m/s; at 3–5 s, the wind speed is kept at 10 m/s. This paper was based on the above simulation. We collected the corresponding wind power output data and branch power data, constructed the corresponding wind power matrix and branch power matrix and analyzed the power grid status.

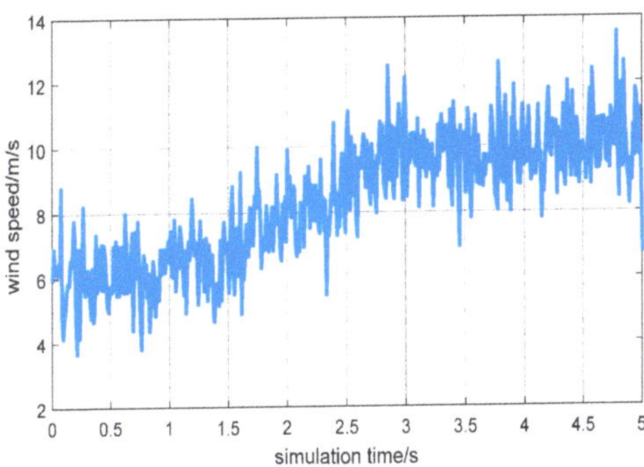

Figure 3. Schematic diagram of wind speed change during simulation.

5.1. Analysis of Power Grid Status Evaluation Index Based on Wind Power Matrix

This section uses the power data of 10 wind turbines at node 10 and 10 wind turbines at node 24 in the simulation stage to construct a wind power matrix with dimension of 20. Let the width of the sliding time window be n = 50 based on $c_1 = p/n \in (0,1)$. The width of the sliding time window in the branch power matrix is also 50. Figure 4 shows the wind power change level index based on the ring law and the wind power change level index based on the M-P law and the comprehensive index of the wind power change rate. Among them, C = 1 in the comprehensive index of the wind power change rate. (The weight preference of MSR and V can be selected according to the actual situation.) Since the sliding time window width is taken as n = 50, the values of various indicators start from t = 0.5.

As shown in Figure 4a, MSR can accurately reflect the change level in wind power. At t = 0–1 s, MSR remained at about 0.9 and was relatively stable in the process. It correctly shows the situation that the wind power changes slightly. When t = 1.01 s, MSR drops obviously. Furthermore, in the stage of t = 1.01–4 s, MSR is in a vibrating state, which corresponds to the wind speed gradually increasing in this stage. It is worth noting that the wind speed remains at 10 m/s from t = 3 s, but the MSR does not start to recover to a relatively stable state until t = 4 s. It shows that the wind speed change at this stage has a great impact on the wind power output. Although the wind speed has been maintained at about 10/s in the stage of t = 3–4 s, the changes in power flow and wind power caused by the changes in wind speed continue until t = 4 s.

The index V describes the variation level of wind power in the grid from the angle of the deviation degree between the spectral distribution and the M-P law. As shown in Figure 4b, when t = 0.5–1, the value of V is kept at about 0.01, because at this stage, the wind speed is kept at 6 m/s, the wind power is kept at a certain level with little change and the value and distribution of the constructed wind power matrix do not change much. Therefore, the corresponding spectral distribution is close to the M-P law, so the V value is small. Starting from t = 1.07 s, the value of V increases abruptly and oscillates, which accurately reflects the change process in wind power level.

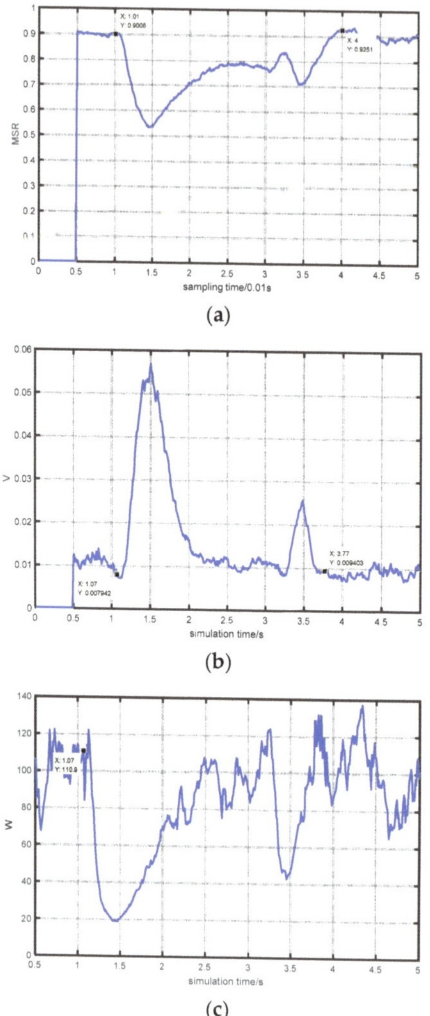

Figure 4. Change trend of power grid status evaluation index based on wind power matrix. (**a**) MSR change trend. (**b**) V change trend. (**c**) W change trend.

The specific wind power change level indicators MSR and V can be found as follows:

(1) MSR suddenly decreased from t = 1.01 s, while V suddenly increased from t = 1.07 s, indicating that the sensitivity of MSR is greater than V. MSR has a continuous value decreasing process during the change in wind speed, while V is similar to the level when the wind power does not change in some periods, such as t = 2.5–3.2 s and t > 3.77 s.

(2) MSR and V have certain correspondence. In addition to sensitivity, both of them fluctuate near the change in wind speed (if MSR is a significant drop and V is a significant increase). Both of them exhibit oscillation at the stage of increasing wind speed and recovered to the level of wind speed of 4 m/s in the later stage of simulation.

The indicator w combines MSR and V to describe the change level in wind power from a comprehensive perspective. That is, when t = 0.07 s, the index began to drop significantly, and it showed a relatively volatile situation in the next stage. It is further explained that the change in wind speed causes the change in wind power, and the change does not stop because of the stopping of the change in wind speed but causes the change in power flow of the power grid and further causes the change in the output of the wind turbine.

5.2. Analysis of Power Grid Status Evaluation Index Based on Branch Power Matrix

In this section, the branch power matrix was constructed with the branch power of 43 branches in the IEEE-33 node, and the width of the sliding time window was n = 50 based on $c_2 = q/n \in (0,1)$. Figure 5 is the entropy index based on this paper.

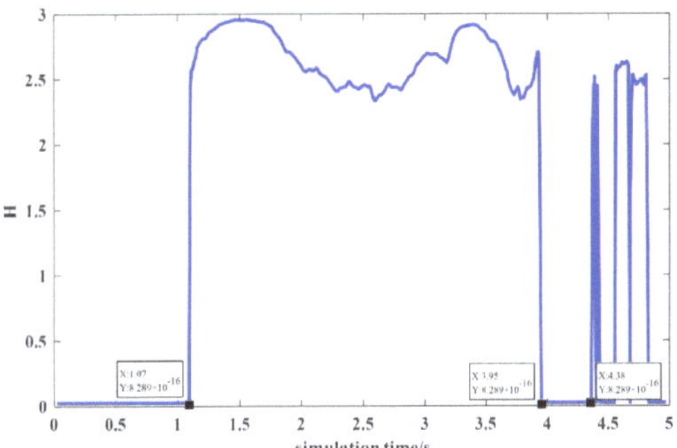

Figure 5. Change trend of power grid status evaluation index based on branch power matrix.

As shown in Figure 5, the proposed entropy index combines the concept of entropy with the high-dimensional random matrix theory, and correctly reflects the change in branch power in the process of wind speed change. When t = 0–1.07 s, the index H remains around 0. This is because at this stage, the wind speed change rate is small, the power flow distribution of the power grid is also small, the power change in each branch is small and the power change distribution is relatively uniform (that is, the power change in each branch is small). At the stage of t = 1.07–3.95 s, the change in wind speed causes the change in wind power, and then causes the change in power flow of the power grid, so that the branch power changes. From t = 3.96–4.37 s, the entropy index gradually decreases to 0, and then increases from t = 4.38 s, which indicates that at the stage of t = 3.96–5 s, the change in wind power makes the state of the power grid change.

In order to further verify the reaction ability of the entropy index proposed in this paper to the change in branch power, Figure 6 describes the change in entropy of each branch power in the whole simulation process (i.e., Equation (14)). It can be seen from Figure 6 that during the period of 0–1 s, the change entropy of the power of each branch is almost zero, which corresponds to the relatively stable wind speed and power grid state at this stage.

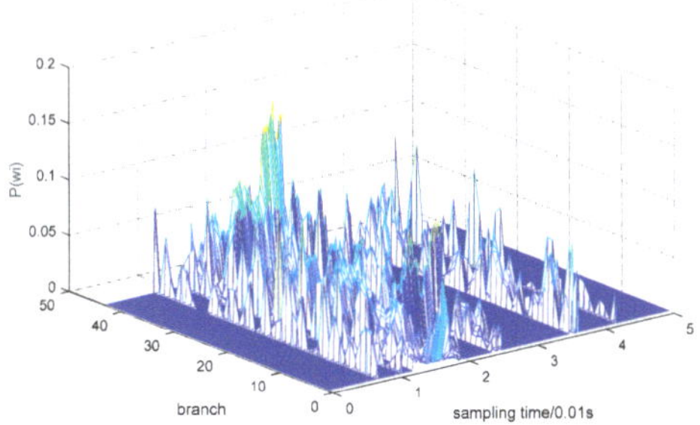

Figure 6. Change in power entropy of each branch.

5.3. Comprehensive Index Analysis of Power Grid Status under Large-Scale Wind Power Access

Based on the calculation results in Sections 4.1 and 4.2, this section analyzes the comprehensive indicators of power grid status under large-scale wind power access, and the results are shown in Figure 7.

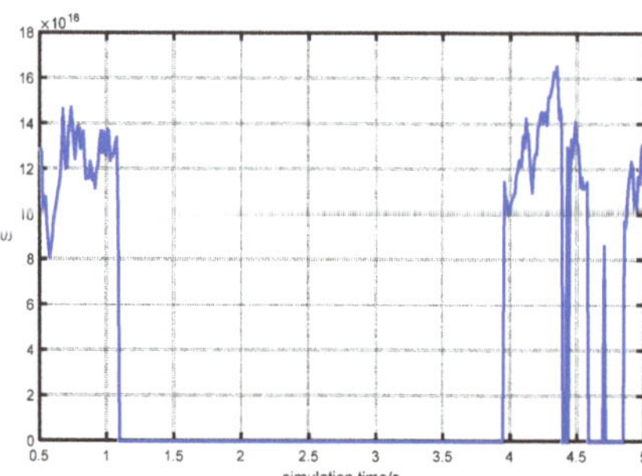

Figure 7. Change trend of comprehensive indicators.

The comprehensive index s proposed in this paper can accurately reflect the change in the power grid state when large-scale wind power is connected. At the stage of t = 0.5–1.1 s and t > 3.9 s, the value of index s is large, which corresponds to the wind speed with a small change rate. At the stage of t = 1.1–3.9 s, the index s is kept at a small value, which corresponds to the wind speed that changed in this stage. It can be seen that the comprehensive index takes into account the changes in wind power and branch power at the same time and describes the power grid status under large-scale wind power access from a more comprehensive perspective.

6. Conclusions

In this paper, the wind power matrix and branch power matrix are, respectively, constructed based on the measured big data, and the corresponding indexes are constructed to analyze the power grid state of large-scale wind power access. The results demonstrated that the proposed indexes can analyze the state of the power grid under large-scale wind power access from different angles. In the application process, the appropriate indexes can be selected according to the needs. Meanwhile, all the indexes have good sensitivity. The wind speed change set in this paper started from t = 1.01 s, and the proposed index changed significantly from 1.1 s at the latest, and the delay time was 0.09 s, which verifies the sensitivity of the proposed index. Combining the proposed index with a sliding time window allows the real-time state analysis of the power grid under large-scale wind power access to be performed.

The work of this paper provides support for power grid state analysis under large-scale wind power access. The follow-up work will further consider the factors affecting the power grid status under large-scale wind power access and construct more comprehensive power grid status analysis indicators to provide decision support for operators.

Author Contributions: Conceptualization, H.M.; Methodology, D.L.; Software, D.L.; Validation, Y.K.; Formal analysis, Y.K.; Investigation, H.L.; Resources, H.L. and X.J.; Data curation, K.C.; Writing—original draft, K.C. All authors have read and agreed to the published version of the manuscript.

Funding: This work was supported by State Grid Corporation, China through the Project No.: 4000-202222070A-1-1-ZN.

Data Availability Statement: This research data is unavailable due to privacy or ethical restrictions.

Conflicts of Interest: The authors declare no conflict of interest.

References

1. Yousefi, H.; Montazeri, M.; Rahmani, A. Techno-economic Analysis of Wind Turbines Systems to Reduce Carbon Emission of Greenhouses: A Case Study in Iran. In Proceedings of the 7th Iran Wind Energy Conference (IWEC2021), Shahrood, Iran, 17–18 May 2021; pp. 1–4. [CrossRef]
2. Yan, J.; Zhang, H.; Liu, Y.; Han, S.; Li, L.; Lu, Z. Forecasting the High Penetration of Wind Power on Multiple Scales Using Multi-to-Multi Mapping. *IEEE Trans. Power Syst.* **2018**, *33*, 3276–3284. [CrossRef]
3. Le, T.; Heo, S.; Kim, H. Toward Load Identification Based on the Hilbert Transform and Sequence to Sequence Long Short-Term Memory. *IEEE Trans. Smart Grid* **2021**, *12*, 3252–3264. [CrossRef]
4. Zheng, X.; Zeng, Y.; Zhao, M.; Venkatesh, B. Early Identification and Location of Short-Circuit Fault in Grid-Connected AC Microgrid. *IEEE Trans. Smart Grid* **2021**, *25*, 145–156. [CrossRef]
5. Ukaszewski, P.; Nogal, U.; Ukaszewski, A. The Identification of Travelling Waves in a Voltage Sensor Signal in a Medium Voltage Grid Using the Short-Time Matrix Pencil Method. *Energies* **2022**, *15*, 456–467.
6. Yuyang, G.; Chao, Q.; Kequan, Z. A Hybrid Method Based on Singular Spectrum Analysis, Firefly Algorithm, and BP Neural Network for Short-Term Wind Speed Forecasting. *Energies* **2016**, *9*, 23–42.
7. Chen, S.; Ai, Y.; Shi, Y.; Zhao, S.; Liu, J. A Novel Online Inductance Identification of Traction Network and Adaptive Control Method for Cascaded H-Bridge Converter. *IEEE Trans. Power Electron.* **2022**, *37*, 33–45. [CrossRef]
8. Wang, N.; Zhao, F. An Assessment of the Condition of Distribution Network Equipment Based on Large Data Fuzzy Decision-Making. *Energies* **2020**, *13*, 197. [CrossRef]
9. He, X.; Qiu, R.C.; Ai, Q.; Chu, L.; Xu, X.; Ling, Z. Designing for Situation Awareness of Future Power Grids: An Indicator System Based on Linear Eigenvalue Statistics of Large Random Matrices. *IEEE Access* **2016**, *4*, 3557–3568. [CrossRef]
10. Gheewala, S.H. Environmental Sustainability of Waste Circulation Models for Sugarcane Biorefinery System in Thailand. *Energies* **2022**, *15*, 9515. [CrossRef]
11. Saldaña-González, A.E.; Sumper, A.; Aragüés-Peñalba, M.; Smolnikar, M. Advanced Distribution Measurement Technologies and Data Applications for Smart Grids: A Review. *Energies* **2020**, *13*, 3730. [CrossRef]
12. Bao, A.R.H.; Liu, Y.; Dong, J.; Chen, Z.P.; Chen, Z.J.; Wu, C. Evolutionary Game Analysis of Co-Opetition Strategy in Energy Big Data Ecosystem under Government Intervention. *Energies* **2022**, *15*, 2066. [CrossRef]
13. Wang, H.; Wang, B.; Luo, P.; Ma, F.; Zhou, Y.; Mohamed, M.A. State Evaluation Based on Feature Identification of Measurement Data: For Resilient Power System. *CSEE J. Power Energy Syst.* **2022**, *8*, 983–992. [CrossRef]
14. Xie, J.; Ma, Z.; Dehghanpour, K.; Wang, Z.; Wang, Y.; Diao, R.; Shi, D. Imitation and Transfer Q-Learning-Based Parameter Identification for Composite Load Modeling. *IEEE Trans. Smart Grid* **2021**, *12*, 1674–1684. [CrossRef]
15. Wang, B.; Wang, H.; Zhang, L.; Zhu, D.; Lin, D.; Wan, S. A data-driven method to detect and localize the single-phase grounding fault in distribution network based on synchronized phasor measurement. *EURASIP J. Wirel. Commun. Netw.* **2019**, *2019*, 195. [CrossRef]
16. Ling, Z.; Qiu, R.C.; He, X.; Chu, L. A New Approach of Exploiting Self-Adjoint Matrix Polynomials of Large Random Matrices for Anomaly Detection and Fault Location. *IEEE Trans. Big Data* **2021**, *7*, 548–558. [CrossRef]
17. Liu, M.; Qiu, X.; Zhang, Z.; Zhao, C.; Zhao, Y.; Zhang, K. Multi-objective reactive power optimization of distribution network taking into account the correlation of wind and solar output. *Grid Technol.* **2020**, *44*, 1892–1899. [CrossRef]
18. Ma, H.; Wang, B.; Gao, W.; Liu, D.; Sun, Y.; Liu, Z. Optimal Scheduling of an Regional Integrated Energy System with Energy Storage Systems for Service Regulation. *Energies* **2018**, *11*, 195. [CrossRef]

Article

Interpretable Hybrid Experiment Learning-Based Simulation Analysis of Power System Planning under the Spot Market Environment

Wei Liao [1], Yi Yang [2,*], Qingwei Wang [1], Ruoyu Wang [1], Xieli Fu [1], Yinghua Xie [1] and Junhua Zhao [2]

[1] Shenzhen Power Supply Co., Ltd., China Southern Power Grid, Shenzhen 518000, China; liaowei@sz.csg.cn (W.L.); wangqw3377@163.com (Q.W.); wangruoyu@sz.csg.cn (R.W.); xlqtzczyyx@163.com (X.F.); xieyinghua@sz.csg.cn (Y.X.)

[2] Shenzhen Institute of Artificial Intelligence and Robotics for Society, Shenzhen 518000, China; zhaojunhua@cuhk.edu.cn

* Correspondence: yangyi2898@gmail.com

Abstract: The electricity spot market plays a significant role in promoting the self-improvement of the overall resource utilization efficiency of the power system and advancing energy conservation and emission reduction. This paper analyzes and compares the potential impacts of spot market operations on system planning, considering the differences between planning methods in traditional and spot market environments through theoretical analysis and model comparison. Furthermore, we conduct research and analysis on grid planning methods under the spot market environment with the goal of maximizing social benefits. Unlike the pricing approach based on historical price data in traditional market simulation processes, a data-driven approach that combines experimental economics and machine learning is proposed, specifically using mixed empirical learning to simulate unit bidding strategies in market transactions. A simulation model for electricity spot market trading is constructed to analyze the performance of the planning results in the spot market environment. The case study results indicate that the proposed planning methods can enable the grid to operate well in the spot market environment, maintain relatively stable nodal prices, and ensure the integration of a high proportion of clean energy.

Keywords: electricity spot market; system planning; social welfare; hybrid experimental learning; data-driven

Citation: Liao, W.; Yang, Y.; Wang, Q.; Wang, R.; Fu, X.; Xie, Y.; Zhao, J. Interpretable Hybrid Experiment Learning-Based Simulation Analysis of Power System Planning under the Spot Market Environment. *Energies* **2023**, *16*, 4819. https://doi.org/ 10.3390/en16124819

Academic Editor: Tomonobu Senjyu

Received: 26 April 2023
Revised: 2 June 2023
Accepted: 12 June 2023
Published: 20 June 2023

1. Introduction

Recently, referring to the experience of electricity market construction in developed countries such as the United States and the United Kingdom, China is also proactively facilitating the reform of power marketization. In the overall reform framework, the spot market, as a key connection part between the medium- and long-term electricity transactions and the real-time operation of the power grid, plays the role of restoring the commodity attributes of electricity, guiding the discovery of power prices, and promoting the optimal allocation of resources [1]. Compared with the power system operating conditions in the traditional centralized dispatching pattern, in the electricity spot market, the power system planning will be affected by several uncertain factors such as node electricity price fluctuation [2] and unit output adjustment [3]. Whether the existing power grid planning scheme can maintain a good operation state in a market-oriented environment needs to be verified urgently.

As a part of electricity market-oriented reformation, the electricity spot market has extensive advantages, such as encouraging demand response [4], promoting market competition [5], and optimizing resource allocation [6]. In recent studies focusing on the spot market, there is a growing interest in the involvement of virtual power plants (VPPs) and the concept of multimarket coupling. VPPs have the potential to generate revenue or mitigate risks through spot price arbitrage or frequency response support. Addressing the coordinated operation challenges of an active distribution network with multiple VPPs participating in a joint energy-reserve market, a two-level model was proposed in [7]. Furthermore, the authors of [8] introduced an optimal bidding strategy for VPPs in the frequency control ancillary service market, along with a joint bidding strategy for VPPs collaborating with wind farms. Employing a game-based approach, the cooperative benefits of both parties are maximized by ensuring fair benefit distribution. Additionally, the authors of [9] presented an analytical method for accurately calculating the cost of VPPs.

Regarding multimarket coupling, this concept can be categorized into various types based on different objectives, such as multilevel market coupling, multi-energy market coupling, or multiregional market coupling. To integrate wholesale and local markets while coordinating transmission system operators, distribution system operators, and distributed energy resources, a decentralized market trading strategy was proposed in [10]. The authors of [11] introduced a game theory-based analysis method to evaluate the impact on the multiarea electricity market through a coordinated transaction scheduling mechanism. To facilitate multi-energy collaborative optimization within local energy markets, the authors of [12] devised an indirect multi-energy transaction approach.

Moreover, market pricing and operating strategies have gained significant attention recently. The authors of [13] proposed an optimization strategy that incorporates thermal energy storage systems and concentrated solar power plants, aiming to stabilize and reduce electricity prices in the spot market. In the Australian national electricity spot and emergency reserve market, the authors of [14] presented a framework for analyzing battery operation and investigating different operating strategies across various states. Additionally, the authors of [15] proposed a selective learning scheme for bidding in the electricity spot market, while the authors of [16] designed a battery operating model that maximizes financial returns while considering battery cycle life costs and capacity degradation.

In the spot market, the issue of generation unit bidding strategy is one of the representative modeling problems in trading behavior. Bidding strategy greatly impacts multiple system state parameters including generation unit output, nodal electricity prices, and system power flows. At the same time, bidding strategy is also constrained by system safety constraints such as generator rated output and transmission line capacity. Traditional models for bidding strategy analysis mainly include experimental-based and machine learning-based methods. Experimental-based bidding strategy analysis models typically use optimization methods or game theory from economics. A stochastic optimization method was proposed in [17] to evaluate the strategies of agents in the market. They proposed a multiagent bidding strategy based on game theory to compute the market Nash equilibrium [18]. In addition, with the development of computer science, machine learning methods, particularly reinforcement learning, have been widely applied to modeling trading behaviors [19]. In the aforementioned methods, experimental-based analysis often relies on many assumptions and may not accurately reflect the true behavioral patterns of participants, while machine learning methods, due to their "black box" nature, have limitations in their application to engineering problems.

In the realm of power system planning, significant attention has been directed toward the transition to decarbonized systems and the optimization of integrated system coupling. The authors of [20] introduced the concept of optimizing emission reduction per unit cost, effectively balancing the tradeoff between cost and emission reduction in the planning process. Meanwhile, the authors of [21] proposed a method for collaborative planning of the electricity system and electric vehicle transportation network, aiming to enhance system flexibility.

Furthermore, in terms of verifying the effectiveness of planning schemes, the literature employs the empirical cumulative distribution function to assess the flexibility of the planning outcomes [22]. Addressing the grid planning process while considering time-varying or spurious electrical loads, the authors of [23] investigated the impact of grids with different sizes. Additionally, the authors of [24] utilized EnergyPLAN to model real-world data and validate planning schemes.

At present, most of the research related to the impact of the spot market on power system planning is still based on theoretical analysis and small-scale simulation. However, in the actual system operation process, more stakeholders are involved, and the system structure is more complicated. In addition, as the scale of the power system continues to enlarge, whether the planning scheme can maintain reliability and an economic requirement under a market-oriented environment still needs to be verified. This paper investigates the differences in planning methods between traditional and market environments through a comparison using mathematical models. Taking the real power system of southern China as an example, we study and analyze the actual performance of the planning schemes obtained through the proposed grid planning method in the electricity spot market environment. Meanwhile, using the hybrid experimental learning (HEL) approach, we simulate the bidding strategies of generating units during the simulation process. Lastly, through market simulation in the spot market environment, we analyze specific indicators, including the economic benefits, environmental benefits, and system reliability of various market entities. Compared to the existing research, the major contributions can be summarized as follows:

1. The analysis of the market environment impact in this paper is grounded on the actual power grid conditions in southern China. Unlike traditional theoretical analysis or benchmark system simulations, this research closely aligns with real system operating conditions. It encompasses more intricate network structures and operational equipment, thereby offering precise guidance for future system planning.
2. This paper utilizes the HEL approach to simulate generator bidding strategies. In contrast to the conventional method of predicting generator bidding prices solely on the basis of historical data, this approach incorporates human irrational factors. Moreover, computer agent quotations driven by data-driven techniques help maintain prices within a reasonable range.
3. Through theoretical analysis and model comparisons, this paper examines and contrasts the disparities between traditional Chinese planning methods and market environmental planning methods. By combining the results of numerical cases, the economic and environmental benefits, as well as the reliability and other indicators for each market participant, are thoroughly analyzed.

The remainder of this paper is organized as follows: Section 2 introduces the possible impact of the spot market on grid planning in detail; Section 3 presents the hybrid experimental learning method; Section 4 constructs the Shenzhen China power grid planning schemes in the PROPHET 2018 software, and then analyzes the possible impact of the implementation of the spot market; Section 5 presents the conclusions.

2. Analysis of Impact on Power System Planning under Spot Market Environment

2.1. Impact Overview

Improving the efficiency and effectiveness of system operation is one of the main motivations for establishing a clean and well-equipped power market. For the current power market dispatch mode used in most countries, the system operator usually takes the power generation marginal cost of each unit as the priority parameter and calls the unit with the lowest marginal cost first. If the unit cannot meet the current demand, the unit with the next lowest marginal cost will be called in sequence. In the process of system optimization, the resources of units with high power generation efficiency will be dispatched preferentially. Therefore, due to the extremely low marginal cost of renewable energy entities, the utilization rate of high-efficiency units and renewable energy can be further improved through market dispatch, achieving the purpose of minimizing the system costs and emissions. On the other hand, these surpluses of discretionary costs can be distributed through the system, and part of them can be used to reduce the retail electricity prices of consumers, thus achieving a win–win situation.

In the spot market environment, locational marginal price (LMP), also named nodal price, serves as a critical indicator for assessing the effectiveness of the system planning scheme. LMP is influenced by various factors, such as supply and demand dynamics, transmission constraints, generation costs, and market regulations. It reflects the real-time equilibrium between electricity supply and demand at a specific location. Within the system, the occurrence of excessively high LMP can indicate the presence of underlying issues, such as escalating energy prices, transmission line congestion, and insufficient power supply capacity. Indeed, the spot market possesses commendable self-regulation capabilities. When consumers are unwilling to bear the high charges associated with the prevailing LMP, the LMP will decrease in response to reduced demand. This self-regulating mechanism ensures that the price aligns with market dynamics. In addition to the inherent self-regulation of the market, there are other crucial factors that can impact market operations. These factors include the retirement of aging generators and the installation of renewable energy sources. The retiring of aging generators plays a vital role in enhancing system efficiency, reducing overall costs, and expediting the transition toward a low-carbon system. By removing outdated and less efficient units from the system, resources can be allocated more effectively, enabling the integration of cleaner and more sustainable energy sources. This transition toward renewable energy is essential for achieving long-term environmental goals and ensuring a more sustainable energy landscape.

To sum up, the introduction of the electricity spot market can effectively improve system efficiency and resource allocation and reduce system costs and overall carbon emissions. However, it should be noticed that the construction of the power spot market will have an impact on the existing system planning methods. As shown in the Figure 1, after the spot market operation, compared with the original planning method, the impact can be summarized in terms of power supply, node price, power flow, emissions, and total system cost. Specifically, the reason can be further attributed to the change of unit bidding strategy, large-scale grid connection of renewable energy, the retirement of old generations, demand response, and grid layout adjustment.

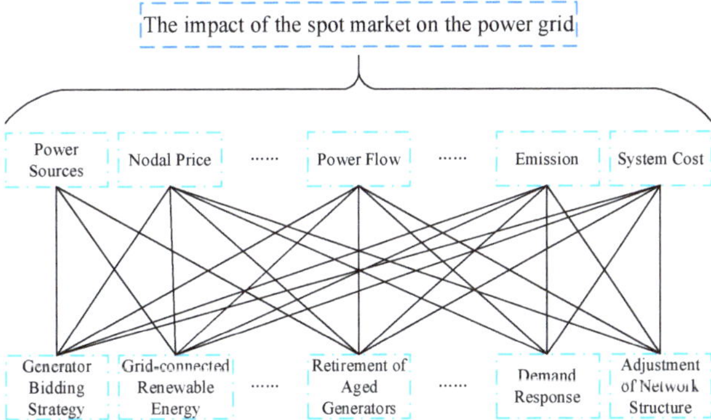

Figure 1. Impact of spot market on power grid planning.

2.2. Power Network Expansion Planning Method

Essentially, network expansion planning methods can be expressed as minimizing the investment cost of adding transmission lines. The mathematical model of the planning objective function is usually a linear equation related to the new link, which can be expressed as follows:

$$Min \sum_{i,j\in\Omega^{ij}} C_{ij}^L L_{ij}, \tag{1}$$

$$s.t. \ S_{ij} + P_i^G = D_i, \tag{2}$$

$$S_{ij} - B_{ij} \cdot \left(L_{ij}^{old} + L_{ij}^{L} \right) \cdot (\theta_i - \theta_j) = 0, \tag{3}$$

$$S_{ij} \leq \left(L_{ij}^{old} + L_{ij}^{L} \right) \overline{S_{ij}}, \tag{4}$$

$$C_{ij}^{L} = \pi^{L} \cdot l_{ij}, \tag{5}$$

$$0 \leq P_i^{G} \leq \overline{P_i^{G}}, \tag{6}$$

$$0 \leq L_{ij}^{L} \leq \overline{L_{ij}^{L}}, \tag{7}$$

$$L_{ij}^{L} \in integer, \tag{8}$$

$$i, j \in \Omega^{ij}, \tag{9}$$

where C_{ij}^{L} denotes the additional investment cost of the new power link ij, which is equal to the transmission line unit price π^{L} (CNY/km) times its length l_{ij}(km), L_{ij}^{L} and L_{ij}^{old} indicate the numbers of newly added lines and existing transmission lines, S_{ij} and B_{ij} represent the power flow and susceptance of the link ij, P_i^{G} and D_i denote the generation output and load of bus i, θ_i and θ_j represent the phase angle of bus i and j, and $\overline{(*)}$ indicates the upper limit of variables. The number of adding lines should be integers, and buses i and j belong to the set Ω^{ij}, which are shown in Equations (8) and (9).

Transmission network expansion models can help system operators plan the power grid efficiently. However, with the increasingly serious environmental problems, this model has been unable to meet the low carbon demand of power system planning. Therefore, under the spot market environment, the grid expansion planning model needs to consider more factors such as unit cost and renewable energy utilization, so as to improve the efficiency of the system and reduce the overall carbon emissions as much as possible.

2.3. Proposed Network Expansion Planning Method under Spot Market Environment

More factors need to be considered in the planning process under spot market operation, such as unit operating costs, external power purchase costs, and installation costs for transmission lines. The introduction of different factors makes the system optimization objectives and constraints more complex.

Objective functions:

$$C_{total} = C^{L} + C^{G} + C^{OS}, \tag{10}$$

$$C^{L} = \sum_{i,j \in \Omega^{ij}} \frac{\pi^{L} \cdot l_{ij} \cdot L_{ij}^{L}}{(1+d)^{Y-1}}, \tag{11}$$

$$C^{G} = \sum_{i \in \Omega^i} \sum_{Y} \frac{\left[a_i \left(P_i^{G} \right)^2 + b_i P_i^{G} + c_i \right] + \left(\lambda_i^{G} P_i^{G} \pi_i^{carbon} \right)}{(1+d)^{Y-1}}, \tag{12}$$

$$C^{OS} = \sum_{i \in \Omega^i} \sum_{Y} \frac{\pi_i^{OS} P_i^{OS}}{(1+d)^{Y-1}}. \tag{13}$$

Constraints:

$$\sum_{ij \in \Omega^{ij}} S_{ij} + P_i^{G} + P_i^{OS} + P_i^{PV} = D_i, \tag{14}$$

$$f(I_i) = \frac{\Gamma(\alpha + \beta)}{\Gamma(\alpha) + \Gamma(\beta)} \cdot \left(\left(\frac{I_i}{I_i^{max}}\right)^{\alpha-1} \left(1 - \frac{I_i}{I_i^{max}}\right)^{\beta-1} \right),$$ (15)

$$P_i^{PV} = \frac{A_i^{PV} \cdot \mu_i^{PV} \cdot I_i}{1000},$$ (16)

$$0 \leq A_i^{PV} \leq \overline{A_i^{PV}},$$ (17)

$$0 \leq P_i^{OS} \leq \overline{P_i^{OS}}.$$ (18)

Under the spot market-oriented environment, some factors (e.g., the line construction cost C^L, the unit generation cost C^G, and the external electricity purchasing cost C^{OS}) need to be further taken into consideration for the objective function, which is shown in Equation (10), where $\frac{1}{(1+d)^{Y-1}}$ indicates the net present value (NPV), referring to the time value of funds, i.e., converting investment costs throughout the planning cycle Y to current prices. The generation relevance cost model is shown in Equation (12), which includes both operation cost and environmental cost, where a_i, b_i, and c_i represent the generator operation cost coefficient. λ_i^G and π_i^{carbon} depict the generation emission factor and emission cost coefficient of generators. The external electricity purchasing cost is equal to the external input power P_i^{OS} multiplied by the spot market electricity price π_i^{OS}. In addition, the constraints such as system power flow equations and unit output upper limits are the same as mentioned above, i.e., Equations (3) and (4), and (6)–(9).

Compared with traditional planning models, the constraints of the new planning method change to the forms in Equations (14)–(18). The solar power output constrains are shown in Equations (15)–(17). For photovoltaic systems, the probability density function of solar radiation can usually be represented by the beta distribution, as shown in Equation (15), where Γ denotes the gramma function, α and β are parameters of Beta distribution, and I_i and I_i^{max} represent the solar radiation intensity and maximum intensity in the corresponding area. The solar power output is equal to the area of installed photovoltaic (PV) panels A_i^{PV} times the conversion efficiency u_i^{PV} times the solar radiation intensity.

Due to the clearing mechanism of the spot market, the main difference between the new planning method and the traditional approach is that the bidding price of various generators and the cost of purchasing external power need to be considered. The cost of the unit will greatly affect its bidding strategy in the spot market. Different bidding strategies lead to changes in the clearing results and eventually affect the network power flow.

Overall, in the traditional grid planning environment, the consideration of generating entities' costs and their market bidding prices is not necessary. The planning scheme primarily relies on assessing the wiring costs associated with different candidate lines. Conversely, in the spot market environment, the system needs to take into account additional market factors such as power generation costs and external power purchase costs. From a policy perspective, the spot market environment introduces a decentralized decision-making process, diminishing the role of the system operators in direct planning and resource allocation. Instead, market participants respond to price signals and make decisions on the basis of their economic interests. The key distinction between planning approaches in traditional and market environments lies in the level of market competition. Traditional approaches prioritize long-term planning, while spot market environments foster market competition and decision making based on prices.

3. Unit Pricing Strategy Model Based on HEL

3.1. HEL

HEL mainly consists of three modules: a human–machine hybrid experiment, a generative model, and an interpretation approach [25]. As shown in Figure 2, participants in hybrid experiments include humans and computer agents. Human participants need to understand how the mechanism works, calculate profits, determine decision ranges, and anticipate potential transaction outcomes. During the experiment, all participants are required to maximize the profits of the roles they represent with a serious attitude. Computer agents are divided into two categories: based on historical data and based on experimental data for training generative models. Agents based on historical data are driven by actual historical data and act as experienced exogenous participants, participating in hybrid experiments with humans in a certain proportion. Generative models generating agents on the basis of experimental data should not join in the first round due to a lack of data. The training data for generative models mainly come from human participants and are adjusted with a small amount of actual data.

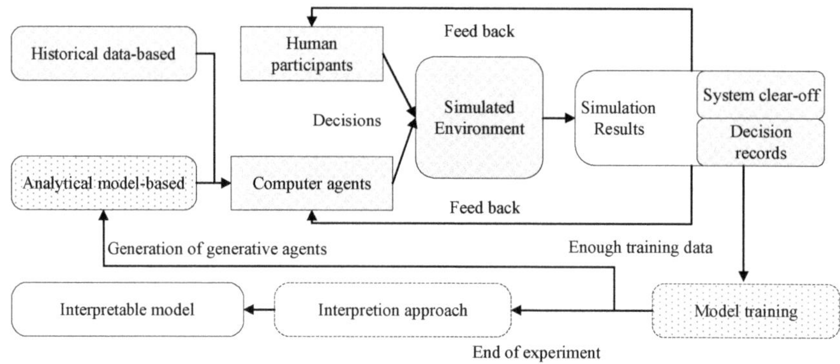

Figure 2. Framework of hybrid experimental learning.

After collecting sufficient data from several rounds of mixed experiments, the generation model is trained using machine learning methods such as generative adversarial networks (GANs). Once trained, the human participants can be replaced by the model. When a specified number of human participants are replaced by the generation model, data-driven behavioral modeling is completed, resulting in a set of transactional behavioral models obtained from the experimental data. The advantages of mixed experimental learning are that the generation model ensures decision-making expertise while reducing the cost of economic experiments. Furthermore, due to the inexplicability of machine learning methods, which cannot be represented by explicit formulas, if some studies pursue the interpretability of behavioral models, interpretative methods can be employed after generation model training. If the purpose of mixed experiments is to simulate markets, explanatory methods and subsequent models can be omitted [25].

3.2. GAN

In this paper, a generative model's machine learning approach using GAN is employed. GAN is an unsupervised learning method that consists of two neural networks competing in a zero-sum game. It can avoid the high-dimensional challenges encountered in classical probability density estimation, such as kernel density estimation. The pricing strategy for the electricity spot market can be generated by GAN, with the following formula:

$$\min_{G}\max_{D} V(G, D) = E_X[D(X)] - E_Z[D(G(Z))], \tag{19}$$

where E_X and E_Z represent the sample expectation of real samples and noise data samples, respectively; $D(\cdot)$ and $G(\cdot)$ denote the discriminator network and the generator network, respectively.

4. Case Study

4.1. Setup

According to the characteristics of the Shenzhen power grid, the Shenzhen regional power transmission network model was constructed according to the 14th Five Year Plan of China. The market clearing model can be solved by PROPHET software. Various parameters such as the load, upper limit of generation output, and solar output are substituted into the established Shenzhen regional grid framework of different years. The 2022–2025 typical daily load of the Shenzhen area is shown in Figure 3, and the load data came from the real data of the nonpublic database. Each unit is optimized and dispatched in the spot market, and then the optimization results, including real-time scheduling strategy, node electricity price, and generation output, can be obtained. Through the simulation analysis of the current planning scheme, it can be verified whether a reliable power system operating state can be maintained under the spot market-oriented environment. The basic parameters of the system are shown in Table 1. In order to verify the effectiveness of the current planning scheme, the following cases are designed:

Case 1.1: The basic case with PV, considering the operation performance of Shenzhen power grid planning results in the spot market environment each year.

Case 1.2: The basic case without PV, considering the operation performance of Shenzhen power grid planning results in the spot market environment each year.

Case 2: The comparison case, where the external power supply is limited.

Case 3: The comparison case, where the energy prices are rising.

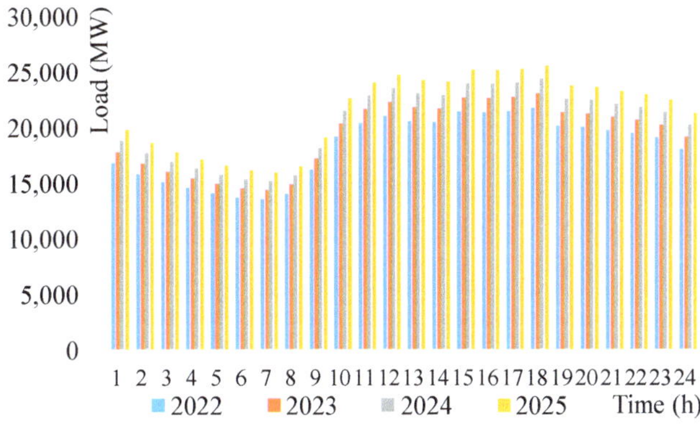

Figure 3. Shenzhen total load bar chart in each year.

Table 1. Basic parameters setting of system.

Parameter	Value	Parameter	Value
π^L	25,000 CNY/KM	π_i^{carbon}	46 CNY/ton
d	0.02	Y	5 years
a_i	0–0.02 CNY/MWh	b_i	50–200 CNY/MWh
c_i	0–500 CNY	π_i^{OS}	200–600 CNY/MWh

Among these cases, Case 1 serves as the base scenario for evaluating the operational performance of the existing system planning scheme in the spot market environment. It analyzes the node power price and various generator outputs to assess system performance according to different indicators. As the specific PV construction plan for the area where the system is situated has not been finalized, Case 1 is divided into two subcases: Case 1.1 and Case 1.2. These subcases compare system operation performance with and without PV, respectively, and analyze the impact of PV construction on system operation. To compare the system operating situations under various conditions, Case 2 and Case 3 are established. Considering the high dependence of the test system on the external power supply, Case

2 investigates system operation when the external power supply is limited. On the other hand, Case 3 considers the recent price fluctuations in energy supply for power generation. It examines system performance when energy prices rise.

Before commencing the experiment, it is essential to establish the basic parameter settings for the HEL approach, with the setting steps referring to [25]. Firstly, concerning the market rule parameterization, considering the operational rules of the specific region's market, each generating unit is permitted to report five price–quantity pairs to construct its own supply curve [25]. This implies that, for five different output levels, corresponding unit electricity prices are quoted. This five-stage quotation strategy forms the foundation for bidding prices at each time point throughout the 24 h day-ahead trading period and cannot be altered during the process. The simulation mandates that the prices for the five-stage strategy must be monotonic, with the upper and lower limits of the unit electricity price set at 1000 and 100 CNY/MWh, respectively. The bidding power associated with each stage of the strategy must be greater than or equal to 0, and the sum of the bidding powers across all five stages must not exceed the upper limit of the generator output capacity, measured in MW.

The clearing rules of the simulated market align with the operational rules of the Guangdong power spot market in China, cleared by LMP. Each simulation simulates a 24 h day-ahead trading period; once completed, the income of each entity is calculated.

Secondly, with regard to the computer agent, all computer agent settings in this study are derived from historical data, specifically from the actual data during the trial operation of the Shenzhen market [25]. The study utilizes GANs to generate agents on the basis of both historical data and generative agents trained using a combination of human experimentation and practice. For the generator of GAN, the training data are divided into two parts; part 1 comprises data accumulated from human participants engaged in repeated mixing experiments, while part 2 consists of historical data recorded from real generating entities. A total of 12,000 five-segment biddings are employed to train the generator. The training bids are generated through a mixture method that combines 120 human participants' experimental bids and 3000 historical bids in a 3:1 mixing ratio. This ratio means that, in this simulation, 9000 samples augment from experimental bids, and 3000 samples derive from historical bids. Furthermore, 20 GANs are trained during the experiments conducted in this study, and the output of the bidding strategy mentioned throughout the paper is randomly selected from the 20 GANs, ensuring a uniform distribution.

4.2. Case Study Results Analysis

The operating results of the system in Case 1.1 in each year are shown in Figure 4. It can be seen that the output of nuclear power and hydropower has remained stable in each year due to relatively low marginal costs. The output of other types of generators is similar to the overall load curve, with a clear peak-to-valley gap. In addition, the output ratios of generators for each year are shown in Figure 5a,b. It can be observed that the output of nuclear and hydro power has remained stable over the years. Moreover, the coal-fired electricity generation will show a decreasing trend from 2022 to 2025, while the trend of the gas generator power output will be opposite. This is due to the coal-to-gas project conducted in Shenzhen and the continuous commissioning of new gas-fired power units. Overall, the Shenzhen power system can still maintain a relatively high proportion of clean energy output in the spot market environment; even with the increasing load, it can still reduce the output ratio of high-energy-consuming coal-fired power units through methods such as coal-to-gas conversion.

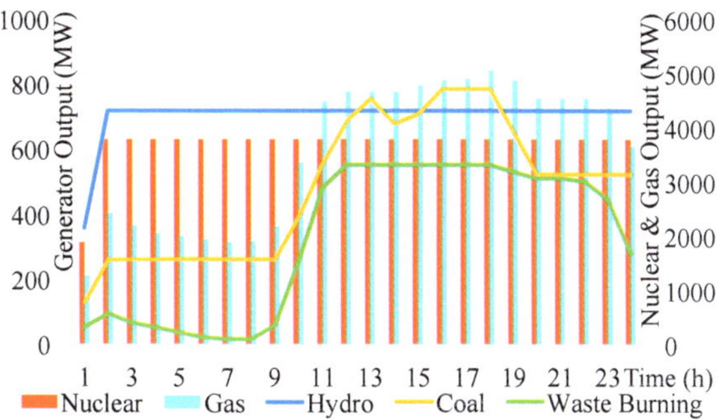

Figure 4. Power output results of various types of generator sets.

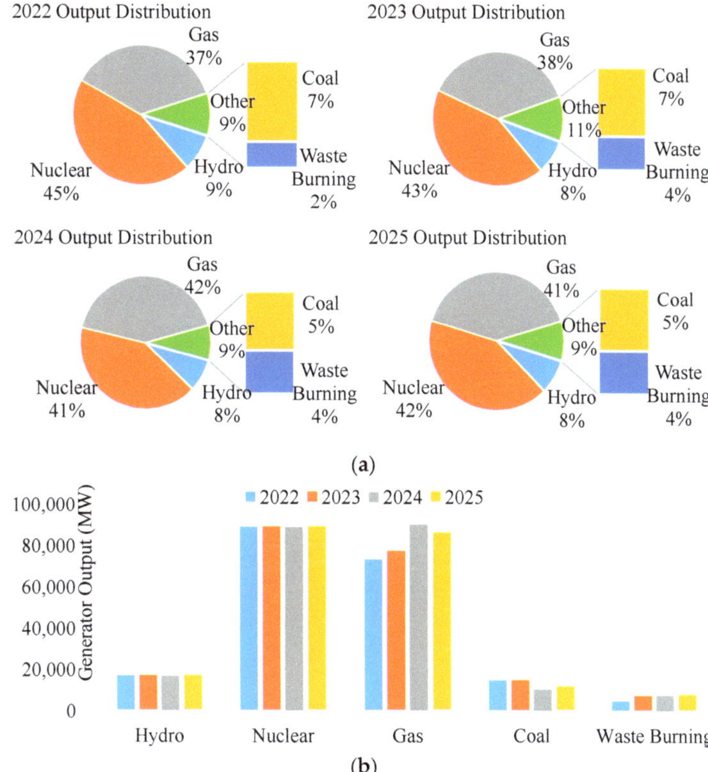

Figure 5. (**a**) Case 1.1 power output proportion of generator sets in each year. (**b**) Case 1.1 power output of generator sets in each year.

In the spot market, nodal electricity prices are an important indicator of system operation. As shown in Figure 6, nodal electricity prices in system simulation results are generally in the range of 400 CNY/MWh to 500 CNY/MWh. There was no significant upward trend in nodal electricity prices in different regions and years, and the average nodal electricity price in Nanshan District even showed a slight decrease over the years. Normally, if the transmission line planning is not adequate, nodal electricity prices in various regions should increase as the load increases year by year. However, the current experimental results reflect that the existing Shenzhen power grid planning for 2022 to 2025 can effectively cope with the increasing load year by year.

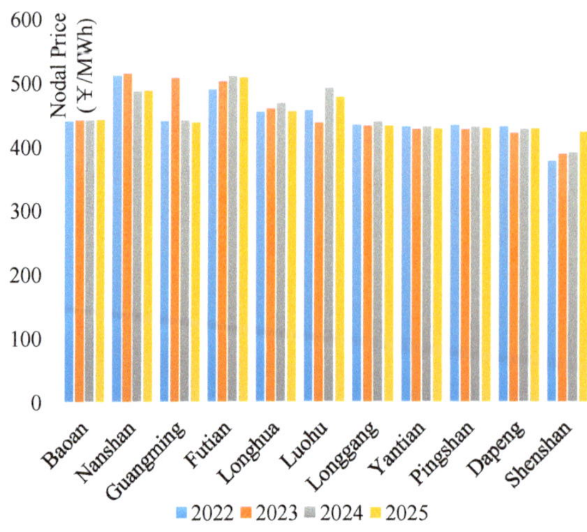

Figure 6. Average electricity nodal price in each district over the years.

Meanwhile, the geographical distribution of nodal electricity prices in different years shown in Figure 7 clearly corresponds to the distribution of loads in Shenzhen. Nanshan, Futian, and Luohu districts have relatively dense populations and concentrated electricity loads, resulting in higher nodal electricity prices in these areas. In addition, average nodal electricity prices in Guangming, Baoan, and Longhua areas are generally higher than those in Longgang, Pingshan, and Yantian areas. Overall, there is a clear trend of higher nodal electricity prices in the southern and eastern areas of Shenzhen compared to the northern and western areas.

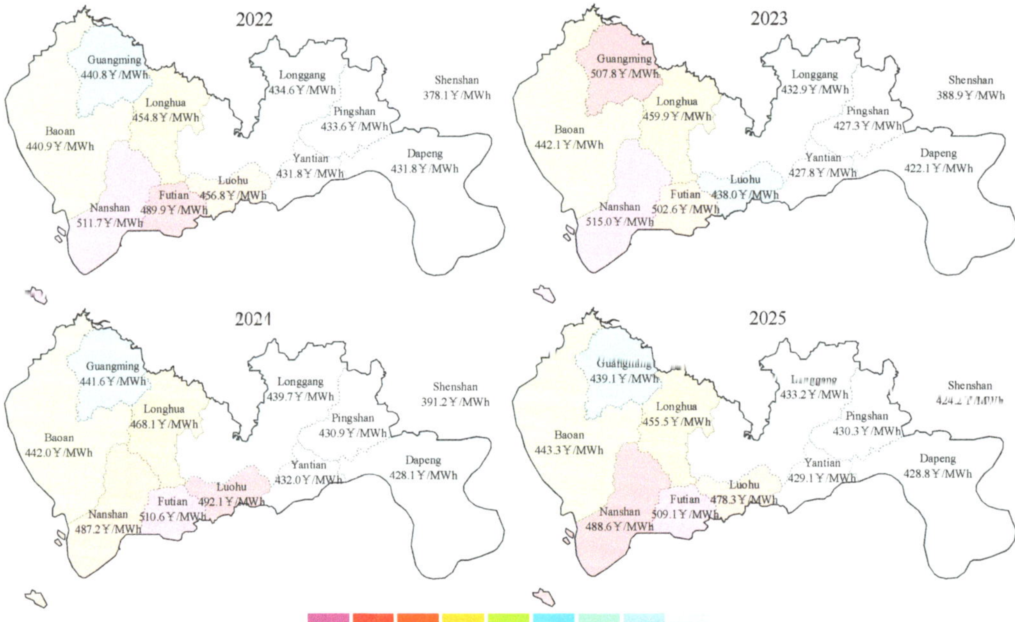

Figure 7. Geographical distribution of electricity nodal prices in different districts and years.

Table 2 presents the node electricity prices in different regions of Shenzhen under different PV distribution scenarios in 2025. Due to the low proportion of PV installed capacity in Shenzhen's total installed capacity and the relatively low level of PV capacity until 2025, the impact of different scenarios on node electricity prices is not significant (about 5 CNY/MWh). The impact of different PV allocation methods on node electricity prices is also negligible. Therefore, this paper adopts the line overload frequency as another indicator to measure PV distribution methods, and the comparison results are shown in Figure 8.

Table 2. Case 1.2 comparison of different PV distribution scenarios in 2025.

Nodal Price (CNY/MWh)	Case 1.1	Case 1.2 Equal Distribution	Case 1.2 Load-Based Distribution
Baoan	443.33	438.73	439.00
Nanshan	488.61	483.01	483.32
Guangming	439.11	437.21	437.07
Futian	509.06	503.74	504.03
Longhua	455.54	452.96	453.04
Luohu	478.34	472.85	473.17
Longgang	433.21	429.54	429.72
Yantian	429.07	425.86	426.00
Pingshan	430.33	425.47	425.70
Dapeng	428.77	424.00	424.19
Shenshan	424.23	419.37	419.64

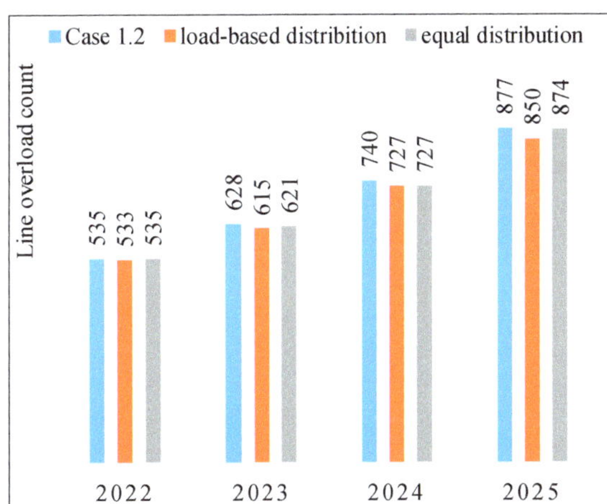

Figure 8. Line heavy load frequency in different cases.

Assuming that a line is considered overloaded when its actual operating power exceeds 80% of its rated capacity, the results show that the occurrence of line overloads is less frequent in cases where there is PV power output compared to cases where there is no PV power. Moreover, when the PV installation is distributed evenly among nodes on the basis of load size, the probability of line overloads is even lower. Therefore, for Shenzhen, allocating the PV installation on the basis of load proportion is a more optimal configuration. Additionally, since the impact of PV installation on system operation results is minimal, the subsequent simulations will no longer consider the influence of PV.

4.3. Simulation Results Analysis of Extreme Conditions of the System

In the pressure test scenario, i.e., Case 2, the impact of the bidding strategy on node electricity prices is not significant when the external restricted input proportion is low. However, when this proportion increases to 40%, the changes in node electricity prices in different regions of Shenzhen in each year are shown in Figure 9, indicating a decreasing trend in the magnitude of changes in node electricity prices with the growth of years. The years 2022 and 2023 show a more noticeable impact of changes in external power transmission on node electricity prices, with changes generally exceeding 4 CNY/MWh. In comparison, the changes in node electricity prices in 2024 and 2025 are mostly below 3 CNY/MWh, indicating that the adjustments in the grid planning scheme play a positive role in effectively reducing the impact of changes in external power transmission on node electricity prices in the Shenzhen grid system, and the stability of node electricity prices also confirms the rationality and safety of the grid planning strategy from the side.

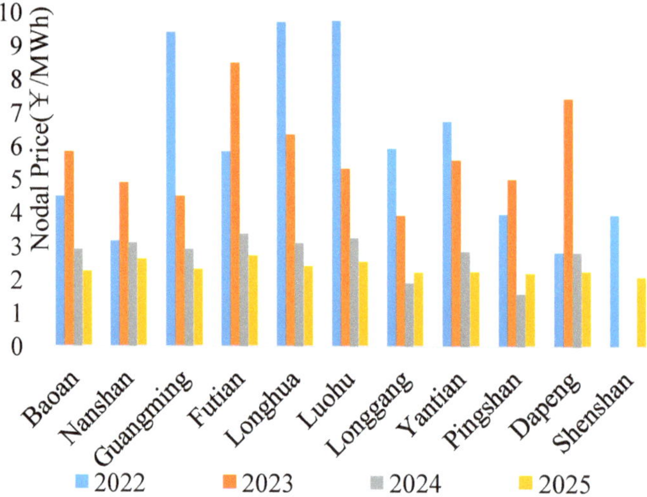

Figure 9. Electricity nodal price growth with 40% restriction on external supply.

In Case 3, as the prices of natural gas and coal energy increase, the bidding prices of the generated units also increase, and the changes in output proportions of different units in different years are shown in Figure 10. Generally, the output proportions of gas-fired and coal-fired units show a decreasing trend as the bidding prices increase, while the output proportion of hydroelectric units shows an increasing trend, and the output proportions of nuclear and waste power plants fluctuate little.

The bidding prices of thermal and gas-fired units generated through HEL are significantly affected by changes in fuel costs, with overall price increases due to cost increases. However, hydroelectric units have lower marginal costs; thus, their cost advantage is reflected when fuel costs rise, resulting in an upward trend in generation proportions. On the other hand, gas-fired and coal-fired units show a decreasing trend in generation proportions. Nuclear power units have extremely low overall bidding prices and are given priority in the dispatch process, resulting in the highest generation proportion with little impact from bidding price fluctuations. Similarly, waste power plants have a relatively low proportion and maintain a low output level, with little noticeable impact from bidding price fluctuations.

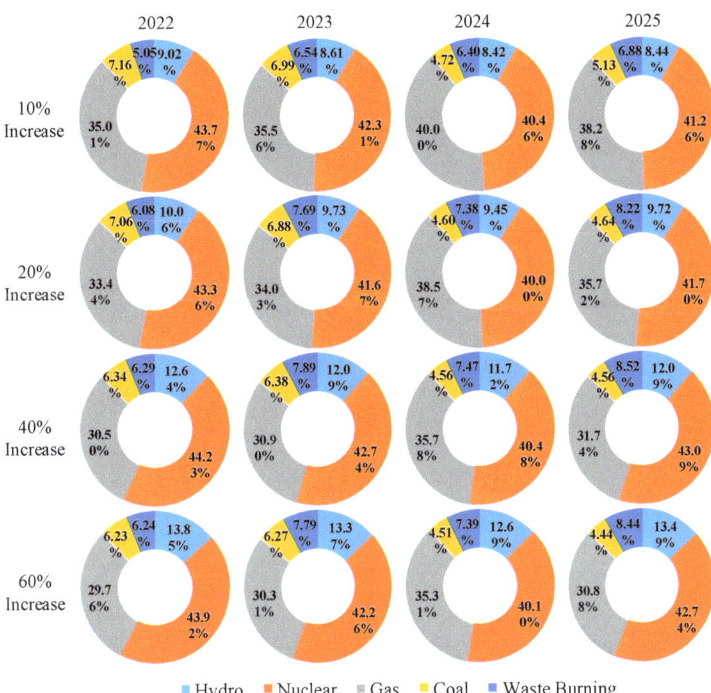

Figure 10. Case 3: the increase in the bidding price of different proportions, and the output ratio of each generator.

The changes in the city-wide average node electricity prices in each year due to bidding price fluctuations are shown in Figure 11. It can be observed that the fluctuations in unit bidding prices have a significant impact on node electricity prices, with a clear increase in city-wide average node electricity prices as the proportion of bidding price increases. This indicates that, under the current planning strategy, it is difficult to maintain stable node electricity prices when bidding prices increase due to changes in unit fuel costs. It is suggested that the situation could be improved by increasing the installation proportion of low-marginal-cost new energy sources such as solar and wind power.

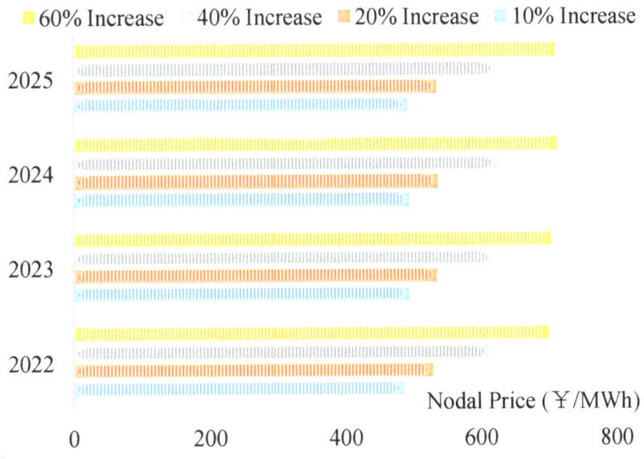

Figure 11. The bidding price growth of different proportions, and the average nodal price of Shenzhen.

5. Conclusions

This paper aimed to analyze and compare the effects of the spot market environment on existing planning strategies. Firstly, mathematical formulas were employed to elucidate the disparities in planning methodologies under varying conditions. Secondly, this paper adopted an HEL approach to generator bidding strategies under different scenarios, thereby simulating the bidding process more accurately according to actual trading prices.

Furthermore, a spot market simulation was designed on the basis of the Shenzhen regional power grid in this paper. From the simulation results, some conclusions can be drawn. Under the spot market environment, the nodal electricity price in the planned area can be approximately maintained within the range of 400–500 CNY/MWh. This result means that there are no significant disparities in nodal prices among different regions attributed to the absence of line congestion, thereby confirming the efficacy of the planning scheme. Additionally, there is almost the same output of various types of generators compared to previous years, and the proportion of clean energy is relatively high. These results indicate that HEL can effectively simulate the bidding strategies of different generator types in a real market environment. Moreover, when the current system faces a limited external power supply, it exhibits minimal impact due to its ample power generation capacity redundancy. Conversely, a high percentage of gas-fired power generation renders the system highly sensitive to energy prices.

This paper holds significant implications for the future development of spot markets in the Shenzhen region. In future research, it is necessary to delve deeper into the bidding strategy of renewable energy in the spot market, aiming to mitigate the potential risk of inadequate system reliability that may arise from the near-term centralized retiring of coal-fired generators. Additionally, as the spot market evolves, conducting further analysis on market simulations encompassing a broader range of generation types and larger system scales becomes crucial.

Author Contributions: Conceptualization, W.L., Q.W., R.W., X.F. and Y.X.; methodology, W.L., J.Z., Q.W., R.W., X.F., Y.X. and Y.Y.; software, Y.Y.; validation, W.L.; formal analysis, W.L., Q.W., R.W., X.F. and Y.X.; investigation, W.L. and Y.Y.; resources, W.L., Q.W., R.W., X.F. and Y.X.; data curation, W.L. and Y.Y.; writing—original draft preparation, W.L. and Y.Y.; writing—review and editing, W.L., J.Z. and Y.Y.; visualization, Y.Y.; supervision, W.L., Q.W., R.W., X.F. and Y.X.; project administration, W.L. and Y.Y.; funding acquisition, W.L., Q.W., R.W., X.F. and Y.X. All authors have read and agreed to the published version of the manuscript.

Funding: This research was funded by a Special Planning Project by China Southern Power Grid Shenzhen Power Supply Co., Ltd.

Data Availability Statement: Not applicable.

Acknowledgments: This paper was supported in part by the Shenzhen Institute of Artificial Intelligence and Robotics for Society (AIRS).

Conflicts of Interest: The authors declare no conflict of interest.

Nomenclature

Abbreviation

VPPs	Virtual power plants
HEL	Hybrid experimental learning
LMP	Locational marginal price
NPV	Net present value
GANs	Generative adversarial networks
PV	Photovoltaic

Indices and Sets

i, j, Ω^{bus}	Index and set of bus locations
Y	Index of years

Parameters

C_{ij}^{L}	Additional investment cost of the transmission lines
π^{L}	Unit price of transmission lines
l_{ij}	Length of transmission lines
L_{ij}^{old}	Existing transmission lines
B_{ij}	Susceptance of the transmission lines
D_i	Load of bus i
$\overline{(*)}$	Upper limit of variables
d	Discount rate
a_i, b_i, c_i	Cost coefficient of generators
λ_i^{G}	Generation emission factor
π_i^{carbon}	Emission cost coefficient of generators
π_i^{OS}	Spot market electricity price
α, β	Beta distribution parameters
I_i, I_i^{max}	Solar radiation intensity and maximum intensity
u_i^{PV}	Conversion efficiency

Variables

L_{ij}^{L}	Numbers of newly added transmission lines
S_{ij}	Power flow on transmission line ij
P_i^{G}	Generation output
θ_i, θ_j	The voltage phase angle of bus i and j
C_{total}	Total planning cost
C^{L}	Lines construction cost
C^{G}	Unit generation cost
C^{OS}	External electricity purchasing cost
P_i^{OS}	External purchasing power
A_i^{PV}, P_i^{PV}	Area of installed pv panels and pv power output
E_X, E_Z	Sample expectation of real samples and noise data samples
$D(\cdot), G(\cdot)$	Discriminator network and the generator network

References

1. Lu, T.; Zhang, W.; Wang, Y.; Xie, H.; Ding, X. Medium- and Long-Term Trading Strategies for Large Electricity Retailers in China's Electricity Market. *Energies* **2022**, *15*, 3342. [CrossRef]
2. Wang, M.; Song, Y.; Sui, B.; Wu, H.; Zhu, J.; Jing, Z.; Rong, Y. Comparative study of pricing mechanisms and settlement methods in electricity spot energy market based on multi-agent simulation. *Energy Rep.* **2022**, *8*, 1172–1182. [CrossRef]
3. Yang, J.; Dong, Z.; Wen, F.; Chen, Q.; Liang, B. Spot electricity market design for a power system characterized by high penetration of renewable energy generation. *Energy Convers. Econ.* **2021**, *2*, 67–78. [CrossRef]
4. Du, Z.; Zhang, C.; Li, Y.; Guo, B.; Sun, Q.; Zhang, Y. Trading Mechanism Design for Demand Response Resources Participating in Electricity Spot Market. In Proceedings of the 2021 International Conference on Power System Technology (POWERCON), Haikou, China, 8–9 December 2021; pp. 639–645. [CrossRef]
5. Green, R.; Newbery, D.M. Competition in the British Electricity Spot Market. *J. Political Econ.* **1992**, *100*, 929–953. [CrossRef]
6. Guan, X.; Wu, J.; Gao, F.; Sun, G. Optimization-Based Generation Asset Allocation for Forward and Spot Markets. *IEEE Trans. Power Syst.* **2008**, *23*, 1796–1808. [CrossRef]
7. Zhang, M.; Xu, Y.; Sun, H. Optimal Coordinated Operation for a Distribution Network With Virtual Power Plants Considering Load Shaping. *IEEE Trans. Sustain. Energy* **2022**, *14*, 550–562. [CrossRef]
8. Chen, W.; Qiu, J.; Zhao, J.; Chai, Q.; Dong, Z.Y. Bargaining Game-Based Profit Allocation of Virtual Power Plant in Frequency Regulation Market Considering Battery Cycle Life. *IEEE Trans. Smart Grid* **2021**, *12*, 2913–2928. [CrossRef]
9. Qu, M.; Ding, T.; Wei, W.; Dong, Z.Y.; Shahidehpour, M.; Xia, S. An Analytical Method for Generation Unit Aggregation in Virtual Power Plants. *IEEE Trans. Smart Grid* **2020**, *11*, 5466–5469. [CrossRef]
10. Ullah, M.H.; Park, J.-D. Transactive Energy Market Operation Through coordinated TSO-DSOs-DERs interactions. *IEEE Trans. Power Syst.* **2023**, *38*, 1978–1990. [CrossRef]
11. Ndrio, M.; Bose, S.; Tong, L.; Guo, Y. Coordinated Transaction Scheduling in Multi-Area Electricity Markets: Equilibrium and Learning. *IEEE Trans. Power Syst.* **2022**, *38*, 996–1008. [CrossRef]
12. Yang, L.; Sun, Q.; Zhang, N.; Li, Y. Indirect Multi-Energy Transactions of Energy Internet With Deep Reinforcement Learning Approach. *IEEE Trans. Power Syst.* **2022**, *37*, 4067–4077. [CrossRef]

13. Borge-Diez, D.; Rosales-Asensio, E.; Palmero-Marrero, A.I.; Acikkalp, E. Optimization of CSP Plants with Thermal Energy Storage for Electricity Price Stability in Spot Markets. *Energies* **2022**, *15*, 1672. [CrossRef]
14. Bayborodina, E.; Negnevitsky, M.; Franklin, E.; Washusen, A. Grid-Scale Battery Energy Storage Operation in Australian Electricity Spot and Contingency Reserve Markets. *Energies* **2021**, *14*, 8069. [CrossRef]
15. Yang, Y.; Ji, T.; Jing, Z. Adaptive learning for strategic bidding in a uniform pricing electricity spot market. *CSEE J. Power Energy Syst.* **2021**, *7*, 1334–1344.
16. Zhai, Q.; Meng, K.; Dong, Z.Y.; Ma, J. Modeling and Analysis of Lithium Battery Operations in Spot and Frequency Regulation Service Markets in Australia Electricity Market. *IEEE Trans. Ind. Inform.* **2017**, *13*, 2576–2586. [CrossRef]
17. Ghorani, R.; Fotuhi-Firuzabad, M.; Moeini-Aghtaie, M. Optimal Bidding Strategy of Transactive Agents in Local Energy Markets. *IEEE Trans. Smart Grid* **2018**, *10*, 5152–5162. [CrossRef]
18. Kanagaraj, A.; Raguru, K.D. Electricity economics for ex-ante double-sided auction mechanism in restructured power market. *Energy Commun. Econ.* **2021**, *3*, 31–37. [CrossRef]
19. Sutton, R.S.; Barto, A.G. *Reinforcement Learning: An Introduction*; MIT Press: Cambridge, MA, USA, 2017.
20. Yang, Y.; Qiu, J.; Ma, J.; Zhang, C. Integrated grid, coal-fired power generation retirement and GESS planning towards a low-carbon economy. *Int. J. Electr. Power Energy Syst.* **2020**, *124*, 106409. [CrossRef]
21. Tao, Y.; Qiu, J.; Lai, S.; Zhang, X.; Wang, G. Collaborative Planning for Electricity Distribution Network and Transportation System Considering Hydrogen Fuel Cell Vehicles. *IEEE Trans. Transp. Electrif.* **2020**, *6*, 1211–1225. [CrossRef]
22. Yang, Y.; Qiu, J.; Zhang, C.; Zhao, J.; Wang, G. Flexible Integrated Network Planning Considering Echelon Utilization of Second Life of Used Electric Vehicle Batteries. *IEEE Trans. Transp. Electrif.* **2021**, *8*, 263–276. [CrossRef]
23. Ruan, J.; Liang, G.; Zhao, J.; Qiu, J.; Dong, Z.Y. An Inertia-Based Data Recovery Scheme for False Data Injection Attack. *IEEE Trans. Ind. Inform.* **2022**, *18*, 7814–7823. [CrossRef]
24. Rana, A.; Gróf, G. Assessment of the Electricity System Transition towards High Share of Renewable Energy Sources in South Asian Countries. *Energies* **2022**, *15*, 1139. [CrossRef]
25. Liu, W.; Zhao, J.; Qiu, J.; Dong, Z.Y. Interpretable hybrid experimental learning for trading behavior modeling in Electricity Market. *IEEE Trans. Power Syst.* **2023**, *38*, 1022–1032. [CrossRef]

Article

Peer-to-Peer Energy Trading among Prosumers with Voltage Regulation Services Provision

Bochun Zhan [1], Changsen Feng [2], Zhemin Lin [3], Xiaoyu Shao [4] and Fushuan Wen [1,*]

1 College of Electrical Engineering, Zhejiang University, Hangzhou 310027, China; zhanbc@zju.edu.cn
2 College of Information Engineering, Zhejiang University of Technology, Hangzhou 310023, China; fcs@zjut.edu.cn
3 Anhui Power Exchange Center Co., Ltd., Hefei 230009, China
4 Economic & Technical Research Institute, State Grid Anhui Electric Power Co., Ltd., Hefei 230071, China
* Correspondence: fushuan.wen@gmail.com

Abstract: The increasing penetration of distributed energy resources (DERs) into distribution networks has changed the energy trading pattern in traditional electricity markets to some degree, and this will possibly cause network congestion and nodal voltage violations. This paper proposes a two-stage modeling framework for peer-to-peer (P2P) energy trading with voltage regulation services provision considered. In the first stage, direct P2P trading among prosumers, considering network congestion management, is enabled. In the second stage, prosumers provide voltage regulation services to address possible voltage violations. Aiming at maximizing social welfare, the alternative direction method of multipliers (ADMM) is applied to solve the two-stage problem. On the basis of the optimal energy solution of the two-stage problem, the energy prices of P2P transactions and the price of voltage regulation services are settled based on the Nash bargaining model. Finally, simulation results of the IEEE 33-bus power system with six prosumers included demonstrate the effectiveness of the proposed models.

Keywords: prosumers; peer-to-peer (P2P) energy trading; voltage regulation; alternative direction method of multipliers (ADMM); Nash bargaining

Citation: Zhan, B.; Feng, C.; Lin, Z.; Shao, X.; Wen, F. Peer-to-Peer Energy Trading among Prosumers with Voltage Regulation Services Provision. *Energies* **2023**, *16*, 5497. https://doi.org/10.3390/en16145497

Academic Editor: Javier Contreras

Received: 26 May 2023
Revised: 15 July 2023
Accepted: 17 July 2023
Published: 20 July 2023

1. Introduction

Consumers equipped with DERs are becoming prosumers with power generation capabilities, and thus, a decentralized and distributed paradigm is more suitable for trading among prosumers than the traditional centralized energy trading pattern [1–3]. The P2P energy trading pattern enables prosumers to trade energy with other prosumers directly [4–6]. It is necessary to design a suitable trading mechanism to promote the prosumers' participation in P2P energy trading.

In order to coordinate emerging prosumers in the energy markets, P2P energy trading models have been studied extensively; these models can be broadly divided into three main categories: auction-based, game theory-based, and consensus-based trading models.

Ref. [7] studied the impacts of auction mechanisms and bidding strategies on the economic efficiencies brought by P2P transactive energy markets. In [8], a discriminatory continuous auction model based on the supply–demand ratio is proposed. In [9], a P2P trading model based on the multi-round double auction and average pricing mechanisms is established. The above auction-based trading models do not take network capacity constraints into account.

Game-theory-based trading models can be divided into non-cooperative game-theory-based models and cooperative game-theory-based ones. A P2P trading model between residential buildings based on non-cooperative games is established in [10]. In [11], a multi-energy trading model based on the Stackelberg game is proposed with various demand-side management measures taken into account. The above trading models that are based on non-cooperative games are characterized as negative-sum games, which may result in extra game costs [12]. In [13], a multi-time-scale trading model of bilateral contracts considering coordination among prosumers is proposed. In [14], a P2P trading model considering the multiple kinds of uncertainties is established based on cooperative games. The P2P energy trading problem is formulated as a centralized optimization problem in the above game-theory-based trading models, and prosumers' privacy may be violated [15].

P2P trading can also be formulated as consensus-based optimization problems with high-efficiency solutions. In [16], the Benders decomposition approach is applied to address the day-ahead scheduling problem of an energy community under a P2P energy trading scheme. ADMM is widely applied to solve the clearing problems of local markets [17,18] and the economic operation problems of power systems [19,20] to protect the prosumers' privacy. In [21], ADMM is applied to solve the clearing problem of local energy markets. The above consensus-based algorithms possess superior convergence properties [22]. In this paper, prosumers share energy within a cooperative alliance to decrease energy costs and energy reliance on the main power grid. The optimization problem of local energy trading is addressed using ADMM.

The existing P2P trading models considering network constraints can be broadly divided into two categories: the constraints of thermal capacity and nodal voltage are incorporated in the optimization model of P2P energy trading [23,24], and ancillary services are needed to address the violations of network constraints after P2P trading [25,26]. In [25,26], active power output regulation of DERs and curtailment of interruptible loads are applied to solve the violations of thermal capacity caused by P2P energy trading. The above models do not take voltage regulation into account.

Fast control actions of transformers and capacitors as well as reactive power regulation of inverters in DERs are widely applied for voltage regulation in distribution networks. In [27], a voltage regulation model based on distribution transformers integrated with a static synchronous compensator is proposed. In [28], a scheduling model of reactive power outputs from photovoltaic (PV) inverters is established. In [29], a muti-step robust model is proposed for distributed generators and energy storage systems (ESSs) to participate in voltage regulation. It should be noted that there is no cost for reactive power generation [30], and the cost of DERs for voltage regulation refers to the cost of inverter losses for the variation of reactive power output [31,32]. Therefore, it is feasible to introduce reactive power regulation of DERs to address the voltage violations caused by P2P trading. In this paper, the reactive power output potentials of PVs and ESSs are explored for voltage regulation.

Extensive research has been conducted on the benefit distribution based on cooperative games. The well-established Shapley value method and nucleolus method are widely used for income distribution in the prosumer alliance [33,34]. An Aumann–Shapley method is proposed for bilateral carbon trading among prosumers in [35]. In [36], a nucleolus-based solution is proposed for a stable and fair payoff distribution scheme of P2P energy trading. The Shapley value method is not applicable for non-convex game models, and the complexity of the nucleolus-based solution increases exponentially with increasing numbers of prosumers. In [37,38], the general Nash bargaining model is utilized to settle the P2P transactions, and the settlement solution is formulated as a centralized optimization problem. This paper applies ADMM to solve the settlement model based on the asymmetric Nash bargaining model in a distributed manner.

In this paper, a two-stage modeling framework is proposed for P2P energy trading with a voltage regulation services provision. Prosumers have the capability to offer voltage regulation services to rectify voltage constraint violations resulting from P2P trading. Energy settlements and voltage regulation services settlements are accomplished using the asymmetric Nash bargaining model. The main contributions of this paper can be summarized as follows:

(1) The reactive power output potentials of PVs and ESSs for voltage regulation are studied. Specifically, the coordinated optimization model of the active power and reactive power in PVs and ESSs are established, respectively.

(2)   P2P energy trading with voltage regulation services provision is characterized as a two-stage optimization problem and is solved based on ADMM to protect prosumers' privacy. On this basis, the optimal active and reactive power outputs of PVs and ESSs as well as the P2P trading scheme are obtained.

(3)   The settlement model for P2P transactions and voltage regulation services are proposed based on the Nash bargaining model, and the settlement solution ensures a fair income distribution for the cooperative alliance.

The rest of this paper is organized as follows. Section 2 introduces the framework of local energy trading with a voltage regulation services provision. Section 3 establishes the local energy trading model considering network congestion management. Section 4 establishes the voltage regulation model. Section 5 solves the two-stage problem based on ADMM. Section 6 establishes the settlement model based on the Nash bargaining model. Section 7 provides simulation results. Conclusions are given in Section 8.

2. Methodological Framework

The methodological framework for P2P energy trading with voltage regulation services provision is illustrated in Figure 1. In the initial stage, direct P2P trading occurs among prosumers, while the responsibility for network congestion management lies with the distribution system operator (DSO). Optimization takes place in this stage for the active power output of PVs, ESSs, and P2P trading energy.

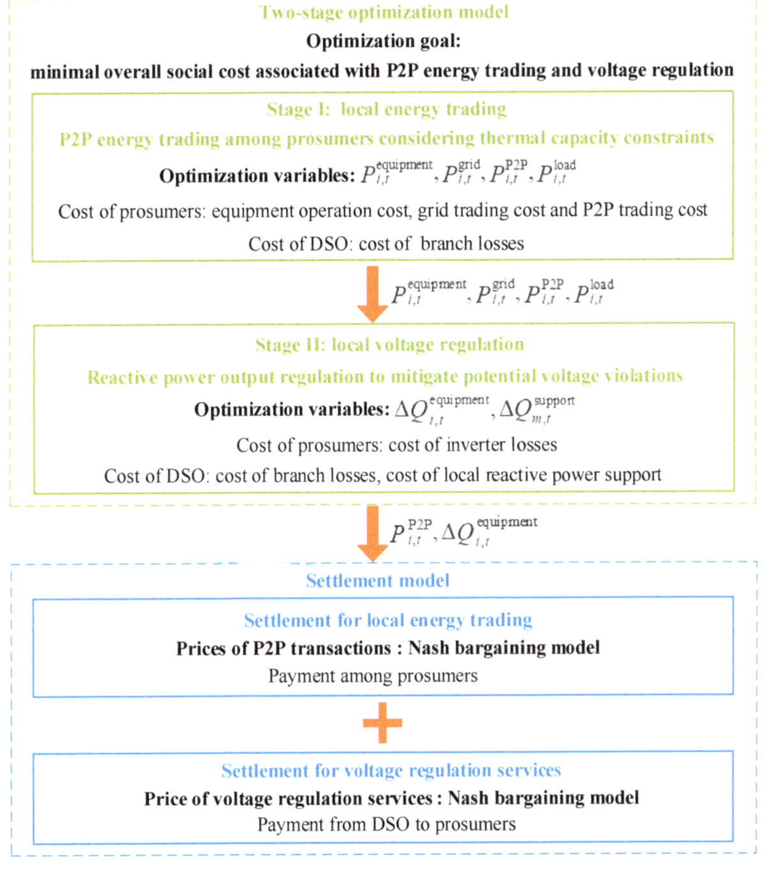

Figure 1. The framework of the proposed two-stage model and settlement model.

In the subsequent stage, the energy traded in P2P transactions and the active power output of the PVs and the ESS remain fixed. In addition to the local reactive power support provided by the DSO, the reactive power output of the PV and the ESS is regulated to mitigate potential voltage violations. The scheduling of reactive power output for the PV system and the ESS occurs in this stage.

The objective of the two-stage model is to minimize the overall social cost associated with P2P energy trading and voltage regulation. The schematic overview of the proposed models is shown in Figure 2. The proposed model's solution is obtained through joint optimization using ADMM. It should be emphasized that the organization and settlement of P2P energy trading and voltage regulation services are distinct.

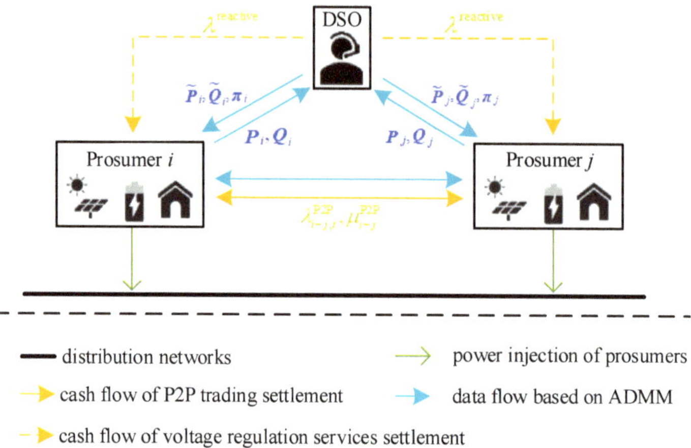

Figure 2. Schematic overview of the proposed models.

3. Modeling of Local Energy Trading

The prosumers in the distribution networks are essentially cooperative. In the first stage, P2P trading among prosumers facilitates energy sharing while simultaneously decreasing energy costs and reducing reliance on the main power grid. The DSO compensates for network losses through the slack bus and ensures that the P2P trading scheme satisfies the thermal capacity constraints of the distribution networks. The primary objective of P2P energy trading optimization is to minimize the social cost.

3.1. Objective Function

The social cost of P2P energy trading is as follows:

$$C_{social}^{total-I} = C_{prosumer}^{total-I} + C_{DSO}^{total-I} = \sum_{i \in A} \left(C_i^{operation} + C_i^{grid} + C_i^{P2P} \right) + C_{DSO}^{loss} \tag{1}$$

where A is the set of prosumers participating in P2P energy trading; $C_{social}^{total-I}$, $C_{prosumer}^{total-I}$, and $C_{prosumer}^{total-I}$ are the social cost, the cost of the prosumers, and the cost of the DSO for P2P energy trading, respectively; $C_i^{operation}$ is the operation cost of equipment for the i-th prosumer; C_i^{grid} is the cost of trading energy with the grid for the i-th prosumer; C_i^{P2P} is the cost of trading energy with other prosumers for the i-th prosumer; and C_{DSO}^{loss} is the cost of network losses for the DSO.

3.1.1. Equipment Operation Cost

The equipment operation cost consists of the operation cost of PV and ESS as follows:

$$C_i^{\text{operation}} = \sum_{t=1}^{T} \left[c_i^{\text{PV,operation}} P_{i,t}^{\text{PV}} + c_i^{\text{ESS,operation}} \left(P_{i,t}^{\text{discharge}} + P_{i,t}^{\text{charge}} \right) \right] \tag{2}$$

where c_i^{ESS} and c_i^{PV} are, respectively, the operation cost coefficient of ESSs and PVs owned by the i-th prosumer; $P_{i,t}^{\text{discharge}}$ and $P_{i,t}^{\text{charge}}$ are, respectively, the charging and discharging power of the ESS at time t; $P_{i,t}^{\text{PV}}$ is the generating power of PV at time t.

3.1.2. Cost of Trading with the Grid

The cost of trading with the grid consists of the cost of buying from the grid and selling to the grid as follows:

$$C_i^{\text{grid}} = \sum_{t=1}^{T} \left[\lambda_t^{\text{purchase,grid}} P_{i,t}^{\text{purchase,grid}} - \lambda_t^{\text{sell,grid}} P_{i,t}^{\text{sell,grid}} \right] \tag{3}$$

where $\lambda_t^{\text{purchase,grid}}$ and $\lambda_t^{\text{sell,grid}}$ are, respectively, the purchase price from the grid and the sell price to the grid at time t; $P_{i,t}^{\text{purchase,grid}}$ and $P_{i,t}^{\text{sell,grid}}$ are, respectively, the power purchased from the grid and sold to the grid at time t.

3.1.3. Cost of Trading with Other Prosumers

The cost of trading with other prosumers consists of the cost of buying from other prosumers and selling to other prosumers:

$$C_i^{\text{P2P}} = \sum_{t=1}^{T} \sum_{j \in A} \left(-P_{i-j,t}^{\text{P2P}} \lambda_{i-j,t}^{\text{P2P}} + P_{j-i,t}^{\text{P2P}} \lambda_{j-i,t}^{\text{P2P}} \right) \tag{4}$$

where $P_{i-j,t}^{\text{P2P}}$ is the power that the i-th prosumer sells to the j-th prosumer at time t, and $\lambda_{i-j,t}^{\text{P2P}}$ is the corresponding trading price.

It should be noted that $\sum\limits_{i \in A} C_i^{\text{P2P}} = 0$.

3.1.4. Cost of Network Losses

The DSO compensates for the power losses of P2P energy trading through the slack bus, and the cost of network losses can be calculated as follows:

$$C_{\text{DSO}}^{\text{loss}} = \phi^{\text{network,loss}} \sum_{t=1}^{T} \sum_{(m,n) \in B} r_{mn} \frac{\left(P_{mn,t}^2 + Q_{mn,t}^2 \right)}{|V_{m,t}|^2} \tag{5}$$

where B is the set of branches in the distribution networks; $\phi^{\text{network,loss}}$ is the transmission loss cost coefficient [17]; $P_{mn,t}$ and $Q_{mn,t}$ are, respectively, the active and reactive power flows in branch (m,n) at time t; r_{mn} is the line resistance of branch (m,n); and $V_{m,t}$ is the voltage magnitude at bus m at time t.

3.2. Constraints

3.2.1. PV Constraint

The active power output of PV is limited as follows:

$$P_{i,t}^{\text{PV,min}} \leq P_{i,t}^{\text{PV}} \leq P_{i,t}^{\text{PV,max}} \tag{6}$$

where $P_{i,t}^{\text{PV,max}}$ and $P_{i,t}^{\text{PV,min}}$ are, respectively, the maximal and minimal active power outputs of PV at time t.

3.2.2. ESS Constraints

(1) State of charge (SOC) constraints.

$$S_{i,t}^{\text{ESS}} = S_{i,t-1}^{\text{ESS}} + \left(\eta_i^{\text{charge}} P_{i,t}^{\text{charge}} - \frac{1}{\eta_i^{\text{discharge}}} P_{i,t}^{\text{discharge}}\right) \frac{\Delta t}{Q_{i,t}^{\text{ESS}}} \tag{7}$$

$$S_i^{\text{ESS,min}} \leq S_{i,t}^{\text{ESS}} \leq S_i^{\text{ESS,max}} \tag{8}$$

where $S_{i,t}^{\text{ESS}}$ is the SOC of the ESS at time t; η_i^{charge} and $\eta_i^{\text{discharge}}$ are, respectively, the efficiency of charging and discharging of ESS; $Q_{i,t}^{\text{ESS}}$ is the capacity of the ESS at time t; Δt is the time step of scheduling.

(2) Charging and discharging constraints.

$$0 \leq P_{i,t}^{\text{charge}} \leq P_i^{\text{charge,max}} \tag{9}$$

$$0 \leq P_{i,t}^{\text{discharge}} \leq P_i^{\text{discharge,max}} \tag{10}$$

where $P_i^{\text{charge,max}}$ and $P_i^{\text{discharge,max}}$ are, respectively, the maximal input and output active powers of the ESS.

Considering that the ESS cannot be charged and discharged at the same time, constraint (11) is required:

$$P_{i,t}^{\text{charge}} P_{i,t}^{\text{discharge}} = 0 \tag{11}$$

3.2.3. Load Constraint

The active load is limited as follows:

$$P_{i,t}^{\text{fix}} \leq P_{i,t}^{\text{load}} \leq P_{i,t}^{\text{fix}} + P_{i,t}^{\text{variable}} \tag{12}$$

where $P_{i,t}^{\text{load}}$ is the load of the i-th prosumer at time t; $P_{i,t}^{\text{fix}}$ and $P_{i,t}^{\text{variable}}$ are, respectively, the uninterruptible load and interruptible load of the i-th prosumer at time t.

3.2.4. Energy Balance Constraints

Considering the traded energy is produced by the DERs of prosumers, the energy balance constraints are as follows:

$$P_{i,t}^{\text{PV}} + P_{i,t}^{\text{discharge}} - P_{i,t}^{\text{charge}} - P_{i,t}^{\text{load}} = P_{i,t}^{\text{sell,grid}} + P_{i,t}^{\text{sell,P2P}} - P_{i,t}^{\text{purchase,grid}} - P_{i,t}^{\text{purchase,P2P}} \tag{13}$$

$$P_{i,t}^{\text{sell,P2P}} = \sum_{j \in A} P_{i-j,t}^{\text{P2P}} \tag{14}$$

$$P_{i,t}^{\text{purchase,P2P}} = \sum_{j \in A} P_{j-i,t}^{\text{P2P}} \tag{15}$$

$$P_{i-j,t}^{\text{P2P}}, P_{j-i,t}^{\text{P2P}}, P_{i,t}^{\text{sell,grid}}, P_{i,t}^{\text{pruchase,grid}} \geq 0 \tag{16}$$

where $P_{i,t}^{\text{sell,P2P}}$ and $P_{i,t}^{\text{purchase,P2P}}$ are, respectively, the total energy that the i-th prosumer sells to others and purchases from others.

Considering that prosumers cannot purchase and sell energy at the same time, constraint (17) is required:

$$\left(P_{i,t}^{\text{sell,grid}} + P_{i,t}^{\text{sell,P2P}}\right)\left(P_{i,t}^{\text{purchase,grid}} + P_{i,t}^{\text{purchase,P2P}}\right) = 0 \tag{17}$$

By accumulating constraints (14) and (15) of all prosumers, constraint (18) is obtained:

$$\sum_{i \in A} \left(P_{i,t}^{\text{sell,P2P}} + P_{i,t}^{\text{purchase,P2P}}\right) = 0 \tag{18}$$

3.2.5. Network Constraints

The DSO is responsible for the management of network congestion. Considering the thermal capacity of lines, constraints (19)–(23) are required:

$$P_{mn,t} = P_{m,t}^{\text{inject}} + \sum_{(n,k) \in B} P_{nk}^t + r_{mn} I_{mn,t}^2 \tag{19}$$

$$Q_{mn,t} = Q_{m,t}^{\text{inject}} + \sum_{(n,k) \in B} Q_{nk}^t + \chi_{mn} I_{mn,t}^2 \tag{20}$$

$$V_{n,t}^2 = V_{m,t}^2 - 2(r_{mn} P_{mn,t} + \chi_{mn} Q_{mn,t}) + (r_{mn}^2 + \chi_{mn}^2) I_{mn,t}^2 \tag{21}$$

$$I_{mn,t}^2 V_{i,t}^2 = P_{mn,t}^2 + Q_{mn,t}^2 \tag{22}$$

$$I_{mn}^{\min} \le I_{mn,t} \le I_{mn}^{\max} \tag{23}$$

where χ_{mn} is the reactance of branch (m,n); $I_{mn,t}$ is the current value of branch (m,n) at time t; $I_{mn}^{\max}$ and $I_{mn}^{\min}$ are, respectively, the upper and lower limits of current value for branch (m,n). $P_{m,t}^{\text{inject}}$ and $Q_{m,t}^{\text{inject}}$ are, respectively, the active and reactive power injections of the i-th prosumer at bus m; the values can be calculated as follows:

$$P_{m,t}^{\text{inject}} = P_{i,t}^{\text{PV}} + P_{i,t}^{\text{discharge}} - P_{i,t}^{\text{charge}} - P_{i,t}^{\text{load}} = P_{i,t}^{\text{inject}} \tag{24}$$

$$Q_{m,t}^{\text{inject}} = Q_{i,t}^{\text{PV,base}} + Q_{i,t}^{\text{ESS,base}} = Q_{i,t}^{\text{inject}} \tag{25}$$

where $Q_i^{\text{PV,base}}$ and $Q_{i,t}^{\text{ESS,base}}$ is the base reactive power output of PV and ESS.

4. Modeling of Local Voltage Regulation

While thermal capacity constraints are considered in the initial stage, P2P energy trading across the distribution networks can lead to undesired violations of nodal voltage. Consequently, the DSO requests voltage regulation services from prosumers during the subsequent stage, aiming to minimize the social cost associated with local voltage regulation.

4.1. Objective Function

The social cost of local voltage regulation is as follows:

$$C_{\text{social}}^{\text{total-II}} = C_{\text{prosumer}}^{\text{total-II}} + C_{\text{DSO}}^{\text{total-II}} = \sum_{i \in A} \left(C_i^{\text{regulation}} - F_i^{\text{service}}\right) + C_{\text{DSO}}^{\text{support}} + C_{\text{DSO}}^{\text{service}} + C_{\text{DSO}}^{\text{loss}} \tag{26}$$

where $C_{\text{social}}^{\text{total-II}}$, $C_{\text{prosumer}}^{\text{total-II}}$, and $C_{\text{DSO}}^{\text{total-II}}$ are the social cost, the cost of the prosumers, and the cost of the DSO for voltage regulation, respectively; $C_i^{\text{regulation}}$ and F_i^{service} are the regulation cost and income of voltage regulation services provision for the i-th prosumer, respectively;

and $C_{DSO}^{regulation}$ and $C_{DSO}^{service}$ are the regulation cost and the payment for the voltage regulation services provision of the DSO.

4.1.1. Regulation Cost for Prosumers

The solution to P2P energy trading provides fixed active power outputs for the PV and the ESS, which are incorporated into the voltage regulation model. The applied strategy for providing voltage regulation services involves regulating the reactive power output of PV and ESS inverters. The cost of voltage regulation for prosumers corresponds to the cost of inverter losses in the PV and the ESS, as characterized below [20,21]:

$$C_i^{regulation} = \sum_{t=1}^{T} \left[c_{i,a}^{PV,loss} \left(S_{i,t}^{PV} \right)^2 + c_{i,b}^{PV,loss} S_{i,t}^{PV} + c_{i,c}^{PV,loss} + c_{i,a}^{ESS,loss} \left(S_{i,t}^{ESS} \right)^2 + c_{i,b}^{ESS,loss} S_{i,t}^{ESS} + c_{i,c}^{ESS,loss} \right] \tag{27}$$

where $c_{i,a}^{PV,loss}$, $c_{i,b}^{PV,loss}$, and $c_{i,c}^{PV,loss}$ are the cost coefficients of inverter losses for the PV; $c_{i,a}^{ESS,loss}$, $c_{i,b}^{ESS,loss}$, and $c_{i,c}^{ESS,loss}$ are the cost coefficients of inverter losses for the ESS. $S_{i,t}^{PV}$ and $S_{i,t}^{ESS}$ are the apparent power of PV and ESS at time t, respectively.

4.1.2. Income of Voltage Regulation Services Provision

The income of the voltage regulation services can be calculaed as follows:

$$F_i^{service} = \lambda^{reactive} \sum_{t=1}^{T} \left(\Delta Q_{i,t}^{PV} + \Delta Q_{i,t}^{ESS} \right) \tag{28}$$

where $\lambda^{reactive}$ is the price for providing voltage regulation services at time t.

Considering that the income of the voltage regulation services for prosumers is paid by the DSO, (26) can be rewritten as follows:

$$C_{social}^{total\text{-}II} = C_{prosumer}^{total\text{-}II} + C_{DSO}^{total\text{-}II} = \sum_{i \in A} C_i^{regulation} + C_{DSO}^{regulation} + C_{DSO}^{loss} \tag{29}$$

4.1.3. Regulation Cost for DSO

The cost of local reactive power support of the DSO can be calculated as follows:

$$C_{DSO}^{support} = \phi^{voltage,support} \sum_{t=1}^{T} \sum_{m \in D} \Delta Q_{m,t}^{support} \tag{30}$$

where D is the set of buses in the distribution system; $\phi^{voltage,support}$ is the cost coefficient of the local reactive power support [22]; and $\Delta Q_{m,t}^{support}$ is the reactive power injection of the DSO at bus m at time t.

4.1.4. Cost of Network Losses

The cost of network losses can be calculated with Equation (5).

4.2. Constraints

4.2.1. PV Constraints

The P–Q plane of an inverter in PV and ESS are illustrated in Figure 3, and the reactive power output of PV should satisfy the constraints as follows [15]:

$$-Q_{i,t}^{PV,max} + Q_i^{PV,base} \leq \Delta Q_{i,t}^{PV} \leq Q_{i,t}^{PV,max} - Q_i^{PV,base} \tag{31}$$

$$Q_{i,t}^{PV,max} = \min \left(\sqrt{\left(S_i^{PV,max} \right)^2 - \left(P_{i,t}^{PV} \right)^2}, P_{i,t}^{PV} \tan \gamma \right) \tag{32}$$

where $Q_{i,t}^{PV,max}$ is the maximal reactive power output of PVs at time t; $S_i^{PV,max}$ is the maximal apparent power of PVs; and γ is the maximal feasible power factor of PVs.

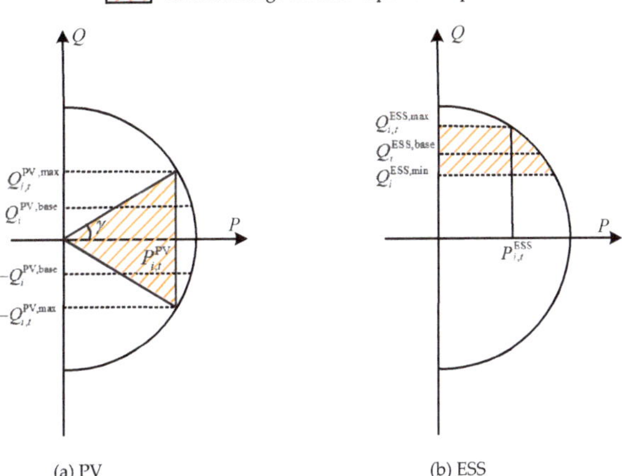

Figure 3. The P–Q plane of an inverter in PV and ESS.

4.2.2. ESS Constraints

The reactive power output of ESS should satisfy the constraints as follows [15]:

$$Q_i^{ESS,base} - Q_i^{ESS,min} \leq \Delta Q_{i,t}^{ESS} \leq Q_{i,t}^{ESS,max} - Q_i^{ESS,base} \tag{33}$$

$$Q_{i,t}^{ESS,max} = \sqrt{\left(S_i^{ESS,max}\right)^2 - \left(P_{i,t}^{discharge}\right)^2 - \left(P_{i,t}^{charge}\right)^2} \tag{34}$$

where $Q_{i,t}^{ESS,max}$ is the maximal reactive power output of the ESS at time t; $Q_i^{ESS,base}$ and $Q_i^{ESS,min}$ are, respectively, the base and minimal reactive power output of the ESS; $S_i^{ESS,max}$ is the maximal apparent power of the ESS.

4.2.3. Network Constraints

The active power injection of the i-th prosumer at bus m can be calculated with Equation (24), and the reactive power injection can be calculated as follows:

$$Q_{i,t}^{injection} = Q_{i,t}^{PV,base} + \Delta Q_{i,t}^{PV} + Q_{i,t}^{ESS,base} + \Delta Q_{i,t}^{ESS} \tag{35}$$

Considering the local reactive power support of the DSO, the total reactive power injection at bus m is as follows:

$$Q_{m,t}^{injection} = Q_{i,t}^{injection} + \Delta Q_{m,t}^{compensation} \tag{36}$$

Besides the constraints of the thermal capacity for distribution networks (19)–(23), the nodal voltage constraints are required:

$$V_m^{min} \leq V_{m,t} \leq V_m^{max} \tag{37}$$

where V_m^{max} and V_m^{min} are, respectively, the upper and lower limits of voltage magnitude at bus m.

5. Joint Optimization of the Two-Stage Model

The optimization goal of P2P energy trading with voltage regulation services provision is to minimize the overall social cost:

$$\begin{cases} \min C_{\text{social}}^{\text{total}} = \min(C_{\text{social}}^{\text{total-I}} + C_{\text{social}}^{\text{total-II}}) = \sum_{i \in A} \left(C_i^{\text{operation}} + C_i^{\text{grid}} + C_i^{\text{P2P}} + C_i^{\text{regulation}} \right) + C_{\text{DSO}}^{\text{loss}} + C_{\text{DSO}}^{\text{support}} \\ \qquad\qquad\qquad \text{subject to}: \ (6)\text{--}(25), (31)\text{--}(37) \end{cases} \tag{38}$$

And the above optimization problem can be rewritten as follows:

$$\begin{cases} \min \sum_{i \in A} C_i \left(P_i^{\text{equipment}}, P_i^{\text{grid}}, P_i^{\text{P2P}}, P_i^{\text{load}}, \Delta Q_i^{\text{equipment}}, P_i^{\text{inject}}, Q_i^{\text{inject}} \right) \\ \quad + C_{\text{DSO}}\left(\widetilde{P}_1^{\text{P2P}}, \widetilde{P}_1^{\text{inject}}, \widetilde{Q}_1^{\text{inject}}, \ldots, \widetilde{P}_i^{\text{P2P}}, \widetilde{P}_i^{\text{inject}}, \widetilde{Q}_i^{\text{inject}}, \Delta Q_1^{\text{support}}, \ldots \Delta Q_m^{\text{support}} \right) \\ \text{subject to}: \ \text{prosumers}: \ (6)\text{--}(17), (24), (25), (31)\text{--}(35) \ \text{DSO}: \ (19)\text{--}(23), (36), (37) \\ \qquad \text{auxilliary}: P_i^{\text{P2P}} = \widetilde{P}_i^{\text{P2P}}, P_i^{\text{inject}} = \widetilde{P}_i^{\text{inject}}, Q_i^{\text{inject}} = \widetilde{Q}_i^{\text{inject}} \end{cases} \tag{39}$$

where $P_i^{\text{equipment}} = \left[P_{i,t}^{\text{PV}}, P_{i,t}^{\text{discharge}}, P_{i,t}^{\text{charge}}, t \in T \right]$ is an active power profile of equipment for the i-th prosumer; $P_i^{\text{grid}} = \left[P_{i,t}^{\text{sell,grid}}, P_{i,t}^{\text{purchase,grid}}, t \in T \right]$ is a grid trading energy profile for the i-th prosumer; $P_i^{\text{P2P}} = \left[P_{i,t}^{\text{sell,P2P}}, P_{i,t}^{\text{purchase,P2P}}, t \in T \right]$ is a P2P trading energy profile for the i-th prosumer; $P_i^{\text{load}} = \left[P_{i,t}^{\text{load}}, t \in T \right]$ is a load profile for the i-th prosumer; $\Delta Q_i^{\text{equipment}} = \left[\Delta Q_{i,t}^{\text{PV}}, \Delta Q_{i,t}^{\text{ESS}}, t \in T \right]$ is a reactive power variation profile of equipment for the i-th prosumer; $P_i^{\text{inject}} = \left[P_{i,t}^{\text{inject}}, t \in T \right]$ is an active power injection profile for the i-th prosumer; $Q_i^{\text{inject}} = \left[Q_{i,t}^{\text{inject}}, t \in T \right]$ is a reactive power injection profile for the i-th prosumer; $\Delta Q_m^{\text{support}} = \left[\Delta Q_{m,t}^{\text{support}}, t \in T \right]$ is a local reactive power support profile for the DSO; and $\widetilde{P}_i^{\text{P2P}}, \widetilde{P}_1^{\text{inject}}$, and $\widetilde{Q}_1^{\text{inject}}$ are, respectively, the auxiliary variables of $P_i^{\text{P2P}}, P_i^{\text{inject}}$, and Q_i^{inject}.

Flowchart of optimal P2P trading with voltage regulation services provision is shown in Figure 4. ADMM is applied to solve the joint optimization problem in the following steps [17–20]:

(1) Set the initial values for the solution variables of the DSO $\left[\widetilde{P}_{i,0}^{\text{P2P}}, \widetilde{P}_{i,0}^{\text{inject}}, \widetilde{Q}_{i,0}^{\text{inject}} \right]$, the dual variables $\left[\pi_{i,0}^{\text{P2P}}, \pi_{i,0}^{\text{active}}, \pi_{i,0}^{\text{reactive}} \right]$ of the auxiliary constraints, and the iteration round to be $k = 0$.

(2) Establish the problem to be solved for the i-th prosumer based on the solution variables of the DSO $\left[\widetilde{P}_{i,k}^{\text{P2P}}, \widetilde{P}_{i,k}^{\text{inject}}, \widetilde{Q}_{i,k}^{\text{inject}} \right]$ and the dual variables $\left[\pi_{i,k}^{\text{P2P}}, \pi_{i,k}^{\text{active}}, \pi_{i,k}^{\text{reactive}} \right]$:

$$\begin{cases} \min \begin{bmatrix} C_i \left(P_i^{\text{equipment}}, P_i^{\text{grid}}, P_i^{\text{P2P}}, P_i^{\text{load}}, \Delta Q_i^{\text{equipment}}, P_i^{\text{inject}}, Q_i^{\text{inject}} \right) + \pi_{i,k}^{\text{P2P}} \left(P_i^{\text{P2P}} - \widetilde{P}_{i,k}^{\text{P2P}} \right) + \\ \pi_{i,k}^{\text{active}} \left(P_i^{\text{inject}} - \widetilde{P}_{i,k}^{\text{inject}} \right) + \pi_{i,k}^{\text{reactive}} \left(Q_i^{\text{inject}} - \widetilde{Q}_{i,k}^{\text{inject}} \right) + \frac{\rho}{2} \| P_i^{\text{P2P}} - \widetilde{P}_{i,k}^{\text{P2P}} \|^2 + \frac{\rho}{2} \| P_i^{\text{inject}} - \widetilde{P}_{i,k}^{\text{inject}} \|^2 + \\ \frac{\rho}{2} \| Q_i^{\text{inject}} - \widetilde{Q}_{i,k}^{\text{inject}} \|^2 \end{bmatrix} \\ \qquad\qquad\qquad \text{subject to}: \ (6)\text{--}(17), (24), (25), (31)\text{--}(35) \end{cases} \tag{40}$$

where ρ is the penalty parameter.

(3) The i-th prosumer solves the optimization problem (40) locally, and the solution variables' updates are labeled as $P_i^{\text{P2P}} = P_{i,k+1}^{\text{P2P}}, P_i^{\text{inject}} = P_{i,k+1}^{\text{inject}}$, and $Q_i^{\text{inject}} = Q_{i,k+1}^{\text{inject}}$; these values are sent to the DSO and used for the DSO update.

(4) Establish the problem to be solved for the DSO based on the solution variables of the prosumers $\left[P_{i,k+1}^{\text{P2P}}, P_{i,k+1}^{\text{inject}}, Q_{i,k+1}^{\text{inject}}, i \in A \right]$ and the dual variables $\left[\pi_{i,k}^{\text{P2P}}, \pi_{i,k}^{\text{active}}, \pi_{i,k}^{\text{reactive}}, i \in A \right]$:

$$\left\{ \min \left\{ \begin{array}{l} C_{\text{DSO}}\left(\widetilde{P}_1^{\text{P2P}}, \widetilde{P}_1^{\text{inject}}, \widetilde{Q}_1^{\text{inject}}, \ldots, \widetilde{P}_i^{\text{P2P}}, \widetilde{P}_i^{\text{inject}}, \widetilde{Q}_i^{\text{inject}}, \Delta Q_{1,t}^{\text{support}}, \ldots \Delta Q_{m,t}^{\text{support}} \right) + \\[2mm] \sum_{i \in A} \left[\pi_{i,k}^{\text{P2P}}\left(P_{i,k+1}^{\text{P2P}} - \widetilde{P}_i^{\text{P2P}} \right) + \pi_{i,k}^{\text{active}}\left(P_{i,k+1}^{\text{inject}} - \widetilde{P}_i^{\text{inject}} \right) + \pi_{i,k}^{\text{reactive}}\left(Q_{i,k+1}^{\text{inject}} - \widetilde{Q}_i^{\text{inject}} \right) + \\[2mm] \frac{\rho}{2}\|P_{i,k+1}^{\text{P2P}} - \widetilde{P}_i^{\text{P2P}}\|^2 + \frac{\rho}{2}\|P_{i,k+1}^{\text{inject}} - \widetilde{P}_i^{\text{inject}}\|^2 + \frac{\rho}{2}\|Q_{i,k+1}^{\text{inject}} - \widetilde{Q}_i^{\text{inject}}\|^2 \right] \end{array} \right\} \right. \tag{41}$$

$$\text{subject to}: (19)\text{–}(23), (36),(37)$$

(5) The DSO solves optimization problem (41) globally, and the solution variables' updates are labeled as $\widetilde{P}_i^{\text{P2P}} = \widetilde{P}_{i,k+1}^{\text{P2P}}$, $\widetilde{P}_i^{\text{inject}} = \widetilde{P}_{i,k+1}^{\text{inject}}$, and $\widetilde{Q}_i^{\text{inject}} = \widetilde{Q}_{i,k+1}^{\text{inject}}$; these values are sent to the i-th prosumer and used for the i-th prosumer update.

(6) Update the dual variables based on the solution variables of the prosumers and the DSO:

$$\left\{ \begin{array}{l} \pi_{i,k+1}^{\text{P2P}} = \pi_{i,k}^{\text{P2P}} + \rho(P_{i,k+1}^{\text{P2P}} - \widetilde{P}_{i,k+1}^{\text{P2P}}) \\[2mm] \pi_{i,k+1}^{\text{active}} = \pi_{i,k}^{\text{active}} + \rho(P_{i,k+1}^{\text{inject}} - \widetilde{P}_{i,k+1}^{\text{inject}}) \\[2mm] \pi_{i,k+1}^{\text{reactive}} = \pi_{i,k}^{\text{reactive}} + \rho(Q_{i,k+1}^{\text{inject}} - \widetilde{Q}_{i,k+1}^{\text{inject}}) \end{array} \right. \tag{42}$$

(7) The termination criteria of the algorithm are as follows:

$$\left\{ \begin{array}{l} \sum_{i \in A} \left[\|P_{i,k}^{\text{P2P}} - \widetilde{P}_{i,k}^{\text{P2P}}\|^2 + \|P_{i,k}^{\text{inject}} - \widetilde{P}_{i,k}^{\text{inject}}\|^2 + \|Q_{i,k}^{\text{inject}} - \widetilde{Q}_{i,k}^{\text{inject}}\|^2 \right] \leq \varepsilon^{\text{primary}} \\[2mm] \sum_{i \in A} \left[\|P_{i,k+1}^{\text{P2P}} - P_{i,k}^{\text{P2P}}\|^2 + \|P_{i,k+1}^{\text{inject}} - P_{i,k}^{\text{inject}}\|^2 + \|Q_{i,k+1}^{\text{inject}} - Q_{i,k}^{\text{inject}}\|^2 \right] \leq \varepsilon^{\text{dual}} \end{array} \right. \tag{43}$$

where $\varepsilon^{\text{primary}}$ and $\varepsilon^{\text{dual}}$ are, respectively, the primary residual value and the dual residual value.

If constraint (43) is respected, terminate the iteration process. Otherwise, update the iteration round to $k = k + 1$; return to step (2); and continue calculating until convergence.

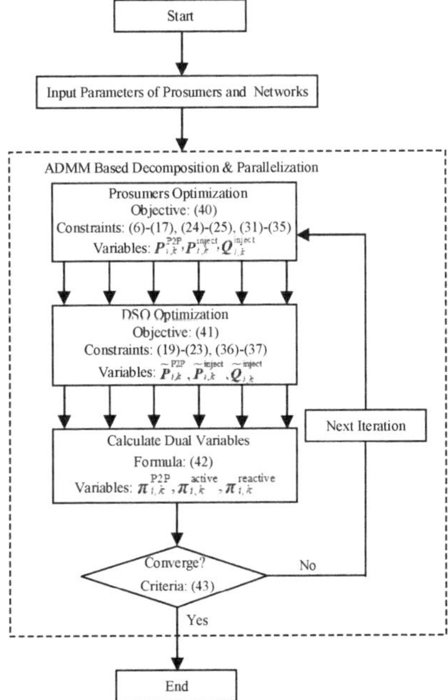

Figure 4. Flowchart of optimal P2P trading with voltage regulation services provision.

6. Settlement Models

Section 5 addresses the two-stage problem and determines the optimal active and reactive power outputs of the PV and the ESS, as well as the energy involved in P2P transactions at each optimization time step. The income distribution based on the Nash bargaining model achieves strong Pareto optimality by considering both the individual interests of the participants and the collective interests of the cooperative alliance [28,29]; the Nash bargaining model is employed in the settlement model for local energy trading and voltage regulation services.

6.1. Settlement model for Local Energy Trading

6.1.1. Objective Function

The energy settlement model among prosumers is established based on the Nash bargaining model as follows:

$$\max \prod_{i \in A} \left[C_i^0 - \left(C_i^{\text{opration}} + C_i^{\text{grid}} + C_i^{\text{P2P}} \right) \right]^{\tau_i} \tag{44}$$

where C_i^0 is the energy cost of the i-th prosumer without P2P energy trading, including the cost of trading with the grid and the cost of equipment operation; τ_i is the bargaining factor of the i-th prosumer as follows:

$$\tau_i = e^{E_i^{\text{sell}}/E_i^{\text{sell,max}}} - e^{-\left(E_i^{\text{purchase}}/E_i^{\text{purcahse,max}} \right)} \tag{45}$$

$$E_i^{\text{sell}} = \sum_{t=1}^{T} \max(0, P_{i-j,t}^{\text{P2P}}) \tag{46}$$

$$E_i^{\text{purchase}} = -\sum_{t=1}^{T} \min(0, P_{i-j,t}^{\text{P2P}}) \tag{47}$$

where E_i^{sell} and E_i^{purchase} are, respectively, the total energy sold and purchased through P2P transactions for the i-th prosumer during the optimization period; $E_i^{\text{sell,max}}$ and $E_i^{\text{purchase,max}}$ are, respectively, the maximum sold energy and the maximum purchased energy among all prosumers.

Equation (45) utilizes the exponential function to quantify the bargaining factors of prosumers. It possesses the following features: (1) any prosumer who has engaged in P2P energy trading has a non-zero bargaining factor value. (2) prosumers who do not participate in P2P energy trading have a bargaining factor value of zero. (3) the bargaining factor value of a prosumer increases with their contribution to the energy sharing.

Objective function (44) is logarithmized to convert it into a convex optimization problem.

$$\min - \sum_{i \in A} \tau_i \left[C_i^0 - \left(C_i^{\text{opr}} + C_i^{\text{grid}} + C_i^{\text{P2P}} \right) \right] \tag{48}$$

6.1.2. Constraints

1. Constraint of price couples.

$$\lambda_{i-j,t}^{\text{P2P}} - \lambda_{j-i,t}^{\text{P2P}} = 0 \tag{49}$$

where $\lambda_{i-j,t}^{\text{P2P}}$ is the price of sell energy from the i-th prosumer to the j-th prosumer.

2. Constraint of income distribution.

$$C_i^0 - \left(C_i^{\text{opr}} + C_i^{\text{grid}} + C_i^{\text{P2P}} \right) > 0 \tag{50}$$

6.1.3. Solving Procedure

The energy settlement model is solved based on ADMM in the following steps:

(1) Calculate the bargaining factor according to (45)–(47).

(2) Set the initial values for $\mu_{i-j,0}^{\text{P2P}}$ and $\lambda_{j-i,t,0}^{\text{P2P}}$, and set the iteration round to be $k = 0$.

(3) Establish the problem to solve for each prosumer based on the augmented Lagrangian function of the settlement model (48) and the values of $\mu_{i-j,k}^{\text{P2P}}$ and $\lambda_{j-i,t,k}^{\text{P2P}}$:

$$\min L_i^{\text{P2P}} = -\tau_i \left[C_i^0 - \left(C_i^{\text{opr}} + C_i^{\text{P2P}} + C_i^{\text{P2P}} \right) \right] + \sum_{j \in A, j \neq i} \sum_{t=1}^{T} \mu_{i-j,k}^{\text{P2P}} \left(\lambda_{i-j,t}^{\text{P2P}} - \lambda_{j-i,t,k}^{\text{P2P}} \right) + \frac{\rho}{2} \sum_{j \in A, j \neq i} \sum_{t=1}^{T} \left(\lambda_{i-j,t}^{\text{P2P}} - \lambda_{j-i,t,k}^{\text{P2P}} \right)^2 \quad (51)$$

where ρ is the penalty parameter.

(4) Each prosumer solves settlement problem (51) locally, and only the updated price is exchanged among prosumers. In the k-th iteration, the following steps are performed:

    a. The i-th prosumer updates the price:

$$\lambda_{i-j,t,k+1}^{\text{P2P}} = \arg\min L_i^{\text{P2P}} \left(\mu_{i-j,k}^{\text{P2P}}, \lambda_{i-j,t,k}^{\text{P2P}}, \lambda_{j-i,t,k}^{\text{P2P}} \right) \lambda_{i-j,t}^{\text{P2P}}, \mu_{i-j}^{\text{P2P}} \quad (52)$$

    b. Other prosumers such as the j-th prosumer receives the updated price of the i-th prosumer and updates the price:

$$\lambda_{j-i,t,k+1}^{\text{P2P}} = \arg\min L_i^{\text{P2P}} \left(\mu_{j-i,k}^{\text{P2P}}, \lambda_{i-j,t,k+1}^{\text{P2P}}, \lambda_{j-i,t,k}^{\text{P2P}} \right) \quad (53)$$

    c. Repeat steps (4)a and (4)b until each prosumer updates the trading price in the k-th iteration.

(5) Update the dual variable:

$$\mu_{i-j,k+1}^{\text{P2P}} = \mu_{i-j,k}^{\text{P2P}} + \rho \left(\lambda_{i-j,t,k+1}^{\text{P2P}} - \lambda_{j-i,t,k+1}^{\text{P2P}} \right) \quad (54)$$

(6) The termination criteria of the algorithm are as follows:

$$\begin{cases} \sum_{t=1}^{T} \sum_{i \in A} \left(\lambda_{i-j,t,k+1}^{\text{P2P}} - \lambda_{j-i,t,k+1}^{\text{P2P}} \right)^2 \leq \varepsilon^{\text{primary}} \\ \sum_{t=1}^{T} \sum_{i \in A} \left(\lambda_{i-j,t,k+1}^{\text{P2P}} - \lambda_{i-j,t,k}^{\text{P2P}} \right)^2 \leq \varepsilon^{\text{dual}} \end{cases} \quad (55)$$

where $\varepsilon^{\text{primary}}$ and $\varepsilon^{\text{dual}}$ are, respectively, the acceptable primary residual value and the dual residual value.

The iteration process terminates if constraint (55) is satisfied. Otherwise, update the iteration round $k = k + 1$, return to step (3), and continue calculating until convergence.

6.2. Settlement Model for Voltage Regulation Services

The price of voltage regulation services paid by the DSO can be determined based on the Nash bargaining model as described below:

$$\max \left(C_{\text{DSO}}^{\text{support},0} + C_{\text{DSO}}^{\text{loss},0} - C_{\text{DSO}}^{\text{support}} - C_{\text{DSO}}^{\text{loss}} - \sum_{i \in A} F_i^{\text{service}} \right) \sum_{i \in A} \left(F_i^{\text{service}} - C_i^{\text{regulation}} \right) \quad (56)$$

where $C_{\text{DSO}}^{\text{support},0}$ and $C_{\text{DSO}}^{\text{loss},0}$ are, respectively, the cost of local reactive power support of the DSO and network losses when prosumers do not provide voltage regulation services. F_i^{service} and $C_i^{\text{regulation}}$ can be calculated with Equations (27) and (28), respectively.

The constraints of income distribution for voltage regulation are as follows:

$$C_{DSO}^{support,0} + C_{DSO}^{loss,0} - C_{DSO}^{support} - C_{DSO}^{loss} - \sum_{i \in A} F_i^{service} > 0 \tag{57}$$

$$F_i^{service} - C_i^{regulation} > 0 \tag{58}$$

7. Simulation Results and Discussion

7.1. Basic Parameters

The IEEE 33-bus system with six prosumers included is used to demonstrate the effectiveness of the proposed model, as shown in Figure 5. The PV generation profile and load profile are shown in Figures 6 and 7, respectively. Prosumer 1 and Prosumer 4 are equipped with an ESS while other prosumers are equipped with a PV and an ESS. The interruptible load constitutes 20% of the total load in each time slot. Figure 8 shows the purchase price from the grid and the sell price to the grid. The technical parameters of the PVs and the ESSs are provided in Tables 1 and 2, respectively. Other parameters are presented in Table 3. The optimization step is set to one hour, and the optimization period is set to one day. The thermal capacity, reactance, and conductance parameters of branches in the IEEE 33-bus power system are referenced from [12].

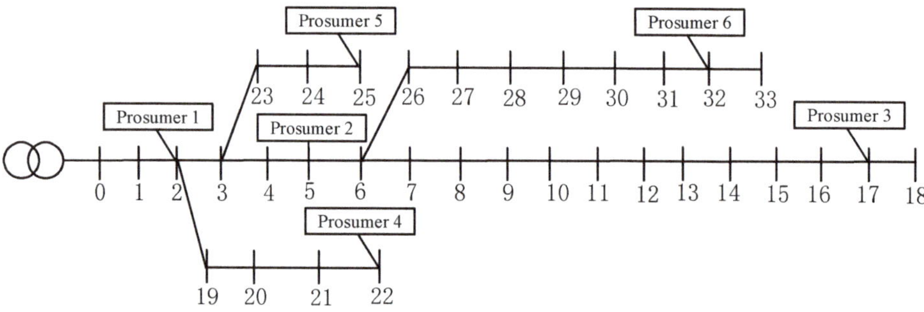

Figure 5. IEEE 33-bus power system with six prosumers.

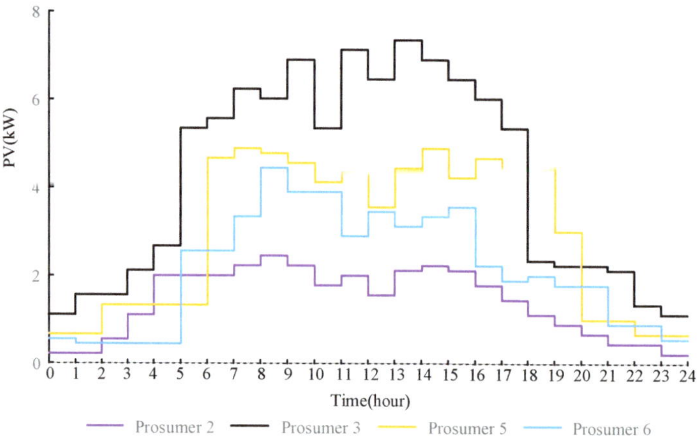

Figure 6. PV generation profile in a typical day.

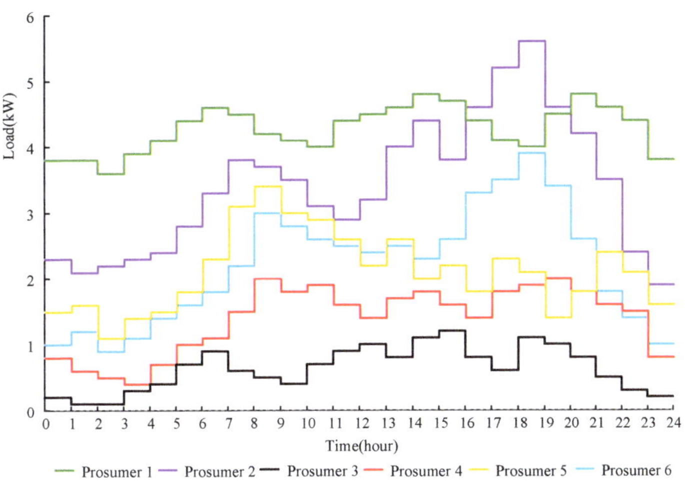

Figure 7. Load profile in a typical day.

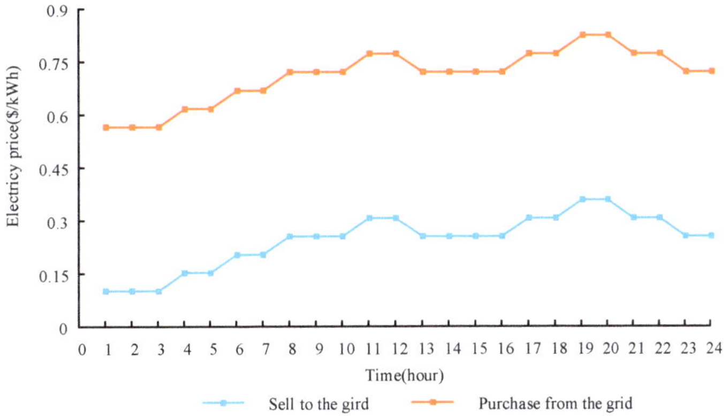

Figure 8. Electricity prices.

Table 1. Technical parameters of PV system.

Parameter	Value	Parameter	Value
Base reactive power	2.0 kVar	Operation cost coefficient	0.04 USD/kW
Maximal apparent power	7.5 kVA	Inverter loss cost coefficient	[0.02, 0.03, 0.03] USD/kVar
Maximal power factor	0.98		

Table 2. Technical parameters of ESSs.

Parameter	Value	Parameter	Value
Base reactive power	2.5 kVar	Minimal SoC	0.10
Maximal apparent power	6.0 kVA	Charge efficiency	0.95
Maximal input active power	3.5 kW	Discharge efficiency	0.85
Maximal output active power	3.0 kW	Operation cost coefficient	0.14 USD/kW
Maximal SoC	0.90	Inverter loss cost coefficient	[0.05, 0.04, 0.01] USD/kVar

Table 3. Other parameters.

Parameter	Value
Cost efficient of network losses	0.13 USD/kW
Cost efficient of voltage support for DSO	0.75 USD/kVar
Primary residual value	10^{-3}
Dual residual value	10^{-3}

7.2. Simulation Results

7.2.1. Convergence Results

Figure 9 depicts the convergence of the two-stage joint optimization model and the settlement model for local energy trading, which is based on the ADMM solution. The two-stage model and the energy settlement model require 22 and 26 iterations, respectively. The solution gap decreases to below 10^{-3} during convergence. The two-stage model takes 412.4 s to compute, while the energy settlement model takes 277.8 s. These results indicate that the ADMM solution demonstrates excellent convergence performance and computational efficiency while ensuring privacy protection for prosumers.

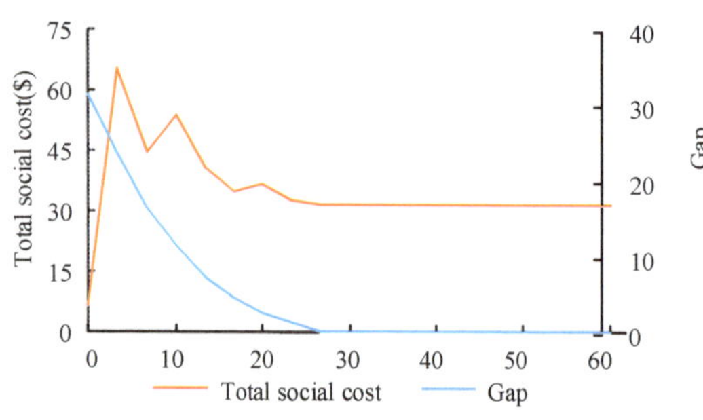

(a) The two-stage joint optimization model

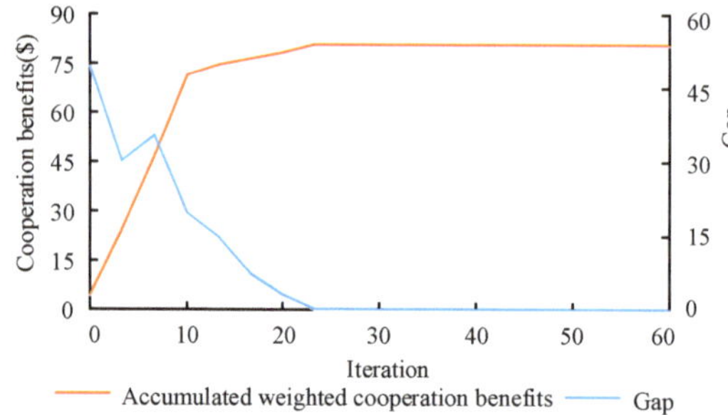

(b) The settlement model for P2P energy trading

Figure 9. Convergence results.

7.2.2. P2P Trading and Settlement Results

Figure 10 presents the electricity consumption structure of prosumers during the optimization period, where the self-consumption energy is defined as the total energy produced and is consumed by the same prosumer. From 1:00 to 6:00, because the PV systems produce a poor output, the proportions of P2P trading energy and self-consumption energy in the total energy consumed are relatively low. From 7:00 to 15:00, the power generation of the PV systems increases significantly, enabling P2P trading energy to meet the load demands of prosumers. Negative self-consumption energy denotes the surplus energy stored in the ESS by prosumers in addition to selling it to the grid. From 16:00 to 24:00, there is a significant increase in the load demand while the PV generation decreases. Prosumers satisfy their load demands by purchasing energy from the grid and scheduling the output of the ESS, and the proportion of energy purchased form the grid is highest in the entire time slots.

Figure 11 presents the average price of P2P transactions for each time slot. During the period of low demand, especially from 1:00 to 6:00, the average prices of P2P transactions are close to the sell price to the grid. However, during the peak demand period, occurring between 17:00 and 22:00, the supply of P2P trading energy falls short of the load demand, resulting in the average prices of P2P transactions being closer to the purchase price from the grid. The average prices of P2P transactions across all time slots are lower than the purchase price from the grid and higher than the sell price to the grid. Consequently, P2P direct trading effectively reduces the total energy cost for prosumers.

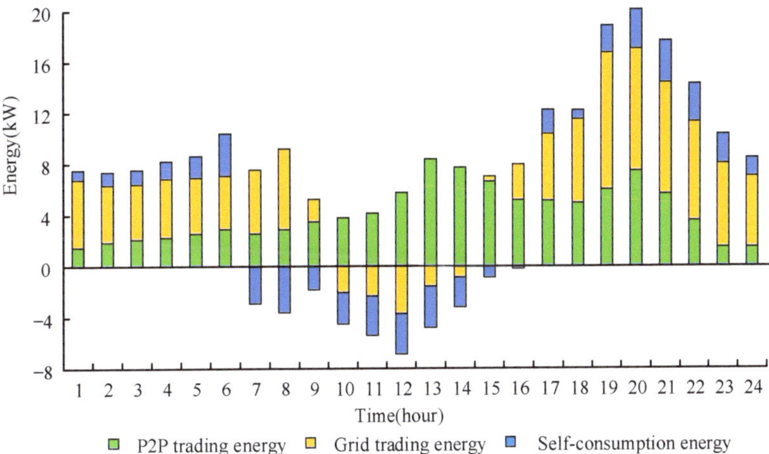

Figure 10. Structure of electricity consumption.

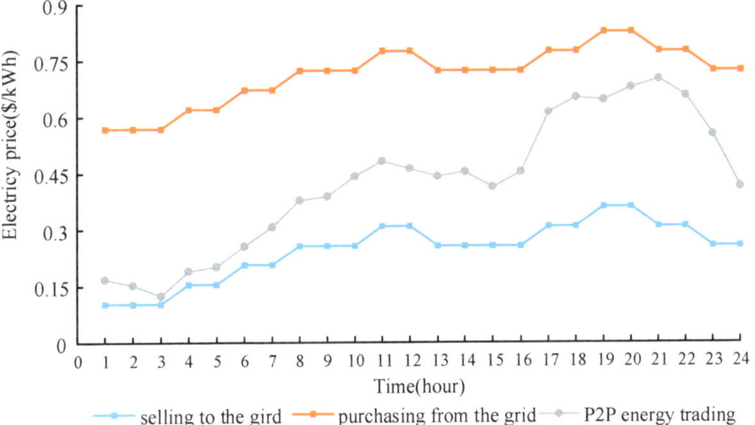

Figure 11. Energy traded with the grid.

Table 4 presents the energy costs of prosumers considering P2P transactions and without considering P2P transactions. The energy settlement cost represents the payment for P2P trading energy. Compared to not participating in P2P trading, the total energy cost for the six prosumers decreased from USD 67.02 to USD 20.7. Prosumer 1 purchases a greater amount of energy than Prosumer 4, and the purchase price for Prosumer 1 is 6.9% lower than that of Prosumer 4. Prosumer 3 sells a greater amount of energy compared to other prosumers and has a higher bargaining factor value. Thus, Prosumer 3 experiences the highest decrease in energy cost, with a reduction of 28.7% compared to other prosumers. The aforementioned analysis demonstrates the effective reduction in energy costs for each prosumer through P2P energy sharing. Additionally, it highlights the attainment of a fair distribution solution for cooperation income based on the contribution of energy sharing.

Table 4. Energy cost.

Prosumer No.	Energy of P2P Trading/kWh	Bargaining Factor	Settlement Cost/USD	Energy Cost without P2P Trading/USD	Energy Cost with P2P Trading/USD	Decreased Energy Cost/USD
Prosumer 1	56.47	1.946	27.45	51.66	36.79	14.87
Prosumer 2	13.98	0.825	−5.28	−2.09	−4.54	2.45
Prosumer 3	31.63	2.371	−19.29	−5.61	−13.33	7.72
Prosumer 4	37.58	1.208	19.62	30.72	25.86	4.86
Prosumer 5	25.14	1.659	−13.32	−4.18	−13.27	9.09
Prosumer 6	23.30	1.110	−11.18	−3.48	−10.81	7.33
Prosumer coalition	94.05	/	/	67.02	20.70	46.32

7.2.3. Voltage Regulation and Settlement Results

The violations of thermal capacity are addressed through congestion management in the local energy trading stage. In the local voltage regulation stage, the violations of nodal voltage are addressed. Figure 12 illustrates the magnitude of each individual nodal voltage throughout the entire time slots, with different colors indicating varying levels of voltage magnitudes. Prior to voltage regulation, overvoltage issues occur at bus 17, with the highest voltage reaching 1.121 p.u at 21:00. Undervoltage problems arise at bus 2, with the lowest observed voltage being 0.942 p.u at 22:00. Solely considering congestion management could lead to undesired voltage violations.

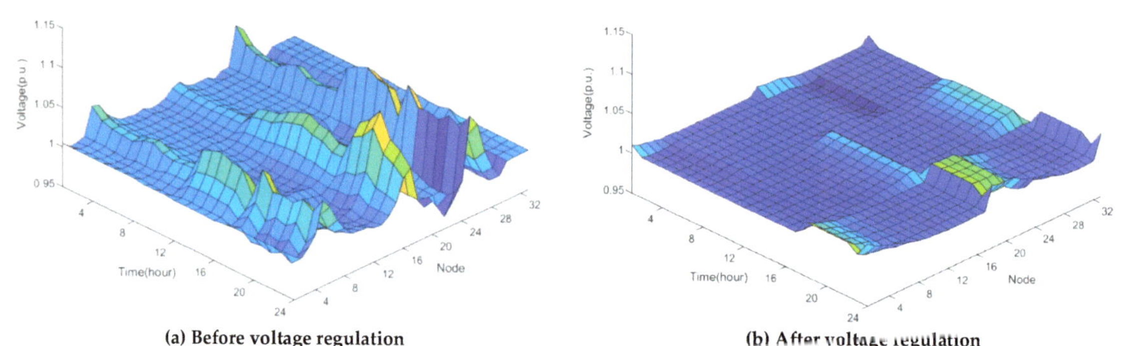

(a) Before voltage regulation (b) After voltage regulation

Figure 12. Nodal voltage magnitudes.

Figure 13 illustrates the regulation of reactive power outputs for Prosumer 3 at bus 17 and Prosumer 1 at bus 2. The reactive power output of the ESS for Prosumer 1 increases by 24.6% at 22:00. The reactive power output of the ESS and the PV for Prosumer 3 decreases by 34.6% and 41.2% at 21:00, respectively. After implementing voltage regulation based on the scheduling of the reactive power output of the PV and the ESS, the highest voltage magnitude at bus 17 decreases to 1.064 p.u, and the lowest voltage magnitude at bus 2 increases to 1.010 p.u. The average deviation of nodal voltage in the networks decreases from 8.4% to 3.7%. The joint optimization results show that local reactive power support from the DSO is unnecessary throughout all time slots. In other words, to minimize the total social cost of local energy trading and voltage regulation, it is not cost-effective to depend on the local reactive power support of the DSO for voltage regulation.

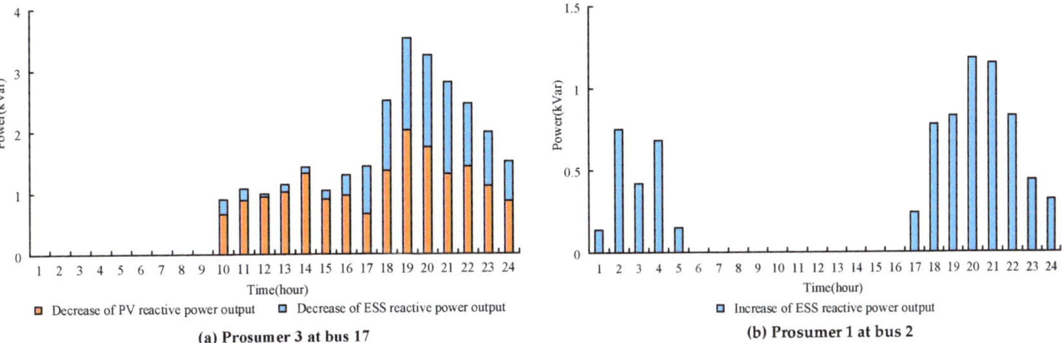

(a) Prosumer 3 at bus 17   (b) Prosumer 1 at bus 2

Figure 13. Reactive power output regulation.

The price of voltage regulation services based on the Nash bargaining model is 0.21 USD/kW. Due to the voltage regulation services provided by prosumers, the voltage regulation cost for the DSO decreases from USD 10.42 to USD 6.28. Despite incurring a cost of USD 4.11 due to inverter losses, prosumers generate an income of USD 6.28 from voltage regulation services. The aforementioned results indicate that the provision of voltage regulation services increases the income for prosumers and reduces the voltage regulation cost for the DSO.

7.2.4. Comparison Analysis

Two patterns are considered for comparison:

(1) Pattern 1: P2P energy trading with voltage regulation services provision is characterized as the two-stage model as proposed in this paper.
(2) Pattern 2: Thermal constraints and nodal voltage constraints are incorporated in the optimization model of P2P energy trading [12].

Figure 14 displays the energy exchanged through P2P transactions among six prosumers during the optimization period of two patterns. The label "3-1" denotes Prosumer 3 purchasing electricity from Prosumer 1. Table 5 presents the energy cost and the total social cost for the two patterns. The self-sufficiency ratio is defined as the ratio of total energy traded among prosumers to the total energy traded, including the energy traded with the grid. Taking into account voltage regulation based on the reactive power output of PVs and ESSs, the voltage constraints are no longer binding constraints in the optimization model that limit the increase in P2P trading energy. Despite inverter losses representing an additional cost, the increase in P2P trading energy among prosumers significantly contributes to reducing energy costs. Consequently, the total cost for prosumers and the total social cost decrease by 46.8% and 34.3%, respectively. Specifically, the DSO compensates prosumers for voltage regulation services, and the settlement fee for these services is not included in the total social cost. The aforementioned results and analysis demonstrate that the two-stage model proposed in this paper can decrease reliance on the grid for energy and foster P2P transactions among prosumers, leading to reduced energy costs for prosumers and reduced overall social cost.

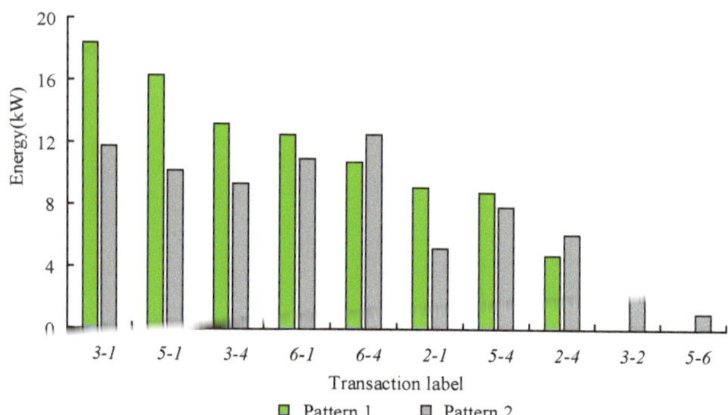

Figure 14. The energy of P2P transactions.

Table 5. Comparison results.

Items	Pattern 1	Pattern 2
Total P2P trading energy/kWh	94.05	76.24
Self-sufficiency ratio/%	77.32	59.83
Total energy cost/USD	20.70	38.91
Network losses cost/USD	6.57	8.82
Inverter losses cost/USD	4.11	0
Total social cost/USD	31.38	47.73

8. Conclusions

In this paper, the reactive power output potentials of the distributed energy resources for voltage regulation are studied. Peer-to-peer energy trading with voltage regulation services provision is characterized as a two-stage optimization problem. The first stage enables peer-to-peer energy trading with network congestion management considered, and the second stage involves requesting reactive power output regulation from distributed energy resources to mitigate nodal voltage violations. The settlement for peer-to-peer energy trading and voltage regulation services is based on the Nash bargaining model. The simulation results show that

(1) The two-stage model and the energy settlement model are solved using the alternative direction method of multipliers, demonstrating excellent convergence performance and computational efficiency while ensuring privacy protection for prosumers.

(2) The peer-to-peer trading scheme in the first stage reduces the overall energy costs for prosumers, and it drops by 69.1%. The prices of peer to peer transactions are determined by the energy traded among prosumers, and the resulting energy settlement solution ensures a fair income distribution.

(3) The potential voltage violations caused by peer-to-peer trading are addressed in the second stage, and the average deviation of nodal voltage in the networks decreases from 8.4% to 3.7%. The income of prosumers grows by 2.8%, and the voltage regulation cost for the distribution system operator slumps by 39.7% due to the provision of voltage regulation services.

(4) In comparison to the optimization model of peer-to-peer trading that incorporates thermal capacity and nodal voltage constraints, the proposed two-stage model explores the reactive power output potentials of distributed energy resources, resulting in a 34.2% decrease in the overall social cost.

Future research can investigate the behavioral models of prosumers and the games between prosumers to control extreme prosumer behaviors.

Author Contributions: Conceptualization, B.Z., C.F. and F.W.; methodology, B.Z., C.F. and F.W.; software, B.Z. and C.F.; validation, B.Z. and C.F.; formal analysis, B.Z. and C.F.; investigation, Z.L. and X.S.; resources, Z.L.; data curation, X.S.; writing—original draft preparation, B.Z. and C.F.; writing—review and editing, B.Z. and F.W.; visualization, B.Z. and C.F.; supervision, F.W.; project administration, Z.L. and F.W.; funding acquisition, F.W. All authors have read and agreed to the published version of the manuscript.

Funding: This work was funded by the National Key Research and Development Project of China entitled "Key technologies for collaborative planning between coal-fired power plants with flexible regulation capability and large-scale renewable energy based generation plants", Project No. 2022YFB2403100.

Data Availability Statement: Not applicable.

Conflicts of Interest: The authors declare no conflict of interest.

Abbreviations

ADMM	alternative direction method of multipliers
DERs	distributed energy resources
DSO	distribution system operator
ESS	energy storage system
P2P	peer-to-peer
PV	photovoltaic

References

1. Paudel, A.; Chaudhari, K.; Long, C.; Gooi, H.B. Peer-to-peer energy trading in a prosumer-based community microgrid: A game-theoretic model. *IEEE Trans. Ind. Electron.* **2019**, *66*, 6087–6097. [CrossRef]
2. Yao, H.; Xiang, Y.; Hu, S.; Wu, G.; Liu, J. Optimal prosumers' peer-to-peer energy trading and scheduling in distribution networks. *IEEE Trans. Ind. Appl.* **2022**, *58*, 1466–1477. [CrossRef]
3. Azim, M.I.; Tushar, W.; Saha, T.K. Cooperative negawatt P2P energy trading for low-voltage distribution networks. *Appl. Energy* **2021**, *229*, 117300. [CrossRef]
4. Zhang, C.; Wu, J.; Zhou, Y.; Cheng, M.; Long, C. Peer-to-peer energy trading in a microgrid. *Appl. Energy* **2018**, *220*, 1–12. [CrossRef]
5. Islam, S.N.; Sivadas, A. Optimisation of buyer and seller preferences for peer-to-peer energy trading in a microgrid. *Energies* **2022**, *15*, 4212. [CrossRef]
6. Yan, M.; Shahidehpour, M.; Paaso, A.; Zhang, L.; Alabdulwahab, A.; Abusorrah, A. Distribution network-constrained optimization of peer-to-peer transactive energy trading among multi-microgrids. *IEEE Trans. Smart Grid* **2021**, *12*, 1033–1047. [CrossRef]
7. Lin, J.; Pipattanasomporn, M.; Rahman, S. Comparative analysis of auction mechanisms and bidding strategies for P2P solar transactive energy markets. *Appl. Energy* **2019**, *255*, 113687. [CrossRef]
8. Chen, K.; Lin, J.; Song, Y. Trading strategy optimization for a prosumer in continuous double auction-based peer-to-peer market: A prediction-integration model. *Appl. Energy* **2019**, *242*, 1121–1133. [CrossRef]
9. Li, Z.; Ma, T. Peer-to-peer electricity trading in grid-connected residential communities with household distributed photovoltaic. *Appl. Energy* **2020**, *278*, 115670. [CrossRef]
10. Cui, S.; Wang, Y.; Shi, Y.; Xiao, J. A new and fair peer-to-peer energy sharing framework for energy buildings. *IEEE Trans. Smart Grid* **2020**, *11*, 3817–3826. [CrossRef]
11. Jing, R.; Xie, M.; Wang, F.; Chen, L. Fair P2P energy trading between residential and commercial multi-energy systems enabling integrated demand-side management. *Appl. Energy* **2020**, *262*, 114551. [CrossRef]
12. Amin, W.; Huang, Q.; Afzal, M.; Khan, A.A.; Umer, K.; Ahmed, S.A. A converging non-cooperative & cooperative game theory approach for stabilizing peer-to-peer electricity trading. *Electr. Power Syst. Res.* **2020**, *183*, 106278.
13. Morstyn, T.; Teytelboym, A.; Mcculloch, M.D. Bilateral contract networks for peer-to-peer energy trading. *IEEE Trans. Smart Grid* **2019**, *10*, 2026–2035. [CrossRef]
14. Li, G.; Li, Q.; Yang, X.; Ding, R. General Nash bargaining based direct P2P energy trading among prosumers under multiple uncertainties. *Int. J. Electr. Power Energy Syst.* **2022**, *143*, 108403. [CrossRef]
15. Kim, H.; Lee, J.; Bahrami, S.; Wong, V.W.S. Direct energy trading of microgrids in distribution energy market. *IEEE Trans. Power Syst.* **2020**, *25*, 639–651. [CrossRef]
16. García-Muñoz, F.; Dávila, S.; Quezada, F. A benders decomposition approach for solving a two-stage local energy market problem under uncertainty. *Appl. Energy* **2023**, *329*, 120226. [CrossRef]
17. Javadi, M.S.; Esmaeel, A.; Jordehi, A.R.; Gough, M.; Santos, S.F.; Catalão, J.P.S. Transactive energy framework in multi-carrier energy hubs: A fully decentralized model. *Energy* **2022**, *238*, 121717. [CrossRef]

18. Nguyen, D.H. Optimal solution analysis and decentralized mechanisms for peer-to-peer energy markets. *IEEE Trans. Power Syst.* **2021**, *36*, 1470–1481. [CrossRef]

19. Maneesha, A.; Swarup, K.S. A survey on applications of alternating direction method of multipliers in smart power grids. *Renew. Sustain. Energy Rev.* **2021**, *152*, 111687. [CrossRef]

20. Zheng, Y.; Song, Y.; Hill, D.J.; Zhang, Y. Multiagent system based microgrid energy management via asynchronous consensus ADMM. *IEEE Trans. Energy Convers.* **2018**, *33*, 886–888. [CrossRef]

21. Babagheibi, M.; Jadid, S.; Kazemi, A. An incentive-based robust flexibility market for congestion management of an active distribution system to use the free capacity of microgrids. *Appl. Energy* **2023**, *336*, 120832. [CrossRef]

22. Feng, C.; Li, Z.; Shahidehpour, M.; Wen, F.; Liu, W.; Wang, X. Decentralized short-term voltage control in active power distribution systems. *IEEE Trans. Smart Grid* **2018**, *9*, 4566–4576. [CrossRef]

23. Guerrero, J.; Chapman, A.C.; Verbič, G. Decentralized P2P energy trading under network constraints in a low-voltage network. *IEEE Trans. Smart Grid* **2019**, *10*, 5163–5173. [CrossRef]

24. Moghadam, A.Z.; Javidi, M.H. Designing a two-stage transactive energy system for future distribution networks in the presence of prosumers' P2P transactions. *Electr. Power Syst. Res.* **2022**, *211*, 108202. [CrossRef]

25. Esmat, A.; Usaola, J.; Moreno, M.A. Distribution-level flexibility market for congestion management. *Energies* **2018**, *11*, 1056. [CrossRef]

26. Zhou, W.; Wang, Y.; Peng, F.; Liu, Y.; Sun, H.; Cong, Y. Distribution network congestion management considering time sequence of peer-to-peer energy trading. *Int. J. Electr. Power Energy Syst.* **2022**, *136*, 107646. [CrossRef]

27. Lai, J.; Yin, X.; Zhang, Z.; Wang, Z.; Chen, Y.; Yin, X. System modeling and cascaded passivity based control for distribution transformer integrated with static synchronous compensator. *Int. J. Electr. Power Energy Syst.* **2019**, *113*, 1035–1046. [CrossRef]

28. Tina, G.; Garozzo, D.; Siano, P. Scheduling of PV inverter reactive power set-point and battery charge/discharge profile for voltage regulation in low voltage networks. *Int. J. Electr. Power Energy Syst.* **2019**, *107*, 131–139. [CrossRef]

29. Tziovani, L.; Hadjidemetriou, L.; Kolios, P.; Astolfi, A.; Kyriakides, E.; Timotheou, S. Energy management and control of photovoltaic and storage systems in active distribution grids. *IEEE Trans. Power Syst.* **2022**, *37*, 1956–1968. [CrossRef]

30. Chen, L.; Liu, N.; Yu, S.; Xu, Y. A stochastic game approach for distributed voltage regulation among autonomous PV prosumers. *IEEE Trans. Power Syst.* **2022**, *37*, 776–787. [CrossRef]

31. Wu, C.; Hug, G.; Kar, S. Smart inverter for voltage regulation: Physical and market implementation. *IEEE Trans. Power Syst.* **2018**, *33*, 6181–6192. [CrossRef]

32. Jay, D.; Swarup, K.S. Game theoretical approach to novel reactive power ancillary services market mechanism. *IEEE Trans. Power Syst.* **2021**, *36*, 1298–1308. [CrossRef]

33. Cremers, S.; Robu, V.; Zhang, P.; Andoni, M.; Norbu, S.; Flynn, D. Efficient methods for approximating the Shapley value for asset sharing in energy communities. *Appl. Energy* **2023**, *331*, 120328. [CrossRef]

34. Chen, Y.; Park, B.; Kou, X.; Hu, M.; Dong, J.; Li, F.; Amasyali, K.; Olama, M. A comparison study on trading behavior and profit distribution in local energy transaction games. *Appl. Energy* **2020**, *280*, 115941. [CrossRef]

35. Nan, J.; Feng, J.; Deng, X.; Wang, C.; Sun, K.; Zhou, H. Hierarchical low-carbon economic dispatch with source-load bilateral carbon-trading based on Aumann–Shapley method. *Energies* **2022**, *15*, 5359. [CrossRef]

36. Feng, C.; Wen, F.; You, S.; Li, Z.; Shahnia, F.; Shahidehpour, M. Coalitional game-based transactive energy management in local energy communities. *IEEE Trans. Power Syst.* **2020**, *35*, 1729–1740. [CrossRef]

37. Zhong, W.; Xie, S.; Xie, K.; Yang, Q.; Xie, L. Cooperative P2P energy trading in active distribution networks: An MILP-based Nash bargaining solution. *IEEE Trans. Smart Grid* **2021**, *12*, 1264–1276. [CrossRef]

38. Cheng, S.; Zuo, X.; Wei, Z.; Ni, K.; Wang, C. Nash bargaining-based cooperative game for distributed economic scheduling of microgrid with charging-swapping-storage integrated station. *Int. J. Electr. Power Energy Syst.* **2022**, *46*, 23927–23938. [CrossRef]

Article

Offshore Wind Power Potential in Brazil: Complementarity and Synergies

Erika Carvalho Nogueira *, Rafael Cancella Morais and Amaro Olimpio Pereira, Jr. *

Energy Planning Program, COPPE, Universidade Federal do Rio de Janeiro (PPE/COPPE-UFRJ),
Rio de Janeiro 21941-914, Brazil; rafaelmorais@ppe.ufrj.br
* Correspondence: erika@ppe.ufrj.br (E.C.N.); amaro@ppe.ufrj.br (A.O.P.J.);
Tel.: +55-21-3938-8777 (E.C.N. & A.O.P.J.)

Abstract: Renewable sources stand out in energy planning due to their contribution to greenhouse gas emission reduction when displacing fossil fuels and the enhancement of energy security through the diversification of the energy matrix. Understanding and optimizing the complementary operative synergy between different energy sources over time and space leads to efficient policies. This article uses an hourly Pearson's correlation coefficient to explore the complementarity between offshore wind and other power generation sources in the Brazilian matrix. An analysis of offshore wind power feasibility in the Brazilian power system will be conducted, considering environmental implications, synergies with the oil industry, costs, and complementarities with other energy sources. The methodology uses an optimization model to minimize costs and optimize the production mix while considering the time series of renewable energy, subject to demand constraints, renewable resource availability, reservoir storage, capacity limitations, and thermal generation. The study concludes that the northeast and southeast electrical subsystems must start offshore wind installation in Brazil due to their complementarity with hydropower production, synergy with the oil and gas industry, and proximity to the largest consumption spots.

Keywords: offshore wind power; complementarity; optimization model; Pearson correlation; energy planning

Citation: Nogueira, E.C.; Morais, R.C.; Pereira, A.O., Jr. Offshore Wind Power Potential in Brazil: Complementarity and Synergies. *Energies* **2023**, *16*, 5912. https://doi.org/10.3390/en16165912

Academic Editors: Fushuan Wen and Xiuli Wang

Received: 22 June 2023
Revised: 13 July 2023
Accepted: 26 July 2023
Published: 10 August 2023

1. Introduction

The importance of renewable sources to energy planning can be attributed to their contribution to reducing greenhouse gas emissions (GHG) as well as maintaining energy security by diversifying the matrix. Energy systems scenario cases from several countries indicate that there are good prospects for the transition toward sustainable energy systems due to significant advances in the development of renewable energy technology, resource assessment, and systems design [1,2]. It is possible to exploit good resources of wind, wave, and solar power in areas that have largely remained untapped [1], such as those in Latin America [3]. Brazil is a continental country with an extensive coastline and an abundance of natural resources, including oil and natural gas. Despite this, there are several opportunities, such as offshore wind power. This work focuses on evaluating the opportunities to use offshore wind resources in Brazil for electricity generation by analyzing the complementarity between offshore wind and the main energy sources. Establishing the most suitable locations for wind offshore power plant installation and contributing to the National Interconnected System's (SIN, in Portuguese) energy security are the desired outputs of this analysis.

Several authors emphasize the benefits of offshore wind, such as job creation, reduced use of fossil resources, improved national energy security, rapid technological improvements, and reduced GHG emissions [4,5]. These authors point out that the main advantage is the quality and resource availability, and the main disadvantage is the high cost of installations in offshore environments. However, technological progress has already enabled a 20% drop in the cost of energy between 2010 and 2018 [6].

In addition, Brazil has great offshore wind potential [7–13]. Silva (2015) estimated the offshore wind potential for the Brazilian exclusive economic zone at 8688 GW and highlighted that the highest values of turbine production are between depths of 0 and 35 m in the northeast region [7]. Further, Silva (2019) analyzed the offshore wind potential in Brazil from different perspectives: theoretical (1687.6 GW), technical (1064.2 GW), and environmental and social (330.5 GW) [8]. The environmental and social potential of offshore wind in Brazil is the most restricted; however, it corresponds to twice the total power capacity currently installed in Brazil and more than 20 times the installed capacity of onshore wind energy [8]. Azevedo et al. (2020) calculated an annual average total power generation of 14,800 TWh based on a Brazilian aerogenerator installable capacity of 3 TW [9].

The Brazilian Wind Potential Atlas [10] provides maps of the annual average wind speed that illustrate what Ortiz and Kampel (2011) state: that winds with greater magnitude in the maritime environment are located on the coast of states that also have a high degree of onshore wind energy use [11]. Pimenta, Kempton, and Garvine (2008) estimated the total potential for offshore wind energy generation in the south and southeast regions of Brazil in shallower waters (up to 50 m) at 102 GW, and they concluded that Brazil has a promising offshore wind resource and it is economically attractive [12]. Vinhoza et al. (2023) also state that Brazil offers a great deal of untapped potential for offshore wind energy, and its costs are competitive internationally [13].

Furthermore, Brazil also has experience using onshore wind resources and offshore oil exploration structures [14–16]. Offshore wind power can be complementary to other energy sources, and when installed near large consumer spots, it reduces the loss in energy transport (transmission and distribution) [7].

In synergy with the oil industry, there is the possibility of extending the useful life of oil and gas fields and reusing data, structures, knowledge, and experience [14] in the wind offshore industry. Moreover, fixed oil platforms in the decommissioning process can be reused for offshore wind power generation [15,17], or floating wind turbines can generate energy for oil platforms as an alternative system. In northeastern Brazil, a region known for its great wind potential, all platforms are nearing their end or have already exceeded their useful lives, presenting low oil productivity [16]. In the next few years, there is no projection for power production from offshore wind in Brazil; however, the reuse of platforms may be the gateway to offshore wind exploration in the country [16].

The aim of this article is to analyze the impact of large-scale offshore wind generation's introduction in the Brazilian electricity system, addressing environmental issues, synergy with the oil industry, costs, and complementarity with other sources of generation. Complementarity was analyzed using Pearson's correlation coefficient between offshore wind resources and power generation of other sources in the Brazilian electrical matrix. Power generation expansion scenarios were built with an optimization model, simulating the generation of offshore wind power in Brazil from reanalysis data and comparing it with other energy sources to analyze its complementarities. These results lead to the identification of the potential costs and benefits of integrating offshore wind sources into the grid. The main outcome is providing information to decision-makers regarding the best measures by achieving a balance between electricity demand, climate change goals, and environmental integrity.

Extensive research has explored the complementarity of multi-energy sources, such as wind, solar, and hydropower [18–24]. Beluco et al. (2003) and Kougias et al. (2016) conducted research on the complementarity between hydropower and solar energy based on different methodologies. The former study proposed dimensionless mathematical indices to evaluate varying degrees of complementarity [18], and the latter one aimed to develop a methodology utilizing an optimization algorithm to assess the complementarity between small hydropower and solar photovoltaic (PV) systems [19]. Peron (2017) analyzed the potential of complementarity between the different sources of power generation based on a statistical approach involving Pearson correlation analysis between the resources and evaluation of combinations and optimization of the generation plots to minimize the energy deficit in the northeast region of Brazil [20]. Silva et al. (2016) identified areas with the possibility of high complementarity between offshore winds and hydrological regimes in Brazil. They used Pearson's correlation coefficient, coherence, and cluster analyses to assess the wind variability and monthly precipitation in Brazil on seasonal and interannual timescales [23]. Naeem et al. (2019) used Pearson's correlation and an optimization model to analyze the spatial and temporal complementary characteristics of solar and wind in order to achieve reliable operation of a grid-connected microgrid in Ireland [21]. Rosa et al. (2017) used the Pearson's correlation coefficient and linear programming to optimize the hydro, photovoltaic, and onshore wind power mix in Rio de Janeiro (Brazil) with a monthly resolution time series [22]. The utilization of Pearson's correlation metric remains prevalent in research papers focused on measuring complementarity among multi-energy sources [24]. However, no study has explored hydro/wind/PV/thermoelectric and specifically wind offshore power output hourly complementarity in Brazil. This article explores the complementarity of already established resources (hydro, onshore wind, and solar PV) with a new energy source in Brazil (offshore wind) using an hourly optimization model and Pearson's correlation coefficient.

Jurasz et al. (2020) reviewed studies that investigated complementarity between renewable energy sources around the world. According to the findings of the study, complementarity assessments should be extended to provide the user with additional information on how these metrics can be applied in practice, as well as the statistical relationship between these energy sources [24]. As part of this article, an analysis of offshore wind power complementarity with other energy sources is provided, enabling the establishment of the most suitable locations for wind farms to be built in the Brazilian electrical system.

2. Offshore Wind and Complementarity with Other Resources

The offshore wind installed capacity worldwide was 64.3 GW in 2022, and the global offshore wind market grew 29% per year between 2013 and 2022, benefiting from rapid technological improvements [25]. The enormous potential of offshore wind makes it a key player in the global future energy mix, highlighting the significance of developing offshore wind in emerging markets. Moreover, largely unexploited areas with good offshore wind resources still exist around the world [3,26–29]. GWEC (2023), for instance, identifies Brazil and India as offshore wind markets that are worth watching [25,29]. Rusu (2019, 2020) points to the Black Sea [26,27] and Baltic Sea [28] as unexploited areas. Although Brazil currently lacks the installed capacity for offshore wind, this renewable energy source holds the potential to serve as a sustainable complement to the country's energy portfolio.

In Brazil, marine wind energy production projects are initiating the environmental licensing process. However, these projects have a short-term, small chance of being installed due to the current economic unviability and the lack of a detailed [30] regulatory framework. As a result, critical factors such as environmental licensing, implementation, or a detailed concession framework remain unsolved, and they are fundamental for the advancement of this new source [31].

Electrical matrix diversification adds resilience to the system to deal with climatic phenomena, supply fuel shocks, or drought. Energy complementarity is the capability of one or more sources to present complementary energy availability and work in a complementary way, that is, to not be positively correlated in time, space, or jointly in both domains [18,24], and this becomes possible when energy resources are combined in extensive regions and over time.

The research exploring the complementarity of renewable resources, including wind, solar, and hydropower [18–24], employs datasets of measured climatic data, or in cases where the weather stations are insufficient, utilizes statistical models to estimate resource availability across various locations over time. In the literature, energy complementarity is usually estimated from the correlation between different regional energy resources, normally using Pearson's correlation coefficient.

Complementarity can be useful for solving wind energy challenges because projects in different locations or wind projects combined with other energy sources in different locations are virtually managed as a complementary solution. Complementarity is an operation strategy that provides better quality of available supplies, a shorter service failure time, and less irregularity in the system generation curve.

In addition to the energy complementarity in time and space, the complementarity of the characteristics and benefits of different energy sources can also be mentioned. No single source has all the necessary features for optimal electrical system operation or is self-sufficient. The best option is complementarity between generation sources, which must operate together, creating a resource mix with different attributes. The aim is to assess the feasibility of offshore wind power integration in the Brazilian electrical system, considering environmental factors, synergy with the oil industry, cost considerations, and complementarity with other energy sources. This analysis will enable the identification of appropriate locations for deploying wind farms.

The Brazilian electricity production and transmission system is a large hydro/thermo/ wind system, with a predominance of hydropower plants. The different regions of the SIN have different hydrological, wind, and solar irradiation regimes that may present seasonal complementarity with each other, so the transmission line interconnection is extremely important. The complementarity between different regions influences reliability improvement. Positive synergy between the sources provides an increase in the percentage guaranteed power index of the integrated generation system [32]. Power demand is concentrated in the southern and eastern regions, and major electricity generation assets (primarily hydropower) are in the northern and western regions. Intermittent resource incorporation in the electrical matrix is based on the interconnection of transmission lines, reserve capacity, and complementarity, increasing capacity for supply, distributing the generation weight between resources, ensuring safety among resources, and ensuring greater safety through decreased variability [33].

The existing methods used to handle fluctuations in electricity demand and manage reserves (including storage, contingency, and response reserves) result in additional expenses or partial loss of energy production. Therefore, it is crucial to explore alternatives [19]. One option is to optimize complementarity between different intermittent sources, specifically addressing the balance between total energy production and its temporal stability.

Offshore wind farms are currently absent in the Brazilian power sector, primarily due to the substantial investment costs involved. Therefore, multiple scenarios were built to analyze the potential competitive conditions for offshore wind power, considering complementarity with other energy sources.

3. Materials and Methods

In Brazil, the Electric System National Operator (ONS) is responsible for coordinating SIN operations and deciding the optimal power generation of each power plant. The medium-term electrical planning is performed using NEWAVE, a dynamic stochastic optimization model, which is a monthly operation planning model with a 5-year horizon up to four Brazilian subsystems. NEWAVE monthly data are disaggregated by other models, such as DECOMP and DESSEM, for weekly and time dispatch [34]. The objective function of NEWAVE, DECOMP, and DESSEM is to minimize operating costs.

In this work, the Climate-based Optimization of Renewable Power Allocation (COPA) model was used [35–37] to optimize the power generation mix, minimize costs, and account for the time series of renewable power production—photovoltaic solar, wind, and hydropower—subject to demand constraints—availability of renewable resources, water storage in reservoirs, capacity restrictions, and thermal generation [38]. It is a linear, deterministic model, with hourly resolution and five regions of discretization. COPA model input data were parametrized according to scenarios provided by the Brazilian Energy Planning Company. For instance, energy exchange constraints and installed capacity of power plants were based on Ten-Year Energy Expansion Plan 2029 [39].

SIN operation models are complex and do not allow various scenario executions or long simulation periods. COPA is a long-term electricity generation model of systems with high renewable source participation. In addition, the COPA model has the advantage of allowing intermittent sources to compete simultaneously with others, aiming to account for the unpredictability of wind and solar photovoltaic production and, thus, create time series considering water flows, wind speeds, and solar radiation [35].

The four SIN subsystems were specified in COPA (north, south, northeast, and southeast/midwest). There is a 5th region for transmission purposes only, where there is no load and no power generation; it is a fictitious node. The model is designed for the Brazilian energy system but can be adapted to other regions if necessary. The scenarios developed involve the introduction of offshore wind as an expansion option to satisfy the future energy load in each of these subsystems. The deployment of additional renewables is limited by the electricity demand that is not satisfied by existing hydropower projects. The system has been optimized for one-year duration.

Electrical system total costs consider the investment, operation, and maintenance costs of all power plants that are necessary to meet demand. Power plants receive cost data individually, according to their technology. There are two categories of power plants: existing, which have zero investment costs and are already installed, and available for expansion, which considers the investment costs and potential in each subsystem. Intermittent renewables have investment costs and a maximum potential that can be invested. For thermal plants, the fuel variable costs are contemplated as well. When the power plant already exists, the investment cost is zero; for example, hydropower plants have no costs in the model because any scenario considers new hydropower plant construction. The objective function is to minimize the system's total costs, which is a simple sum of production costs throughout the period [36].

Since COPA is a deterministic model, there are no hydrological, load, or planning uncertainties. It has perfect resource foresight. This limitation may result in underestimating the costs of integrating uncertain resources. The model chooses between the different renewable energy capacities in different regions and optimizes the production mix represented by the hourly time series of wind, solar, and water production for each plant (existing or new) in each subsystem, depending on its capacity factor. The model also manages hydropower reservoirs and thermal backup dispatch using energy hourly series. The results indicate optimal regions to expand the system through powerplant installation, which in this study's case is an offshore wind farm settlement.

3.1. Correlation Analyses

Resource and energy production complementarity in Brazil was investigated using Pearson's correlation coefficient (r). The coefficient r indicates the intensity and association direction between two variables, given by the following:

$$r = \frac{1}{n-1} \Sigma \left(\frac{x_i - \bar{x}}{S_x} \right) \left(\frac{y_i - \bar{y}}{S_y} \right) \tag{1}$$

here x_i represents the observed value of the x series ($x_1, \ldots, x_n$), $\bar{x}$ represents the average, and S_x is the standard deviation. The same logic can be applied to y. This coefficient was calculated by comparing resource and power generation over time (hourly series in one year) and space (regions of Brazil). Pearson's correlation coefficient, which ranges from -1 to $+1$, provides information on the relationship between variables. A positive signal indicates that the variables vary in the same direction and have similar behavior, while negative signal suggests that one variable decreases when the other increases. Furthermore, the magnitude of the coefficient reflects the strength between the variables [23]. Thus, if it is closer to -1, the more complementary the resources are; that is, when one tends to increase, the other tends to decrease. When the value approaches 1, the series are more similar, and the features are more correlated.

As these extreme values (0 or |1|) are hardly found in practice, this article considers values of r (+ or −), ranging from 0.10 to 0.29 indicating weak correlation, from 0.30 to 0.49 moderate correlation, and values exceeding 0.50 representing -strong correlation. The complementarity was investigated using Pearson's correlation coefficient between power production obtained from COPA over time and space.

3.2. Input Data

The model expansion capacity relies on hourly time series data of renewable energy production obtained from reanalysis data. These data are employed to establish the optimal combination of capacity generation, assuming perfect forecasting, in addition to simulating the power plant's hourly dispatch.

Climate input variables used in this study are derived from the MERRA-2 global reanalysis data set. The reanalysis data combine historical observational data into a numerical weather forecast model to reconstruct the climate, generating a complete set of data in space–time [40]. For this analysis, data were retrieved from the Renewables Ninja website [41], a calibrated model that simulates the MERRA-2 hourly power produced by wind farms located anywhere in the world and has been validated in 23 European countries [40]. MERRA-2 reanalysis data were selected due to their convenient accessibility, high spatial and temporal resolution, and consistent stability across different time scales.

Meteorological parameters change significantly from year to year, which is essential for the interannual complementarity study.

It is important to note that climate change affects the climatic conditions that are crucial to renewable resources. Due to climate change, renewable energies in Brazil are vulnerable [42,43]. In Brazil's poorest regions, climate change may negatively affect electricity generation (particularly hydropower) and biofuel production [43]. Brazil's energy matrix is sensitive to climate variations due to hydroelectricity and wind power [44].

However, to represent a long period of data, one must choose a year with the typical characteristics of the site. The typical meteorological year allows us to apply the knowledge of local climatology to the procedures for evaluation, design, planning, and operation of generation plants from renewable sources [44]. Since 2014, Brazil has been going through a water crisis scenario that is still being regularized, so 2013 was chosen to be the typical meteorological year. The most important input data for the operational model are power capacity factors from onshore and offshore wind, photovoltaic solar energy, and hydropower power production (Table 1).

Table 1. Summary of data sources.

Data	Reference
Offshore wind resource	[41]
Flow	[45]
Maximum capacity and the initial and final levels of the reservoir	[46]
Solar radiation	[41]
Onshore wind resource	[47]
Onshore wind and solar installed capacity	[48]
Maximum onshore wind power capacity	[10]
Maximum offshore wind capacity	[31]
Capacity of transmission lines and power exchange; installed capacity of thermal power plants	[49]
Brazilian energy demand in 2030	[50]
Offshore wind energy investment cost	[51]
Investment costs of other sources	[51]

Source: Authors.

The offshore wind capacity factors obtained from the Renewables Ninja website are from 2013, with a 7 MW turbine model at a height of 140 m. This turbine model was chosen because it represents the average nominal power of 6.8 MW in 2018 worldwide [29].

A limitation of the study in using only MERRA-2 reanalysis data without a calibration specifically for Brazil is that data from numerical simulation models add uncertainty to energy resource assessment estimates. The use of experimental offshore wind measurement data through measurement campaigns and offshore weather stations is necessary; however, there is a high cost in obtaining these types of data.

Offshore wind power production is generated using hourly data from the capacity factor of 45 points in Brazil, 10 km, 100 km, and 150 km (15 points each) far from coast, distributed across the north, northeast, southeast, and south regions (Figure 1). The hourly capacity factor of each subsystem was calculated by averaging the hourly capacity factors of the points in each subsystem.

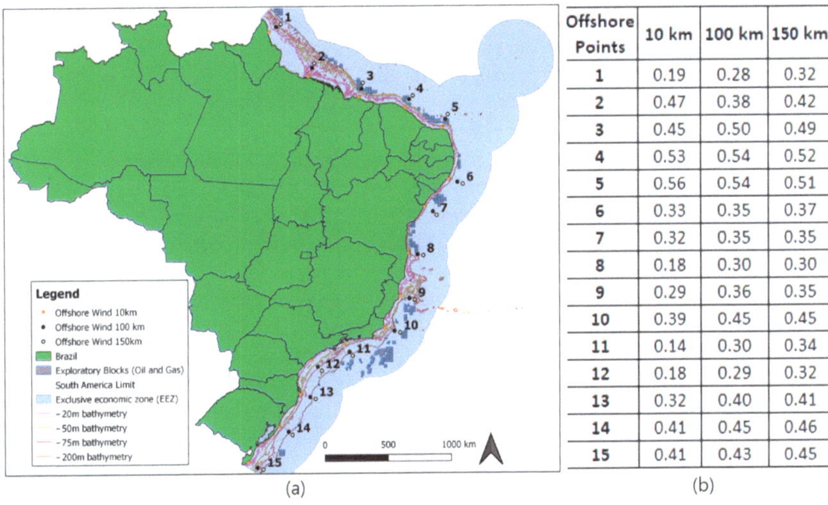

Offshore Points	10 km	100 km	150 km
1	0.19	0.28	0.32
2	0.47	0.38	0.42
3	0.45	0.50	0.49
4	0.53	0.54	0.52
5	0.56	0.54	0.51
6	0.33	0.35	0.37
7	0.32	0.35	0.35
8	0.18	0.30	0.30
9	0.29	0.36	0.35
10	0.39	0.45	0.45
11	0.14	0.30	0.34
12	0.18	0.29	0.32
13	0.32	0.40	0.41
14	0.41	0.45	0.46
15	0.41	0.43	0.45

(a) (b)

Figure 1. Data location, bathymetry lines, and exploratory blocks (oil and gas) map: (**a**) Brazil map; (**b**) offshore wind hourly capacity factor 10 km, 100 km, and 150 km far from coast points.

Points 4 and 5 (on the coast of Ceará and Rio Grande do Norte) show the best locations for wind generation in Brazil, contemplating only capacity factors and depth. These points have the highest average capacity factors at all distances from the coast, where hourly peaks can be found with up to 96% capacity factor. In addition, the ocean depth in the northeast subsystem is slight, especially in Ceará and Rio Grande do Norte, where it is possible to find maximum depths from 50 m up to 70 km from the coast [52].

The 10 km distance was used to represent the wind farms with a fixed foundation near the coast, as they do not interfere so much in touristic regions and economic activities along the coast. In addition, all 15 points chosen 10 km from the coast have depths less than 50 m, which makes the installation of a fixed foundation viable. The location of offshore wind farms has a minimum distance from the coast between 2 and 15 km [8]. A minimum distance of approximately 10 km from the coast produces a 48% reduction in the public rejection rate, as the turbines would have no effect on recreational use of the beach [53]. UK and Germany currently have the farthest wind farms from shore, at over 100 km each [29]. The 100 km distance was chosen to represent the farthest from shore, and in the case of Brazil, floating wind farms, because the average depth at that distance is greater than 60 m. The 150 km distance is an analysis of future floating wind farms further from the coast.

3.3. Scenarios

To assess offshore wind complementarity with other energy sources, we have developed 10 scenarios to simulate offshore wind generation and determine impacts of this energy type in different subsystems, installed capacities, and system performance. In each scenario, we defined an offshore wind capacity specified by subsystem at a distance of 10 km from the Brazilian shore, because when offshore wind enters the market, it will likely have a fixed foundation.

Scenarios were created with the installed capacities of 400 MW and 800 MW offshore wind in different regions. Nowadays, Brazil has at least 800 MW of offshore wind in environmental licensing process in the northeast, with a single park of 400 MW [31]. In addition, the average size of wind farms in 2018 was 561 MW worldwide.

The reference scenario does not have offshore power plants. In this scenario, offshore wind was an option for expansion; however, the model was not chosen due to the high installation cost. In the other 9 scenarios, we defined an offshore wind development to analyze the SIN behavior—zero investment costs. The location and combination of regions in the scenarios were chosen based on data of greater resource complementarity, proximity to demand spots, and synergy with the oil and gas sector.

4. Results

We analyzed the changes in the Brazilian power mix incorporating a range of offshore wind expansion capacities and verifying the fact that the complementarity between hydro reservoirs and offshore wind plants is able to replace thermal power plants in some cases. The hourly Pearson's correlation coefficients indicated a low complementarity between solar and offshore wind production in the north, south, and southeast/midwest. In different subsystems, offshore wind plants are correlated. The same occurs with onshore and offshore winds; they are correlated. Coastal areas in the northeast and southeast have more positive points for the deployment of offshore wind farms. There is a greater complementarity between offshore wind generation in the northeast and southeast and hydropower in the southeast/midwest, northeast, and north regions. In addition, there is synergy with the oil industry in those two regions. The southeast is the nearest region with the highest energy consumption. With those results, it is possible to minimize decision-makers' uncertainties in investment allocations. Brazil's offshore winds can diversify the electricity matrix, stabilize water fluctuations, and reduce thermal plant use, which increases the cost of production and the emissions of polluting gases.

4.1. Complementarity Analysis of Resources in Brazil

The hourly resource time series in Brazil covering 2013 was used to calculate the Pearson's correlation coefficient. The results are divided between different subsystems and energy resources, indicating the cross-correlations between resources. There is a significant complement between marine wind resources and water sources (Table 2), mainly in the southeast/midwest, and north regions, with offshore winds from the northeast and southeast (lowest grade). The results here are similar to those found by Silva (2015) and Borba (2023). There is a moderate seasonal complementarity between the flow in the southeast/midwest, and northeast, 10 km from shore winds ($r = -0.48$).

Regarding solar onshore and wind offshore resources, there is low complementarity. This relationship is slightly more significant when comparing the offshore wind resources in the southeast with solar radiation in the southeast/midwest, and northeast. Nascimento et al. (2022) examined offshore wind and offshore solar energy sources, and their findings also indicated an annual and hourly complementarity between these two sources. The offshore solar complements offshore wind up to 40% in the northeast region within a water depth of up to 50 m [54]. Additionally, offshore wind plants at varying distances from the shoreline are correlated, primarily within the subsystem. The same occurs when comparing winds at sea and on land, as the resources are more correlated than complementary.

Table 2. Correlation coefficient between different regions and energy resources: south (S), southeast/midwest (SE/MW), northeast (NE), and north (N) subsystems and the normalized values of solar radiation, flow, and the offshore wind speed at 10, 100, and 150 km far from shore and onshore winds. Green colors are more complementary and reds more correlated and bold numbers are moderate/strong correlation.

		1.00	0.80	0.60	0.40	0.20	0.00	−0.20	−0.40	−0.60	−0.80	−1.00
		Correlated					No correlation			Complementary		

		Offshore Wind (10 km)				Offshore Wind (100 km)				Offshore Wind (150 km)			
		S	SE/MW	NE	N	S	SE/MW	NE	N	S	SE/MW	NE	N
Offshore wind (10 km)	S	1.00											
	SE/MW	0.34	1.00										
	NE	0.04	0.21	1.00									
	N	0.10	0.17	0.30	1.00								
Offshore wind (100 km)	S	0.94	0.32	0.05	0.06	1.00							
	SE/MW	0.40	0.90	0.25	0.17	0.42	1.00						
	NE	0.06	0.19	0.93	0.24	0.06	0.27	1.00					
	N	0.09	0.08	0.07	0.65	0.05	0.11	0.10	1.00				
Offshore wind (150 km)	S	0.89	0.30	0.04	0.05	0.99	0.41	0.05	0.03	1.00			
	SE/MW	0.40	0.84	0.28	0.17	0.44	0.98	0.30	0.11	0.44	1.00		
	NE	0.05	0.18	0.91	0.24	0.06	0.27	0.99	0.10	0.05	0.30	1.00	
	N	0.05	0.04	0.10	0.86	0.01	0.07	0.09	0.76	−0.01	0.07	0.09	1.00
Onshore wind	S	0.86	0.22	0.06	0.13	0.76	0.26	0.13	0.12	0.72	0.26	0.12	0.09
	SE/MW	0.14	0.27	0.06	0.14	0.13	0.34	0.22	0.23	0.12	0.34	0.22	0.17
	NE	0.15	0.12	0.45	0.21	0.12	0.22	0.59	0.19	0.11	0.23	0.56	0.14
	N	0.11	0.19	0.18	0.45	0.11	0.21	0.15	0.24	0.10	0.20	0.14	0.29
Solar	S	−0.16	−0.13	−0.02	−0.07	−0.10	−0.17	−0.23	−0.18	−0.08	−0.15	−0.20	−0.10
	SE/MW	−0.15	−0.21	0.00	−0.13	−0.08	−0.21	−0.18	−0.20	−0.06	−0.19	−0.15	−0.15
	NE	−0.14	−0.22	0.01	−0.09	−0.06	−0.20	−0.17	−0.15	−0.05	−0.17	−0.14	−0.09
	N	−0.13	−0.10	0.02	−0.11	−0.07	−0.16	−0.20	−0.25	−0.06	−0.14	−0.18	−0.18
Hydro generation	S	0.08	0.15	0.02	−0.14	0.13	0.17	0.03	−0.14	0.15	0.17	0.03	−0.25
	SE/MW	−0.11	−0.16	**−0.48**	0.02	−0.16	−0.18	**−0.49**	0.20	−0.17	−0.20	**−0.49**	0.25
	NE	−0.12	−0.16	−0.27	0.02	−0.17	−0.18	−0.30	0.14	−0.19	−0.19	−0.31	0.24
	N	−0.09	−0.24	**−0.54**	−0.03	−0.12	−0.26	**−0.56**	0.14	−0.13	−0.27	**−0.57**	0.25

Offshore wind behavior varies in comparison to onshore winds depending on the Brazilian geographic region, and there is a significant variability of winds along the coast during the seasons. The southeast has the lowest average offshore capacity factor in Brazil, and in the autumn and winter, it is slightly lower than in the spring and summer. The onshore and offshore values are to some degree correlated, and the marine capacity factor varies less than the land resource.

The northeast subsystem offshore capacity factor value is more constant than the great variation that exists onshore. These values are moderately correlated. This subsystem has the largest average capacity factor in Brazil, and its highest values occur in the winter and spring.

In the north subsystem, offshore wind capacity factors are slightly higher during the summer and autumn, while in the winter and spring, these values are lower. These results are similar to those found by Silva (2015). In this subsystem, the onshore capacity factor is negligible, and the offshore suffers great variations throughout the year.

The south subsystem has the second-largest average capacity factor in Brazil. Offshore wind is highly correlated with onshore wind. In addition, the subsystem has the highest rate of oscillations throughout the year, with the most significant values in the winter and spring.

4.2. Results of Offshore Wind Power Integration into the Brazilian Electrical System (COPA)

The first scenario considers the current installation cost of offshore wind power plants, resulting in no offshore wind generation in Brazil. This technology was not competitive compared to other available options for expansion. For this reason, scenarios were created to simulate offshore wind generation, considering already-installed power plants (zero installation cost) to evaluate their complementarity with other sources. The share of each source in the Brazilian electrical matrix indicates that offshore wind power shifts thermoelectric generation into the electricity mix (Table 3).

Table 3. Brazilian electric matrix scenarios with the insertion of offshore wind (OW) source (% in energy).

Scenarios	Offshore Wind (%)	Onshore Wind (%)	Total Wind (%)	PV (%)	Hydro (%)	Thermo (%)
Without OW	0.00	26.68	26.68	1.99	58.95	12.38
OW northeast (800 MW)	0.37	26.43	26.80	1.99	58.96	12.25
OW southeast (800 MW)	0.24	26.56	26.81	1.99	58.96	12.25
OW southeast (400 MW)	0.12	26.37	26.50	1.99	59.16	12.35
OW south (400 MW)	0.18	26.60	26.79	1.99	58.95	12.28
OW north (400 MW)	0.15	26.64	26.79	1.99	58.92	12.30
OW southeast (400 MW) and northeast (800 MW)	0.50	26.28	26.78	1.99	58.99	12.25
OW southeast and northeast (800 MW each)	0.61	26.31	26.92	1.99	58.95	12.14
OW south and northeast (800 MW each)	0.73	26.24	26.98	1.99	58.96	12.08
OW south, southeast, and northeast (800 MW each)	0.98	26.10	27.07	1.99	58.93	12.01

The hourly time series resulting from COPA covering 2013 was used to calculate the Pearson's correlation coefficient. These coefficients between offshore wind generation in different regions and energy production from different sources (Table 4) in all scenarios presented a similar correlation coefficient; that is, the power generation behavior is the same when the offshore wind is installed in a certain subsystem, varying only in the amount of energy produced.

4.2.1. Wind Offshore and Hydropower

Hydropower energy represented 64.9% of the country's electricity generation in 2019 [55], and grid stabilization in the medium and long term is necessary. In order to provide more stable and reliable electrical generation, the complementarity of these sources also allows for a gain in water storage during critical periods of drought where hydraulic generation is compromised.

With the increase in wind energy penetration, the aim is to bring benefits to the SIN taking advantage of the water storage (energy) in hydropower plants, the opportunity to use the existing water and transmission resources more efficiently, and the potential to improve hydrological operations, as well as to develop a more diversified, robust, and clean general energy supply portfolio [56].

Northeast Subsystem

Offshore wind generation has a strong seasonality in the northeast. Although the annual behavior of the accumulated hourly generation of hydropower dams and offshore wind power in the northeast subsystem is complementary, there is no correlation between the electricity production values ($r = -0.03$). However, when offshore wind generation in the northeast is compared to hydropower in the north, there is a moderate complementarity ($r = -0.38$), which benefits the system. In the north subsystem, hydropower plants have no reservoir due to environmental issues and are run-of-the-river hydroelectric powerplant (without seasonal or multiannual regularization). In other words, the hydropower plant's potential is only reached during the wet period because, in the dry period, the generation is much lower. Therefore, the complementary generation of energy by offshore wind in the northeast is beneficial for the system, and the seasonality of the winds (greatest generation between June and November) is inverse to the seasonality of the rain regime (greatest generation between December and May).

Table 4. Summary correlation coefficient between different regions and energy production by different sources in relation to offshore wind power. Green colors are more complementary and reds more correlated.

| | | COPA Offshore Wind Power Generation | | | |
		South	Southeast/Midwest	Northeast	North
Offshore wind power generation	South	1.00			
	Southeast/midwest	0.30	1.00		
	Northeast	0.01	0.16	1.00	
	North	-	-	-	-
Onshore wind power generation	South	0.84	0.20	−0.01	0.10
	Southeast/midwest	0.13	0.29	−0.08	0.05
	Northeast	0.15	0.11	0.24	0.21
	North	-	-	-	-
Photovoltaic	South	−0.15	−0.10	0.07	−0.02
	Southeast/midwest	−0.13	−0.16	0.10	−0.07
	Northeast	−0.12	−0.18	0.11	−0.05
	North	−0.12	−0.06	0.11	0.02
Hydropower	South	−0.40	−0.06	0.17	0.07
	Southeast/midwest	−0.26	−0.06	0.01	0.02
	Northeast	−0.11	0.10	−0.01	0.04
	North	−0.08	−0.16	−0.38	−0.21
Thermal	South	−0.21	−0.11	0.05	0.00
	Southeast/midwest	−0.16	−0.09	0.07	0.01
	Northeast	−0.08	0.00	0.05	−0.03
	North	0.05	0.18	0.46	0.19

Southern Subsystem

There is a moderate ($r = -0.4$) complementarity between hydropower in the southern subsystem and offshore wind generation in the same subsystem. Although the resources are not complementary (Table 2), the hourly series of hydropower energy production is strongly influenced by storage in the reservoirs.

The complementarity between offshore wind power in the south subsystem and water generation in the southeast/midwest subsystem, although weak ($r = -0.26$), can bring benefits to the system, as the southeast subsystem has the largest hydro storage capacity and also has the higher energy demand of the SIN. In other words, the south subsystem exports energy to the southeast to meet its demand. Additionally, the south subsystem contributes to reduce the depletion of reservoirs in the southeast.

Southeast/Midwest Subsystem

The hourly correlation between hydropower and offshore wind generation in the southeast is insignificant ($r = -0.06$); however, the seasonal complementarity between these two sources in the subsystem can be highlighted. Offshore wind generation is greatest between June and November, when the dry season occurs.

North Subsystem

Offshore wind production in the northern subsystem presents a weak complementarity ($r = -0.21$) in relation to hydro generation in the same subsystem. When comparing this new energy production with the generation in hydropower plants in other regions, there is no correlation between the sources.

Electrical system operation planning expects natural source complementarity, so that the secondary source (offshore wind) complements the main source (hydraulic) in periods of low hydro availability. One of the benefits of complementarity is that offshore wind meets the modulation and seasonality demanded by hydropower plants. The hydro/wind complementarity identified in some regions can optimize the use of hydropower energy reservoirs in the country.

In Brazil, wind generation has played a fundamental role in times of low hydro generation [57]. There is a great generation synergy between hydropower and wind sources, with wind energy production being generally stronger during the dry season (from May to October).

Brazil

Offshore wind and hydro generation have greater seasonal complementarity in the northeast and southeast (lowest grade) when assessing the energy production from these sources along Brazil (Figure 2). When offshore wind farms are implemented in the northeast and southeast regions, the largest offshore wind production occurs in the second half of the year, while for hydro, it occurs in the first. Thus, there is complementarity in these subsystems between offshore and hydropower wind sources. That is, during the dry period, when the levels of the reservoirs and the energy production of the hydropower plants are reduced, it occurs precisely when the offshore wind farms have their greatest generation. Consequently, the results show a lower marginal operating cost and less need for thermoelectric drive. In addition, Borba et al. (2023) also demonstrated that as the wind power capacity increases, dynamic dispatch is shifted, and natural gas becomes less important during the dry seasons. Using existing reservoirs, offshore wind farms can be integrated into highly renewable scenarios, but they are not sufficient in a complete phaseout of fossil fuels, which requires other storage sources [58].

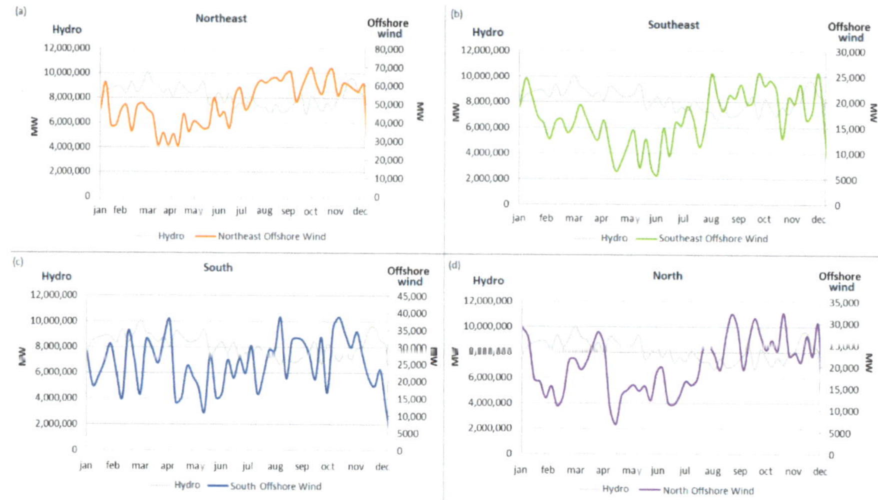

Figure 2. Hydro generation throughout Brazil compared to offshore wind generation in each subsystem: (**a**) northeast offshore wind—scenario with 800 MW; (**b**) Ssutheast offshore wind—scenario with 400 MW; (**c**) south offshore wind—scenario with 400 MW; (**d**) north offshore wind—scenario with 400 MW.

4.2.2. Wind Offshore and PV

Offshore wind power plants installed in the south, southeast, and north are negatively (weak) hourly correlated with photovoltaic farms (Table 2). When analyzing the correlation between sources on a typical day (correlation between the average generation of each source at one hour of the day), the magnitude of this complementarity increases (Table 5).

Table 5. Correlation coefficient between solar PV and offshore wind generation in one day. Green colors are more complementary and reds more correlated.

		Offshore Wind Power Generation			
		South	Southeast/Midwest	Northeast	North
	South	−0.80	−0.48	0.56	−0.20
Photovoltaic	Southeast/midwest	−0.86	−0.54	0.59	−0.21
Power	Northeast	−0.88	−0.68	0.66	−0.24
	North	−0.80	−0.34	0.49	−0.20

The northeast subsystem has a strong correlation between solar generation and offshore wind power. Meanwhile, the most negative correlation occurs between offshore wind in the south and photovoltaic in the northeast ($r = -0.88$), indicating that greater offshore wind generations are observed at night, when there is no photovoltaic generation.

During the year, there is a weak correlation between wind offshore and solar PV. While offshore wind generation is greater during the second half of the year, photovoltaic generation does not change sharply over the year (Figure 3). However, in terms of the correlation between the average generation in daily hours, there is a strong complementation between these two sources since wind offshore plants are installed in the south and southeast (Table 5). In addition, there is a benefit of these two sources' integration complementarity: the combined generation curve has a more subtle variation than the one with only PV production (Figure 3). That is, the integration of offshore wind and solar resources can contribute to minimize the intermittent resource's electricity variability.

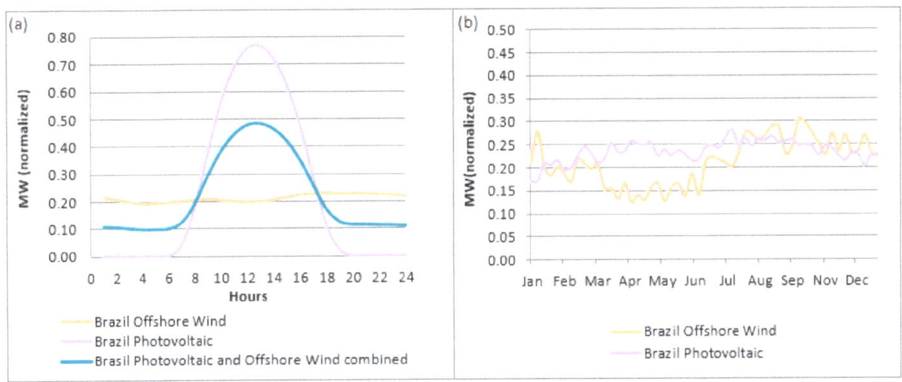

Figure 3. Sum of offshore wind generation and standardized photovoltaics, in the "southeast (400 MW) and northeast (800 MW) offshore wind" scenario; (**a**) hourly average generation on a typical day normalized to their maximum value; and (**b**) weekly generation throughout the year normalized to their maximum value.

4.2.3. Wind Offshore and Thermal Plants

Thermopower generation in the electric matrix is a strategic factor for the Brazilian electricity sector, as it can meet the base load, complement renewable sources, or meeting cutting-edge demands. With major hydropower generation of electricity in Brazil, the thermoelectric plants have acted significantly in periods of hydrological scarcity. However, they emit GHG during energy production.

Offshore wind generation in Brazil displaces a certain share of the thermal generation of the electrical mix (Table 3), making it even more based on renewable resources. Although small, thermoelectric generation in the northeast subsystem decreases with the highest offshore wind generation share entering the matrix (Figure 4). As a first exercise, offshore wind penetration in the matrix in each scenario is modest, but there is a trend that the greater the entry of offshore wind in the matrix, the smaller the use of thermal generation, which can be replaced by hydro/wind complementary.

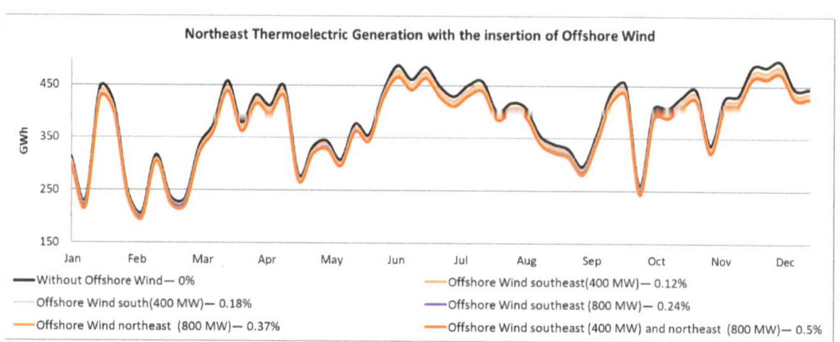

Figure 4. Thermoelectric generation in the northeast with several percentages of insertion of offshore wind in different scenarios. Source: Authors.

4.2.4. Offshore Wind Farm's Location

Each subsystem in Brazil has favorable and unfavorable points for offshore wind farm installation. However, the regions with the most favorable points are the coastal areas of the northeast and southeast.

In the north subsystem, the average capacity factor is higher than in the southeast. However, there is no resource complementarity, no synergy with oil, and it is far from the energy demand spot in Brazil, requiring large investment in the transmission lines. The south subsystem has one of the largest capacity factors in Brazil, but the complementarity of wind and water resources is less than in the northeast and southeast subsystems, and there is no synergy with the oil industry.

In the southeast, the wind offshore capacity factor is the lowest one, but this resource is somewhat complementary with the water source. However, an offshore wind farm installed in this subsystem is located close to the largest demand spot in Brazil, which would reduce the energy transport loss and investments in new transmission line construction. In addition, there is synergy with the oil industry, as there are several active fields that offshore wind energy can be used to supply the platform itself or deployed on decommissioned fixed platforms.

In the northeast, there are several favorable characteristics that could make this subsystem the main gateway to this new source of wind energy in Brazil. The biggest capacity factors are present in this subsystem, and these wind resources are quite complementary with the water source. There is synergy with the oil industry and decommissioning of platforms, as in the southeast. In addition, the depth of the ocean in the region is relatively slight.

Because of their favorable points, the northeast and southeast regions were chosen for the installation of offshore wind energy and representation of the Brazilian electric mix projected for 2030 through the scenario "northeast offshore wind (800 MW) and southeast (400 MW)" (Figure 5).

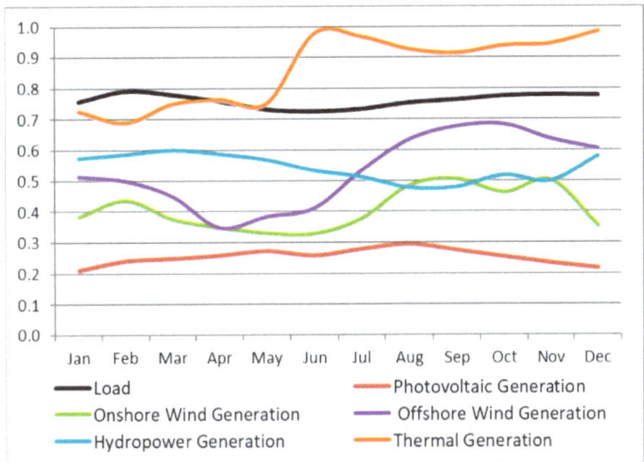

Figure 5. Brazil annual performance of electric generation by source and consumption of electricity (the load curve) in the scenario "Offshore wind power southeast (400 MW) and northeast (800 MW)". Monthly variation of the parameters normalized to their max value.

To represent the beginning of offshore wind power production in Brazil, we selected the scenario "Offshore Wind southeast (400 MW) and northeast (800 MW)" due to the benefits already mentioned in each subsystem and the system's reliability index increment with two distinct regions. The resource complementarity between different regions influences the improvements to the system's reliability [32]. These complementarities, whether separately or combined, provide flexibility and constancy for SIN energy production.

From the annual behavior of electricity generation by source and consumption of electricity (load curve) in Brazil (Figure 5), it is observed that the consumption of electricity and photovoltaic generation remain practically constant; they do not vary much during the year, while hydropower generation reduces due to rain reduction in the dry period (from May to November). However, onshore and offshore wind generation increases, especially in the second half of the year, when the reservoirs water levels are low. The highest wind intensity months are those with the lowest rainfall intensity and vice versa.

5. Discussion and Conclusions

Brazil has a significantly growing energy demand, and energy planning should focus on defining an electrical matrix with fewer environmental and social impacts without preventing the expansion of installed capacity in an economically viable way.

The COPA model encompasses historical offshore wind availability profiles per region using a typical meteorological year. Future work can include the effect of climate change on wind availability to address the complementarity between offshore wind power plants and the other sources of electricity conversion.

Brazil has a great offshore wind power potential (330.5 GW—more restrictive potential considering social and environmental aspects [8]), diversified in time and space, and complementary with hydropower and solar resources. The Brazilian electric matrix is based on hydropower generation and medium- and long-term grid energy stabilization is necessary. Commonly, hydropower is complemented by conventional sources, such as thermoelectric. However, offshore wind energy can assist hydropower generation storing water for future power generation or other uses. In addition, the operational system cost is reduced since with greater offshore wind penetration, there is a reduction in thermoelectric use, which has the most expensive operation costs due to fuel use.

In addition, there is the complementarity of the characteristics and benefits of hydropower and wind energy. Hydropower reservoir flexibility will help in wind energy integration through storage [56,58]. At the same time, wind resources will help hydropower reservoirs in the long term, as their seasonal variability and drought impacts are expected to decrease [23,58]. Brazilian coast offshore winds can diversify the electrical matrix, stabilize water fluctuations, avoid rationing and blackouts, and reduce thermal plant use, which increases production costs and pollutes gas emissions [7].

The offshore wind source has important advantages in Brazil: the greatest potential is the possibility of installation close to load spots, synergy with oil, and the possibility of complementarity with water and solar sources. Coastal areas in the northeast and southeast have more favorable points for the deployment of offshore wind farms. There is a greater complementarity between offshore wind generation in the northeast and southeast and hydropower in the southeast/midwest, northeast, and north regions. In addition, there is synergy with the oil and gas industry in those two regions, and the southeast is the nearest region with the highest energy consumption. Although the south subsystem has a greater offshore wind capacity factor than the southeast, it does not have these last two factors that are positive to the location of the offshore wind.

With those results, it is possible to minimize decision-makers' uncertainties in investment allocations. Brazil's offshore winds can diversify the electricity matrix, stabilize water fluctuations, and reduce thermal plant use, which increases electricity production costs and the emission of polluting gases.

Since the complementarity between wind offshore resources and further sources of electricity conversion has been proven in this paper, future works should encompass the development of power regulation to accommodate hydrogen production from offshore wind and its environmental impacts.

Author Contributions: Conceptualization, E.C.N. and A.O.P.J.; methodology, E.C.N., A.O.P.J. and R.C.M.; software, E.C.N. and R.C.M.; validation, E.C.N., A.O.P.J. and R.C.M.; formal analysis, E.C.N. and R.C.M.; investigation, E.C.N. and R.C.M.; resources, E.C.N.; data curation, E.C.N.; writing—original draft preparation, E.C.N.; writing—review and editing, A.O.P.J. and R.C.M.; visualization, E.C.N., A.O.P.J. and R.C.M.; supervision, A.O.P.J. All authors have read and agreed to the published version of the manuscript.

Funding: This work was supported by the National Council for Scientific and Technological Development (CNPq) through scholarships awarded to Erika Nogueira (Grant number 141198/2020-6) and Amaro Pereira (303432/2022-5).

Data Availability Statement: Not applicable.

Conflicts of Interest: The authors declare no conflict of interest.

Abbreviations

GHG	greenhouse gas
SIN	National Interconnected System (in Portuguese)
PV	photovoltaic solar energy
GWEC	Global Wind Energy Council
ONS	Electric System National Operator (in Portuguese)
NEWAVE	Long- and Medium-Term Interconnected Hydro-Thermo-Wind Systems Operation Planning Model (Modelo de Planejamento da Operação de Sistemas Hidro-termo-eólicos Interligados de Longo e Médio Prazo in Portuguese)
DECOMP	Short-term planning model for the operation of interconnected hydrothermal systems (Modelo de Planejamento de Curto Prazo da Operação de Sistemas Hidrotérmicos Interligados in Portuguese)
DESSEM	Short-Term Hydrothermal Dispatch Model (Modelo de Despacho Hidrotérmico de Curto Prazo in Portuguese)
COPA	Climate-based Optimization of Renewable Power Allocation
S	south
SE/MW	southeast/midwest
NE	northeast
N	north

References

1. Østergaard, P.A.; Duic, N.; Noorollahi, Y.; Mikulcic, H.; Kalogirou, S. Sustainable Development Using Renewable Energy Technology. *Renew. Energy* **2020**, *146*, 2430–2437. [CrossRef]
2. Østergaard, P.A.; Duic, N.; Noorollahi, Y.; Kalogirou, S. Latest Progress in Sustainable Development Using Renewable Energy Technology. *Renew. Energy* **2020**, *162*, 1554–1562. [CrossRef]
3. Rusu, E.; Onea, F. A Parallel Evaluation of the Wind and Wave Energy Resources along the Latin American and European Coastal Environments. *Renew. Energy* **2019**, *143*, 1594–1607. [CrossRef]
4. IEA. *Offshore Wind Outlook 2019*; IEA: Paris, France, 2019.
5. Esteban, M.D.; Diez, J.J.; López, J.S.; Negro, V. Why Offshore Wind Energy? *Renew. Energy* **2011**, *36*, 444–450. [CrossRef]
6. IRENA. *Renewable Power Generation Costs in 2018*; IRENA: Abu Dhabi, United Arab Emirates, 2019.
7. Silva, A.R. Energia Eólica em Alto Mar: Distribuição dos Recursos e Complementaridade Hídrica. Ph.D. Thesis, Universidade Federal do Rio Grande do Norte, Natal, Brazil, 2015.
8. De Carvalho Silva, A.J.V. *Potencial Eólico Offshore No Brasil: Localização de Áreas Nobres Através de Análise Multicritério*; Universidade Federal do Rio de Janeiro: Rio de Janeiro, Brazil, 2019.
9. De Azevedo, S.S.P.; Pereira, A.O.; da Silva, N.F.; de Araújo, R.S.B.; Júnior, A.A.C. Assessment of O Shore Wind Power Potential along the Brazilian Coast. *Energies* **2020**, *13*, 2557. [CrossRef]
10. CEPEL. *Atlas Do Potencial Eólico Brasileiro: Simulações 2013*; CEPEL: Rio de Janeiro, Brazil, 2017.
11. Ortiz, G.P.; Kampel, M. *Potencial de Energia Eólica Offshore na Margem do Brasil*; V Simpósio Brasileiro de Oceanografia: Santos, Brazil, 2011.
12. Pimenta, F.; Kempton, W.; Garvine, R. Combining Meteorological Stations and Satellite Data to Evaluate the Offshore Wind Power Resource of Southeastern Brazil. *Renew. Energy* **2008**, *33*, 2375–2387. [CrossRef]
13. Vinhoza, A.; Lucena, A.F.P.; Rochedo, P.R.R.; Schaeffer, R. Brazil's Offshore Wind Cost Potential and Supply Curve. *Sustain. Energy Technol. Assess.* **2023**, *57*, 103151. [CrossRef]
14. De Carvalho, L.P. *A Potencial Sinergia Entre a Exploração e Produção de Petróleo e Gás Natural e a Geração de Energia Eólica Offshore: O Caso Do Brasil*; Universidade Federal do Rio de Janeiro: Rio de Janeiro, Brazil, 2019.
15. Da Costa, K.M. *Reutilização de Plataformas Fixas Para Geração de Energia Eólica Offshore*; UFRJ: Rio de Janeiro, Brazil, 2018; Volume 53.
16. Barros, J.C.; Fernandes, G.C.; Silva, M.M.; Da Silva, R.P.; Santos, B. Fixed Platforms at Ageing Oil Fields—Feasibility Study for Reuse to Wind Farms. In Proceedings of the Annual Offshore Technology Conference; Offshore Technology Conference, Houston, TX, USA, 1–4 May 2017; Volume 6, pp. 4564–4581.
17. Braga, J.; Santos, T.; Shadman, M.; Silva, C.; Assis Tavares, L.F.; Estefen, S. Converting Offshore Oil and Gas Infrastructures into Renewable Energy Generation Plants: An Economic and Technical Analysis of the Decommissioning Delay in the Brazilian Case. *Sustainability* **2022**, *14*, 13783. [CrossRef]
18. Beluco, A.; Krenzinger, A.; Souzza, P. A Complementariedade No Tempo Entre as Energias Hidrelétrica e Fotovoltaica. *Rev. Bras. Recur. Hídricos* **2003**, *8*, 99–109. [CrossRef]
19. Kougias, I.; Szabó, S.; Monforti-Ferrario, F.; Huld, T.; Bódis, K. A Methodology for Optimization of the Complementarity between Small-Hydropower Plants and Solar PV Systems. *Renew. Energy* **2016**, *87*, 1023–1030. [CrossRef]
20. Peron, A. Análise da Complementaridade das Gerações Intermitentes No Planejamento da Operação Eletro-Energética da Região Nordeste Brasileira. Master's Thesis, Unicamp, Campinas, Brazil, 2017.
21. Naeem, A.; Hassan, N.U.; Yuen, C.; Muyeen, S.M. Maximizing the Economic Benefits of a Grid-Tied Microgrid Using Solar-Wind Complementarity. *Energies* **2019**, *12*, 395. [CrossRef]
22. Rosa, C.; Costa, K.A.; Christo, E.; Bertahone, P.B. Complementarity of Hydro, Photovoltaic, and Wind Power in Rio de Janeiro State. *Sustainability* **2017**, *9*, 1130. [CrossRef]
23. Silva, A.R.; Pimenta, F.M.; Assireu, A.T.; Spyrides, M.H.C. Complementarity of Brazils Hydro and Offshore Wind Power. *Renew. Sustain. Energy Rev.* **2016**, *56*, 413–427. [CrossRef]
24. Jurasz, J.; Canales, F.A.; Kies, A.; Guczgouz, M.; Beluco, A. A Review on the Complementarity of Renewable Energy Sources: Concept, Metrics, Application and Future Research Directions. *Sol. Energy* **2020**, *195*, 703–724. [CrossRef]
25. GWEC. *Global Wind Report 2023 | GWEC*; GWEC: Brussels, Belgium, 2023.
26. Rusu, L. The Wave and Wind Power Potential in the Western Black Sea. *Renew. Energy* **2019**, *139*, 1146–1158. [CrossRef]
27. Rusu, E. A 30-Year Projection of the Future Wind Energy Resources in the Coastal Environment of the Black Sea. *Renew. Energy* **2019**, *139*, 228–234. [CrossRef]
28. Rusu, E. An Evaluation of the Wind Energy Dynamics in the Baltic Sea, Past and Future Projections. *Renew. Energy* **2020**, *160*, 350–362. [CrossRef]
29. GWEC. *Global Offshore Wind Report 2020*; GWEC: Brussels, Belgium, 2020.
30. González, M.O.A.; Santiso, A.M.; de Melo, D.C.; de Vasconcelos, R.M. Regulation for Offshore Wind Power Development in Brazil. *Energy Policy* **2020**, *145*, 111756. [CrossRef]
31. EPE. *Roadmap Eólica Offshore Brasil*; EPE: Brasilia, Brazil, 2020; Volume 141.
32. Haydt, G.; Guerreiro, A.; Rosa, F.; Lopes, L.; Cunha, S. Avaliação da Confiabilidade de Parques Eólicos na Região Nordeste do Brasil. In Proceedings of the XVI ERIAC-16° Encuentro Regional Iberoamericano de Cigré; XVI Eriac Decimosexto Encuentro Regional Iberoamericano de Cigré, Iguazú Port, Argentina, 17–21 May 2015; pp. 4–12.

33. Pimenta, F.M.; Assireu, A.T. Simulating Reservoir Storage for a Wind-Hydro Hydrid System. *Renew. Energy* **2015**, *76*, 757–767. [CrossRef]
34. CEPEL. Energy Operation Planning. Available online: http://www.cepel.br/pt_br/produtos/programas-computacionais-por-categoria/planejamento-da-operacao-energetica.htm (accessed on 2 January 2021).
35. Schmidt, J.; Cancella, R.; Pereira, A.O. The Role of Wind Power and Solar PV in Reducing Risks in the Brazilian Hydro-Thermal Power System. *Energy* **2016**, *115*, 1748–1757. [CrossRef]
36. Schmidt, J.; Cancella, R.; Pereira, A.O. An Optimal Mix of Solar PV, Wind and Hydro Power for a Low-Carbon Electricity Supply in Brazil. *Renew. Energy* **2016**, *85*, 137–147. [CrossRef]
37. Schmidt, J.; Cancella, R.; Junior, A.O.P. The Effect of Windpower on Long-Term Variability of Combined Hydro-Wind Resources: The Case of Brazil. *Renew. Sustain. Energy Rev.* **2016**, *55*, 131–141. [CrossRef]
38. Schmidt, J. COPA Model. Available online: https://homepage.boku.ac.at/jschmidt/COPA/index.html (accessed on 10 August 2020).
39. MME/EPE. *PDE 2029—Ten-Year Energy Expansion Plan- Executive Plan*; MME/EPEL: Brasília, Brasil, 2019.
40. Staffell, I.; Pfenninger, S. Using Bias-Corrected Reanalysis to Simulate Current and Future Wind Power Output. *Energy* **2016**, *114*, 1224–1239. [CrossRef]
41. Staffell, S.; Pfenninger, I. Renewables Ninja Downloads. Available online: https://www.renewables.ninja (accessed on 19 June 2019).
42. Lucena, A.F.P.; Szklo, A.S.; Schaeffer, R.; de Souza, R.R.; Borba, B.S.M.C.; da Costa, I.V.L.; Júnior, A.O.P.; da Cunha, S.H.F. The Vulnerability of Renewable Energy to Climate Change in Brazil. *Energy Policy* **2009**, *37*, 879–889. [CrossRef]
43. Ruffato-Ferreira, V.; da Costa Barreto, R.; Oscar Júnior, A.; Silva, W.L.; de Berrêdo Viana, D.; do Nascimento, J.A.S.; Freitas, M.A.V. A Foundation for the Strategic Long-Term Planning of the Renewable Energy Sector in Brazil: Hydroelectricity and Wind Energy in the Face of Climate Change Scenarios. *Renew. Sustain. Energy Rev.* **2017**, *72*, 1124–1137. [CrossRef]
44. Machado, R.D.; Bravo, G.; Starke, A.; Lemos, L.; Colle, S. Generation of 441 Typical Meteorological Year from INMET Stations—Brazil. In Proceedings of the ISES Solar World Congress 2019 and IEA SHC International Conference on Solar Heating and Cooling for Buildings and Industry 2019, Santiago, Chile, 4–7 November 2019; pp. 2189–2200.
45. ONS. Dados Hidrológicos-Vazões. Operador Nacional do Sistema Elétrico. Available online: http://www.ons.org.br/Paginas/resultados-da-operacao/historico-da-operacao/dados_hidrologicos_vazoes.aspx (accessed on 15 September 2019).
46. ONS. Energia Armazenada. Operador Nacional do Sistema. Available online: http://www.ons.org.br/Paginas/resultados-da-operacao/historico-da-operacao/energia_armazenada.aspx (accessed on 15 September 2019).
47. Gruber, K.; Klöckl, C.; Regner, P.; Baumgartner, J.; Schmidt, J. Assessing the Global Wind Atlas and Local Measurements for Bias Correction of Wind Power Generation Simulated from MERRA-2 in Brazil. *Energy* **2019**, *189*, 116212. [CrossRef]
48. ANEEL. Sistema de Informações de Geração da Agência Nacional de Energia Elétrica—SIG. Available online: https://www.aneel.gov.br/siga (accessed on 21 June 2023).
49. CCEE. Todos os Dados Resultantes do Processamento do Modelo. Available online: https://www.ccee.org.br (accessed on 21 June 2023).
50. La Rovere, E.L.; Wills, W.; Pereira, A.O., Jr.; Dubeux, C.B.S.; Cunha, S.H.F.; Oliveira, B.C.P.; Moreira, M.M.R.; Watanabe, S.; Loureiro, S.; Moreira, L.S.S.C.; et al. *Economic and Social Implications of GHG Mitigation Scenarios in Brazil until 2030*; International Economics: Rio de Janeiro, Brazil, 2016.
51. EPE. *Estudos Para Expansão da Geração. Custo Marginal de Expansão Do SetorElétrico Brasileiro Metodologia e Cálculo—2019*; EPE: Rio de Janeiro, Brazil, 2019.
52. Pimentel, J.; Andrade, F. *Brazilian Wind Generation Guidebook*; BTG Pactual S.A Bank- BTG Pactual Affiliate Research: São Paulo, Brazil, 2019; p. 30.
53. Parsons, G.R.; Firestone, J. *Atlantic Offshore Wind Energy Development: Values and Implications for Recreation and Tourism*; BOEM: Sterling, VA, USA, 2018.
54. de Souza Nascimento, M.M.; Shadman, M.; Silva, C.; de Freitas Assad, L.P.; Estefen, S.F.; Landau, L. Offshore Wind and Solar Complementarity in Brazil: A Theoretical and Technical Potential Assessment. *Energy Convers. Manag.* **2022**, *270*, 116194. [CrossRef]
55. EPE. *Balanço Energético Naciona 2020*; EPE: Rio de Janeiro, Brazil, 2020.
56. Acker, T. *IEA Wind Task 24 Final Report: Issues, Impacts, and Economics of Wind and Hydropower Integration*; IEA: Paris, France, 2011; Volume 1.
57. Oliveira, C.; Zulanas, C.; Kashiwagi, D. A Long-Term Solution to Overcome the Problems Caused by Droughts in the Brazilian Power Systems. *Procedia Eng.* **2016**, *145*, 948–955. [CrossRef]
58. Borba, P.C.S.; Sousa, W.C.; Shadman, M.; Pfenninger, S. Enhancing Drought Resilience and Energy Security through Complementing Hydro by Offshore Wind Power—The Case of Brazil. *Energy Convers. Manag.* **2023**, *277*, 116616. [CrossRef]

 *energies*

MDPI

Article

Optimal Scheduling Strategy for Distribution Network with Mobile Energy Storage System and Offline Control PVs to Minimize the Solar Energy Curtailment

San Kim [1] and Jinyeong Lee [2],*

1   School of Electrical Engineering, Anam Campus, Korea University, 145 Anam-ro, Seoul 02841, Republic of Korea; kimsan011@korea.ac.kr
2   Electricity Policy Research Center, Korea Electrotechnology Research Institute (KERI), Uiwang 16029, Republic of Korea
*   Correspondence: jinyeong@keri.re.kr; Tel.: +82-31-420-6148

Abstract: As offline control photovoltaic (PV) plants are not equipped with online communication and remote control systems, they cannot adjust their power in real-time. Therefore, in a distribution network saturated with offline control PVs, the distribution system operator (DSO) should schedule the distributed energy resources (DERs) considering the uncertainty of renewable energy to prevent curtailment due to overvoltage. This paper presents a day-ahead network operation strategy using a mobile energy storage system (MESS) and offline control PVs to minimize power curtailment. The MESS model efficiently considers the transportation time and power loss of the MESS, and models various operating modes, such as the charging, discharging, idle, and moving modes. The optimization problem is formulated based on mixed-integer linear programming (MILP) considering the spatial and temporal operation constraints of MESSs and is performed using chanced constrained optimal power flow (CC-OPF). The upper limits for offline control PVs are set based on the probabilistic approach, thus mitigating overvoltage due to forecasting errors. The proposed operation strategy was tested in the IEEE 33-node distribution system coupled with a 15-node transportation system. The test results show the effectiveness of the proposed method for minimizing curtailment in offline control PVs.

Keywords: offline control photovoltaic; mobile energy storage system; renewable curtailment mitigation; distribution system operator; chanced constrained optimal power flow

Citation: Kim, S.; Lee, J. Optimal Scheduling Strategy for Distribution Network with Mobile Energy Storage System and Offline Control PVs to Minimize the Solar Energy Curtailment. *Energies* **2024**, *17*, 2234. https://doi.org/10.3390/en17092234

Academic Editors: Fushuan Wen and Xiuli Wang

Received: 9 April 2024
Revised: 30 April 2024
Accepted: 2 May 2024
Published: 6 May 2024

1. Introduction

Over the last decade, the penetration of renewable energy resources (RESs) in a distribution grid has dramatically increased in many countries across the world. However, integration of the distributed and highly uncertain generation from RESs into the distribution grid poses several challenges for grid planning and operation. System operators (SOs) have faced several system security and quality problems, such as voltage rise, congestion, stability, and low flexibility [1–4]. These problems have inevitably caused a percentage of renewable energy generation to be cut off.

Fast-developing energy storage technology could offer carbon-free power system operations with a fast response, improving flexibility and stability of the system, while enabling high efficiency, a long life cycle, and low maintenance. Energy storage types are classified into four technologies for grid-scale applications including electrical, mechanical, chemical, and thermal [5,6]. Mechanical storage systems, such as pumped-hydro and compressed-air, are the most widely adopted type of energy storage system, accounting for 99% of the worldwide storage capacity [7]. However, these systems have strict locational requirements, which hinder their widespread development. Battery-based energy storage technologies have emerged as the most promising technologies for grid-scale storage with the increasing penetration of renewable energy [8,9]. It can help to mitigate system uncertainties and ensure the stable and reliable operation of power grids with a high share of renewable energy in the electricity generation portfolio. However, an intelligent operation strategy is required to maximize the benefits offered by battery energy storage systems (BESSs), as such devices are expensive and have limited energy resources. Different operation strategies perform the scheduling of different objective functions (e.g., maximizing profit and system reliability, minimizing power loss and operating costs), subject to the system operation constraints (e.g., system constraints, charging/discharging limit, energy limit, and power rating) [10–14].

In recent years, the application of battery-based energy storage for transportation in power systems has been introduced and studied extensively. Mobile energy storage systems (MESSs) are a mobile and transportable storage technology, consisting of battery cells and a power converter carried on a truck [15]. This resource is flexible both spatially and temporally, being free from spatial constraints unlikely in traditional energy storage systems. It is a powerful tool that can enhance system reliability and respond flexibly and rapidly to digesters or system uncertainties. MESSs can be positioned prior to a disaster for emergency power supply and can be re-positioned after the disaster to maximize the system restoration time [16]. The researchers in [17] and [18] used the MESSs to realize optimal energy allocation in emergency situations. The distribution system is reconfigured to change the network topology using MESS. In [7], system restoration was further considered along with the routing and scheduling of MESSs. MESS investment and relocating strategies were proposed in [19] to minimize the load shedding under emergency conditions. The critical load restoration method was proposed in [20] to select the optimal position of MESSs along with a network topology and a load switching sequence for power grids. The researcher in [21] used the mobile energy resource to perform energy management in the island group.

The spatial and temporal operation of MESSs can also offer various benefits for grid operation, such as congestion management, grid security enhancement, profit improvement, expansion deferral, and uncertainty mitigation of RES. System operators have sought to directly own and operate MESS under various operational conditions within distribution and transportation systems. In [22], MESSs were treated as the alternative to reduce transmission congestion in power systems with high renewable penetration, proposing a transmission planning using both mobile and stationary storage resources. The reliability assessment of distribution system with MESSs was developed using an analytic approach based on Markov models [23]. In [24], MESSs were used to alleviate the impact of photovoltaic fluctuations on distribution networks and to postpone equipment upgrades. An optimization framework that can be performed under the realistic operation conditions of distribution system was proposed in [25]. This framework proposes an optimal scheduling algorithm for MESS to minimize the EV waiting at a charging station. A day-ahead optimization method for MESS operation was introduced in [26] to improve the profit of the DSO while maintaining a normal voltage profile along the distribution feeder. Stochastic scheduling of power systems with wind farms was proposed in [27], which can alleviate wind energy curtailment and reduce operating costs. The author in [28] focused on the MESS optimal scheduling in a distribution grid with wind and PV units to mitigate renewable energy curtailment.

The curtailment of RES in the distribution grid is expected to worsen as reinforcement and investment in the distribution network are delayed. In particular, offline-controlled PVs can be easily tripped to solve overvoltage issues due to the uncertain production from RESs because these resources are not equipped with online communication and remote control systems. The MESS is an effective energy storage technology that can contribute to solving the energy curtailment problems of RESs caused by voltage rises during different periods at different locations. In many previous research works, MESS-based operation strategies have been studied to minimize RES curtailment in distribution grids. However, based on the current literature, optimal scheduling and energy management approaches that can minimize RES curtailment in distribution grids with high penetration offline control PVs have not yet been developed. This paper presents a day-ahead operation strategy using MESS and offline control PVs to minimize PV energy curtailment in distribution grids with a high penetration of offline control units. The spatial and temporal operation model for the MESS efficiently accounts for power loss during transportation was developed. This model includes various operating modes, such as the charging, discharging, idle, and moving mode. The reactive power of the MESS was linearized using the circular constraint linearization method. Day-ahead offline control for PVs was performed considering forecasting errors. The proposed operation strategy was integrated into the linearized DistFlow algorithm to minimize curtailment and was solved using the MILP optimization model. The contributions of this paper can be summarized as below:

- Day-ahead offline control for PVs is implemented to mitigate the PV trips resulting from uncertainties in renewable energy. This day-ahead offline control is predetermined based on a probabilistic approach to respond to unexpected voltage rises because offline control units cannot adjust the upper limit in real-time.
- CC-OPF is used to minimize energy curtailment considering soft chance constraints related to voltage, which can ensure that the system conditions are not below the confidence level. This can lead to moving the MESS to nodes with overvoltage risk to minimize energy curtailment due to overvoltage.
- The energy and transportation scheduling is performed to minimize total operating costs considering the cooperative operation of MESS and offline control PVs. The proposed method can significantly reduce PV production curtailment in forecasting error scenarios, while hardly increasing the operating costs of DSO in the base case scenario without forecasting errors.

The rest of this paper is organized as follows: Section 2 introduces the mathematical model for offline control PV and MESS. In Section 3, the proposed optimization model is formulated considering spatial and temporal operating conditions. Section 4 presents case studies on the IEEE 33-node distribution system coupled with a 15-node transportation system. Finally, the conclusions are drawn in Section 5.

2. Mathematical Modeling

2.1. Offline Control PV Model

In this study, offline control PVs are defined as solar PVs that cannot adjust their power output in real-time to satisfy the electricity demand or system limits. They are not equipped with an online communication and remote control system; thus they cannot change the upper limit in real-time. Therefore, the system operator should request day-ahead curtailment to adjust offline control PV owners to manually change the upper limit if PV curtailment is expected to be required [29]. Figure 1 shows a day-ahead offline control of PVs to prevent generator trips due to overvoltage. The black dashed line is a day-ahead forecasted output and a red dashed line is the upper limit of PV output to prevent overvoltage due to an oversupply at the node. In a network with high output uncertainty, day-ahead offline control may cause curtailment due to the upper limit of the output, but it can alleviate total energy curtailment because it can prevent PV trips due to voltage rises. If the upper limit is lower than the power output before energy curtailment, the PV output is reduced by the upper limit. On the other hand, if the upper limit is higher than the power output before energy curtailment, the PV output does not curtail. The PV output with the upper limit can be modeled in Equation (1).

$$P_{i,t}^{PV} = \begin{cases} PV'_{i,t} & PV'_{i,t} < PV_{i,t}^{Upper} \\ PV_{i,t}^{Upper} & PV'_{i,t} \geq PV_{i,t}^{Upper} \end{cases}, \quad \forall i,t \tag{1}$$

where $P_{i,t}^{PV}$ is the power output of PV at node i at time t, $PV_{i,t}^{Upper}$ is upper limit offline control PV output at node i at time t, $PV_{i,t}'$ is the power output of offline PV before energy curtailment at node i at time t.

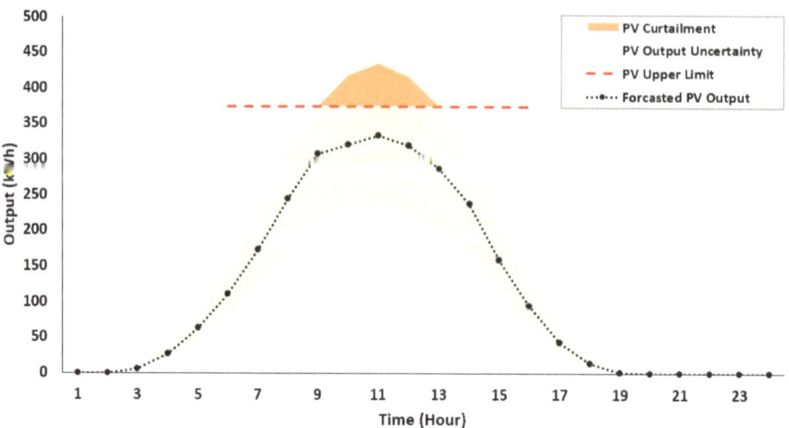

Figure 1. Spatial–temporal movement of MESS.

If the upper limit is lower than the forecasted output, the energy abandonment can be computed by the forecasted output from the upper limit. Second, energy abandonment due to overvoltage is equal to total energy at a node with overvoltage because offline control PVs cannot adjust their output in real-time.

2.2. The MESS Model

MESSs can be described as the mobilized grid-scale battery systems employed in power systems [4]. They consist of an energy storage system carried on a truck. MESSs can be operated in moving mode, as well as in charging, discharging, or idle mode, which are operating modes of traditional stationary energy storage systems (SESSs) [30,31]. Figure 2 illustrates the characteristics of spatiotemporal MESS mobility in a transportation network. The vertical axis in Figure 2 denotes the MESS station, and the horizon axis represents the time span. S1, S2, S3, and S4 are locations with MESS charging stations, and the MESS can only connect nodes with charging stations. There are four possible mobility states: S1→S1, S1→S2, S1→S3, and S1→S4 during the initial period. The solid lines represent the determined path during each time period, while the dashed lines are the possible mobility paths during each time period.

The spatiotemporal status of the MESS can be described from the binary variables. The binary variables in this paper are classified into four statuses, such as the charging, discharging, idle, and moving modes. The idle mode means that the MESS waits at the connected node before charging, discharging, or moving. The location of the MESS can be can be expressed as Equation (2). It means that MESS can only connect to one position. The MESS initially operates based on a pre-determined spatiotemporal status. It should be relocated to this status at the end of the scheduling period. This paper assumes that the initial and final locations of the MESS should be connected to node 1, as expressed in Equations (3) and (4).

$$\sum_i \left(o_{i,t}^3 \right) \leq 1, \quad \forall\, t \tag{2}$$

$$o_{1,t_0}^3 = 1 \tag{3}$$

$$o_{1,T}^3 = 1 \tag{4}$$

where $o^3_{i,t}$ are the binary variables indicating the connection statues of MESS at node i at time t, t_0 are the initial scheduling period, and T is the final period.

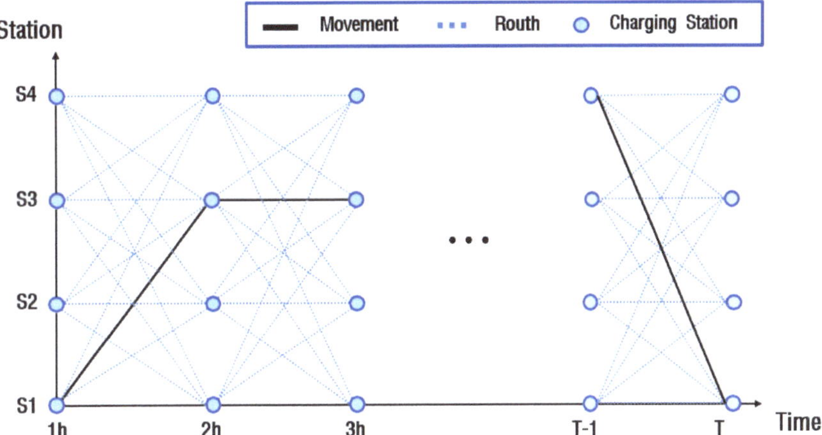

Figure 2. Spatial–temporal movement of MESSs.

The charging and discharging power of MESS can only be provided if MESS is connected to nodes, as expressed in Equation (5).

$$o^1_{i,t} + o^2_{i,t} \leq o^3_{i,t}, \quad \forall\, t \tag{5}$$

where $o^1_{i,t}$ and $o^2_{i,t}$ are the binary variables describing the charging and discharging statues of MESS at node i at time t.

MESS can travel between nodes with a charging station, and the transportation of MESS from node i to node j can be modeled by Equations (6) and (7). Equation (6) means that the total travel time $TT_{i,j}$ is required to transport between i to node j. Equation (7) means that a non-connection time of MESS cannot be longer than the $TT_{i,j}$.

$$o^3_{i,t} + o^3_{j,u} \leq 1, \quad \forall\, i,j,t, i \neq j, u = \{t+1, \cdots, t + TT_{i,j}\} \tag{6}$$

$$o^3_{i,t} \leq \sum_{u=t+1}^{t+TT_{i,j}} o^3_{j,u} \quad \forall\, i,t \tag{7}$$

The transportation path is modeled by Equations (8) and (9). This means that the MESS is located at node i at time t and has moved to node j at time u. The minimum number N^{moving} of moving modes is limited by Equation (10).

$$o^4_{i,j,t} \geq o^3_{i,t} \quad \forall\, i,t \tag{8}$$

$$o^4_{i,j,t} \geq o^3_{j,u} \quad \forall\, i,t \tag{9}$$

$$\sum_{t}\sum_{i}\sum_{j} \left(o^4_{i,j,t} \right) \geq N^{moving} \tag{10}$$

where $o^4_{i,j,t}$ means the transportation path, which MESS departures at node i at time t to node j.

3. Problem Formulation

3.1. Objective Function

The proposed mathematical operation model for MESS is integrated into the energy scheduling of the distribution network [32–34]. The problem formulation is described using the DistFlow algorithm and is solved using mixed integer linear programming [19,25]. The objective of the proposed approach is to minimize the total operation costs of the distribution network. The operation costs include the electricity purchase cost from upstream grids, the cost of the energy supplied by PV, and the charging/discharging costs of the MESS, which is stated as:

$$\min(OC) = \min \left(\sum_{t \in T} \lambda_t P_t^{Sub} + \sum_{t \in T} \sum_{i \in I} \lambda_t (1 - z_{i,t}) P_{i,t}^{PV} \right.$$
$$+ \sum_{t \in T} \sum_{i \in I} \lambda_t \left(PV_i^{Upper} - PV'_{i,t} \right) + \sum_{t \in T} \sum_{i \in I} \lambda_t P_{i,t}^{ch} \tag{11}$$
$$\left. + + \sum_{t \in T} \sum_{i \in I} \lambda_t P_{i,t}^{dch} \right)$$

where P_t^{Sub} is active power purchased from upstream gird at time t, $z_{i,t}$ is the binary variable indicating PV trip at node i at time t, $P_{i,t}^{ch}$ is charging power of MESS at node i at time t, $P_{i,t}^{dch}$ is discharging power of MESS at node i at time t, and λ_t is the electric price at time t.

3.2. Charging and Discharging of the MESS

The power of the MESS can be described using two mutually exclusive terms, charging and discharging. This means that the power of the MESS cannot be charged and discharged simultaneously. The active power is limited by Equations (12) and (13), and they are defined using two binary variables such that the two conditions cannot occur at the same time. The reactive power is limited from Equation (14). The net active power injected from the battery same as the difference between its charging and discharging power and is represented in Equation (15). The sum of active and reactive power in both the charge and discharge modes cannot be greater than the rated power of the battery, and this condition is imposed in Equation (16). Equations (17) and (18) are the sum of the charging and discharging power of MESS s at time t.

$$P_m^{ch,min} o_{i,t}^1 \leq P_{i,t}^{ch} \leq P_m^{ch,max} o_{i,t}^1 \quad \forall\, i, t, m \tag{12}$$

$$P_m^{dch,min} o_{i,t}^2 \leq P_{i,t}^{dch} \leq P_m^{dch,max} o_{i,t}^2 \quad \forall\, i, t, m \tag{13}$$

$$-Q_m^{min} \left(o_{i,t}^1 + o_{i,t}^2 \right) \leq Q_{i,t}^{Net} \leq Q_m^{max} \left(o_{i,t}^1 + o_{i,t}^2 \right) \quad \forall\, i, t, m \tag{14}$$

$$P_{i,t}^{Net} = P_{i,t}^{ch} + P_{i,t}^{dch} \quad \forall\, i, t \tag{15}$$

$$\sqrt{\left(P_{i,t}^{Net} \right)^2 + \left(Q_{i,t}^{Net} \right)^2} \leq S_m^{MESS} \quad \forall\, i, t, m \tag{16}$$

$$P_{m,t}^{ch} = \sum^{i} P_{m,i,t}^{ch} \quad \forall\, i, t, m \tag{17}$$

$$P_{m,t}^{dch} = \sum^{i} P_{m,i,t}^{dch} \quad \forall\, i, t, m \tag{18}$$

where $P_m^{ch,max}$ and $P_m^{ch,min}$ are maximum and minimum charging active power of MESS m, $P_m^{dch,max}$ and $P_m^{dch,min}$ are maximum and minimum discharging active power of MESS m, Q_m^{min} is maximum charging reactive power of MESS m, $P_{i,t}^{Net}$ and $Q_{i,t}^{Net}$ are net active and net reactive power of MESS at node i at time t, and S_m^{MESS} is maximum allowable MESS power rating.

To avoid the non-linearity of the MESS model based on Equation (16), this constraint is substituted by Equations (19) and (20). Two constraints are approximated using the circular constraint linearization method, as shown in Figure 3 [35]. The vertical axis 'P' in Figure 3 represents the active power, and the horizontal axis 'Q' represents the reactive power. The radius of the circle 'S' represents the maximum allowable power rating.

$$-\sqrt{2}S_m^{MESS} \leq P_{i,t}^{Net} + Q_{i,t}^{Net} \leq \sqrt{2}S_m^{MESS} \quad \forall\, i,t,m \tag{19}$$

$$-\sqrt{2}S_m^{MESS} \leq P_{i,t}^{Net} - Q_{i,t}^{Net} \leq \sqrt{2}S_m^{MESS} \quad \forall\, i,t,m \tag{20}$$

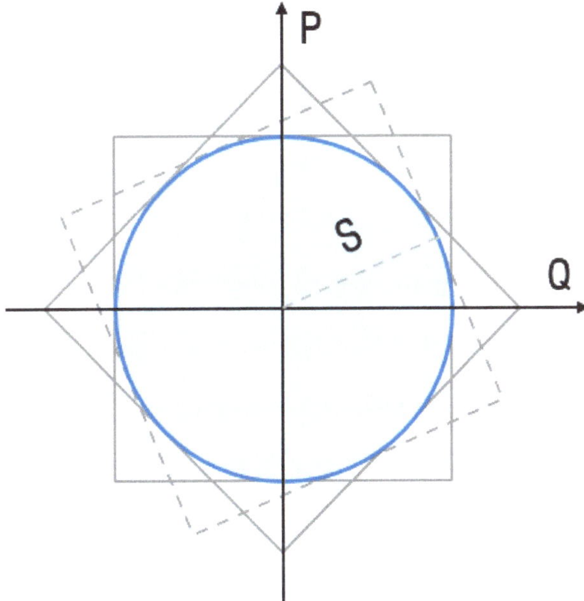

Figure 3. Circular constraint linearization method.

The stored energy in the battery during any period is a function of the previously stored energy and the net energy transacted in the current period. It is equal to the stored energy during the previous period, the energy charge and discharge during the present period, and the energy loss due to movement during the present period. The stored energy should be limited between the upper and lower bounds. Finally, the state of charge (SOC) at the end of the time periods must equal the initial conditions. These constraints are modeled in Equations (21)–(23).

$$SOC_{m,t}^{MESS} = SOC_{m,t-1}^{MESS} + \eta_m^{ch} P_{s,t}^{ch} - \frac{P_{m,t}^{dch}}{\eta_m^{dch}} - o_{i,j,t}^4 P_{i,j}^{tr} \quad \forall\, i,j,t,m \tag{21}$$

$$SOC_m^{MESS,min} \leq SOC_{m,t}^{MESS} \leq SOC_m^{MESS,max} \quad \forall\, t,m \tag{22}$$

$$SOC_{m,t_0}^{MESS} = SOC_{m,T}^{MESS} \quad \forall\, t,m \tag{23}$$

where $SOC_{m,t}^{MESS}$ is the SOC of MESS m at time t, $SOC_m^{MESS,max}$ is maximum energy content of MESS m, $SOC_m^{MESS,min}$ is minimum energy content of MESS m, η_m^{ch} is charging efficiency of MESS m, η_m^{dch} is discharging efficiency of MESS m, $P_{i,j}^{tr}$ is energy loss due to travel of MESS between different nodes.

92

3.3. The System Constraints

The power flow in a radial distribution network is derived from a set of recursive equations, called the DistFlow method [19,25]. The active power, reactive power, and voltage magnitude at the sending end of the branch are used to represent the same quantities at the receiving end of a branch, as below:

$$\sum_{ij} P_{ij,t} = P_t^{sub} - P_{i,t}^L - P_{i,t}^{ch} + P_{i,t}^{dch} + P_{i,t}^{PV} - r_{ij}\frac{P_{ij,t}^2 + Q_{ij,t}^2}{V_{i,t}^2} \quad \forall\, i, t, ij \tag{24}$$

$$\sum_{ij} Q_{ij,t} = Q_t^{sub} - Q_{i,t}^L - Q_{i,t}^{ch} + Q_{i,t}^{dch} + Q_{i,t}^{compensator} - x_{ij}\frac{P_{ij,t}^2 + Q_{ij,t}^2}{V_{i,t}^2} \quad \forall\, i, t \tag{25}$$

$$V_{j,t}^2 = V_{i,t}^2 - 2\left(r_{ij}P_{ij,t} + x_{ij}Q_{ij,t}\right) + \left(r_{ij}^2 + x_{ij}^2\right)\frac{P_{i,t}^2 + Q_{i,t}^2}{V_{i,t}^2} \quad \forall\, i, j, t, ij \tag{26}$$

where $P_{ij,t}$ is active power flow on branch ij at time t, $Q_{ij,t}$ is reactive power flow on branch ij at time t, $Q_{s,i,t}^{compensator}$ is reactive power of compensator at node i at time t, r_{ij} is resistance of branch ij, x_{ij} is reactance of branch ij, $V_{i,t}$ is voltage at node i at time t, $P_{i,t}^L$ is active load at node i at time t, $Q_{i,t}^L$ is reactive load at node i at time t.

Equations (24)–(26) should be approximated to overcome the quadratic terms. Some tolerable assumptions and simplifications are applied in the linearization process. First, the branch losses, which are non-linear terms, can be ignored in the DistFlow equations when the non-linear terms are much smaller than the branch flows. Second, it is assumed that the voltage deviation at each bus is very small and can be neglected. Then, the linearized DistFlow equations are expressed in Equations (27) and (28). The lower and upper voltages are limited in Equation (30). The active and reactive power supplied from the upstream grid is represented by Equations (31) and (32), respectively. The voltage magnitude of the substation bus is always valued as 1.0 p.u. according to Equation (33). The active power supplied from PVs can be modeled as Equation (34) and can be tripped to prevent overvoltage at the node by a binary variable.

$$\sum_{ij} P_{ij,t} \cong P_t^{sub} - P_{i,t}^L - P_{i,t}^{ch} + P_{i,t}^{dch} + P_{i,t}^{PV} \quad \forall\, i, t, ij \tag{27}$$

$$\sum_{ij} Q_{ij,t} \cong Q_t^{sub} - Q_{i,t}^L - Q_{i,t}^{ch} + Q_{i,t}^{dch} + Q_{s,i,t}^{compesator} \quad \forall\, i, t, ij \tag{28}$$

$$V_{j,t} \cong V_{i,t} - r_{ij}P_{ij,t} - r_{ij}Q_{ij,t} \quad \forall\, i, t, ij \tag{29}$$

$$V_i^{min} \leq V_{i,t} \leq V_i^{max} \quad \forall\, i, t \tag{30}$$

$$P^{sub,min} \leq P_t^{sub} \leq P^{sub,max} \quad \forall\, t \tag{31}$$

$$Q^{sub,min} \leq P_t^{sub} \leq Q^{sub,max} \quad \forall\, t \tag{32}$$

$$V_t^{sub} = 1.0 \quad \forall\, t \tag{33}$$

$$P_{i,t}^{PV} = PV_{i,t}z_{i,t} \quad \forall\, i, t \tag{34}$$

where V_i^{max} is the maximum voltage limit at node i, V_i^{min} is the minimum voltage limit at node i, $P^{sub,max}$ is the maximum active power supplied from upstream grid, $P^{sub,min}$ is the minimum active power purchased from upstream grid, $Q^{sub,max}$ is the maximum reactive power supplied from upstream grid, $Q^{sub,min}$ is the minimum reactive power supplied from upstream grid, V_t^{sub} is a voltage of substation node at time t.

3.4. Chance-Constraints

Chance-constrained optimal power flow (CC-OPF) is used to deal with the uncertainties of power injections [36,37]. Chance constraints are used to ensure that the system risks are not below the confidence level. This paper only considers soft constraints related to voltage, and the remaining constraints are reflected as deterministic constraints. The confidence level can be determined using analysis to minimize total energy curtailment. Then, the voltage constraints in Equations (35) and (36) can be rewritten using acceptable violation probability of voltage ϵ_V as:

$$Pr(V_{i,t} \leq V_i^{max}) \geq 1 - \epsilon_V \quad \forall\, i,t \tag{35}$$

$$Pr\left(V_{i,t} \geq V_i^{min}\right) \geq 1 - \epsilon_V \quad \forall\, i,t \tag{36}$$

4. Proposed Algorithm

In distribution networks with a high penetration of offline control PVs, the optimal operation must be performed considering the day-ahead curtailment request, as offline control PVs cannot adjust the upper limit in real-time. The PV output can be excessively curtailed if a strict upper limit of the output is applied, while the generator can be tripped due to overvoltage if the upper limit of the output is relaxed. This paper computes an upper limit determined by the probabilistic method that considers output uncertainty. First, the state of the MESS is pre-determined by the optimization problem based on forecasting data. Second, Monte Carlo simulation (MCS) is used to generate a large number of scenarios [38]. Third, optimal power flow (OPF) without voltage constraints in each scenario is performed to determine the voltage distribution. If there is a node with a voltage violation in each scenario, MESS charge to manage voltage rising using the voltage sensitivity in Equation (37). After this, if overvoltage still occurs, the PV is curtailed to bring the voltage back down to the reference voltage level, as outlined in Equation (38). Fourth, the day-ahead curtailment requirement of offline control PVs is computed using the forecasted output and curtailment to alleviate voltage to the reference level in each scenario, and it is expressed as in Equation (39). Finally, optimal operation with day-ahead curtailment is implemented to consider renewable energy uncertainty [39]. Figure 4 summarizes the overall process of the proposed scheduling strategy, taking into account the MESS and offline control PVs.

$$V'_{s,i,t} = \begin{cases} V^{intial}_{s,i,t} - \Delta P^{MESS}_{s,i,t} \frac{\partial V_{s,i}}{\partial P_{s,i}}, & if\ V^{intial}_{0,i,t} > V^{max} \\ V^{intial}_{s,i,t} & if\ V^{intial}_{s,i,t} \leq V^{max} \end{cases} \quad \forall\, s,i,t \in T_{V'} \tag{37}$$

$$P^{curtail}_{s,i,t} = \begin{cases} (V^{max} - V'_{s,i,t}) \frac{\partial P_{s,i}}{\partial V_{s,i}}, & if\ V'_{s,i,t} > V^{max} \\ 0 & if\ V'_{s,i,t} \leq V^{max} \end{cases} \quad \forall\, s,i,t \in T_v \tag{38}$$

$$PV^{upper}_i = mean\left(PV'_{s,t} - P^{curtail}_{s,i,t}\right), \quad \forall\, s \in S_v, i \tag{39}$$

where $V^{intial}_{s,i,t}$ is an initial voltage at node i at time t in scenario s, $P^{curtail}_{s,i,t}$ is PV energy curtailment to alleviate the voltage to the reference voltage level at node i at time t in scenario s, T_v is the time when curtailment occurs most frequently, S_v is the scenario where overvoltage occurs. N_s and N_i are the number of scenarios and nodes, respectively.

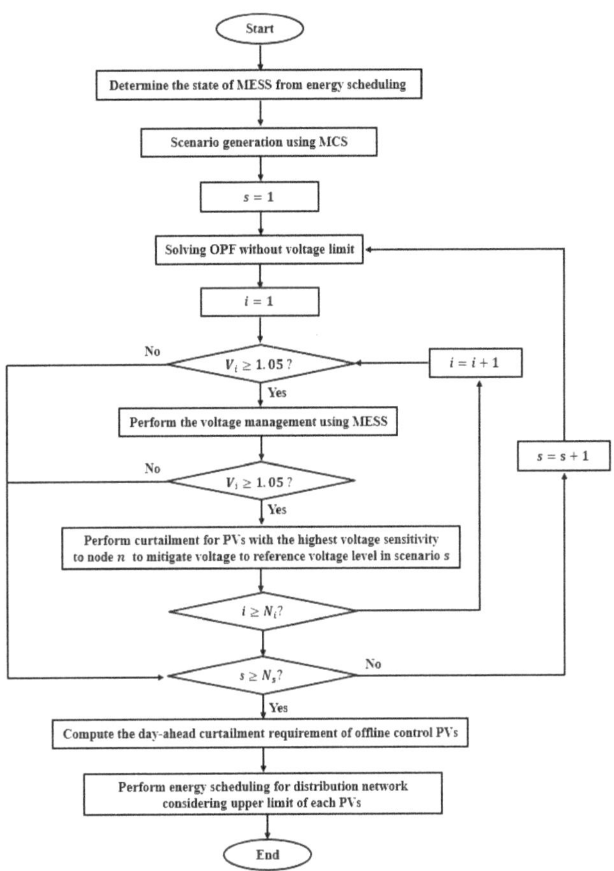

Figure 4. The overall process of proposed scheduling strategy.

5. Case Study

Simulation was performed on an IEEE 33-bus distribution system and a 15-node transportation system, as shown in Figure 5 [25]. These coupled systems include 33 nodes, 6 offline control PVs, 34 lines, 4 reactive power compensators, 32 loads, and 6 charging stations. The line, load, and system data can be found in [40]. The offline control PVs were located at buses 8, 10, 18, 25, 30, and 33. The hourly profile of the offline control PVs is shown in Figure 6. The details of the reactive power compensator are presented in Table 1. The system was equipped with 500 kW and 1000 kWh MESS with charging stations at buses 1, 3, 6, 12, 20, 24, and 31 [26]. The battery charging and discharging efficiencies were equal to 0.95 and the initial location was bus 1. The initial, minimum, and maximum SOCs were 0.5, 0.2, and 0.8, respectively. The traveling efficiency was set to $\eta_s^{tr} = 2\text{kW}/\Delta t$ and the traveling time was assumed to be 3 min/km. MCS was used to generate 1000 random scenarios to compute the day-ahead curtailment. The simulations were implemented on an Intel(R) Core(TM) I7-970 CPU with 8 GB memory. The optimization problem was solved using the MILP solver within MATLAB R2020b. This study considers four cases to compare the results of the operating costs and expected curtailment:

- Case 1: OPF without day-ahead curtailment;
- Case 2: OPF with day-ahead curtailment;
- Case 3: CC-OPF without day-ahead curtailment;
- Case 4: CC-OPF with day-ahead curtailment.

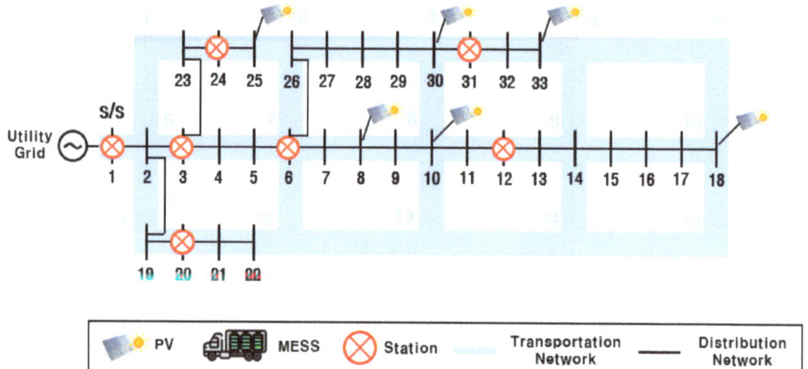

Figure 5. Distribution and transportation network for the case studies.

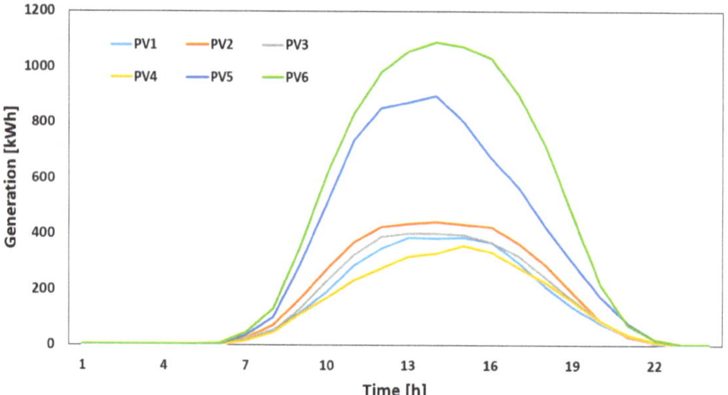

Figure 6. Hourly PV generation forecasting profiles in the distribution network.

Table 1. Reactive power compensators data.

Case	Reactive Power (MVAR)
6	0.3
14	0.3
18	0.4
33	0.6

The main results for the transportation schedule of the MESS are presented in Figure 7. The initial and final positions of the MESS were located on bus 1 in all four cases. The MESS in Case 1 moved to station 3 at hour 6, station 2 at hour 16, and station 1 at hour 21. In Case 2, the MESS moved to station 3 at hour 11, station 2 at hour 16, station 1 at hour 22. The MESS in these cases was mainly placed around bus 1 to minimize the power loss due to movement of the MESS because they did not consider the overvoltage risk due to output uncertainty. On the other hand, in Cases 3 and 4, it can be confirmed that MESS was located at station 4, near the node at the overvoltage risk due to high PV output between hours 12 and 15. The MESS in Case 3 moved to station 2 at hour 2, station 3 at hour 9, station 4 at hour 22, and station 1 at hour 14. The MESS in Case 4 moved to station 4 at hour 12, station 3 at hour 16, station 2 at hour 21, and station 1 at hour 23. They used CC-OPF for distribution network scheduling and set strict voltage limits to keep the overvoltage risk below a certain probability. Therefore, the MESS was located at station 4, which has high voltage sensitivity for voltage management during high PV output periods. Table 2 shows the hourly operating mode of the MESS. The operating mode of the MESS was classified into charging, discharging, idle, and moving modes. The operating mode pattern in all four cases is similar because the MESS operated to obtain revenue from SMP arbitrage. This means that the two optimization problems had a greater impact on the MESS's travel paths rather than the operating mode.

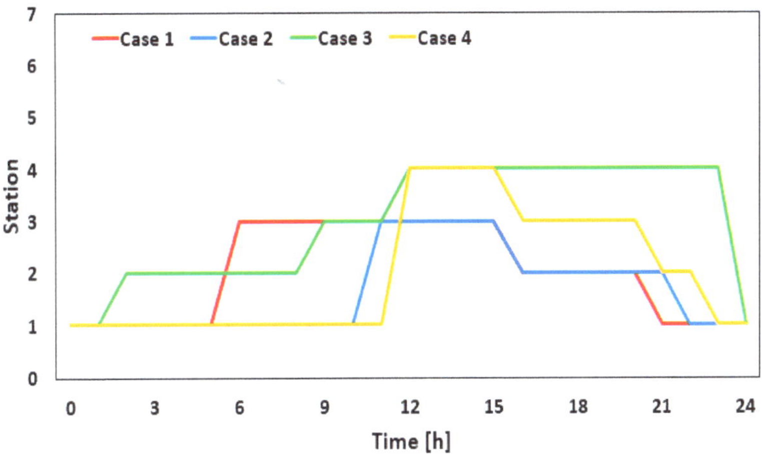

Figure 7. The results of the transportation schedule of the MESS.

Table 2. The hourly operating mode of the MESS for the four cases.

Hour	Case 1	Case 2	Case 3	Case 4
1	Discharging	Discharging	Discharging	Discharging
2	Idle	Idle	Moving	Idle
3	Idle	Idle	Idle	Idle
4	Charging	Charging	Charging	Charging
5	Charging	Charging	Charging	Charging
6	Moving	Idle	Idle	Idle
7	Discharging	Idle	Idle	Moving
8	Discharging	Discharging	Discharging	Discharging
9	Idle	Idle	Moving	Idle
10	Discharging	Discharging	Discharging	Discharging
11	Idle	Moving	Idle	Idle
12	Idle	Idle	Moving	Idle
13	Charging	Charging	Charging	Charging
14	Charging	Charging	Charging	Charging
15	Charging	Charging	Charging	Charging
16	Moving	Moving	Idle	Idle
17	Charging	Idle	Idle	Idle
18	Idle	Discharging	Idle	Idle
19	Idle	Idle	Idle	Discharging
20	Discharging	Discharging	Discharging	Discharging
21	Moving	Discharging	Idle	Moving
22	Charging	Moving	Idle	Idle
23	Charging	Idle	Idle	Moving
24	Idle	Idle	Moving	Idle

Detailed results of the MESS operating schedule, such as the charge, discharge, and SOC, are given in Figure 8. It can be seen that the SOC of the MESS increased as the MESS charged power, and decreased as the MESS discharged power and movement loss. The initial and final SOCs of the MESS were fixed at 50% by the SOC constraints. The optimization problem was solved to minimize the operating costs of the distribution network. In all four cases, the active power was mainly charged at hours 4–5 and 13–15, which were low SMP periods, and discharged at hours 1, 8, 10, and 20, which were high SMP periods. The reactive power was mainly charged and discharged to maintain the voltage. The MESS mainly charged reactive power during daytime hours when the voltage was high due to the PV output, and discharged reactive power mainly during dawn hours when the voltage was low owing to demand. However, Cases 3 and 4 tended to charge more reactive power than Cases 1 and 2 from hours 13 to 15 when overvoltage was expected. This was the result of strict voltage constraints in the optimization problems using CC-OPF.

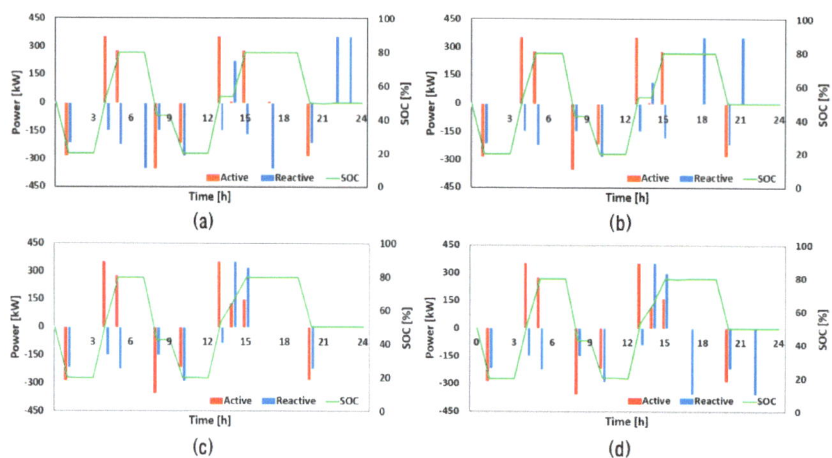

Figure 8. Results of charging and discharging schedule of the MESS. (**a**) Case 1, (**b**) Case 2, (**c**) Case 3, and (**d**) Case 4.

To analyze the impact of renewable energy uncertainty on the distribution network, the results of the voltage profile in the four cases were compared. The base scenario was forecasting data without forecasting error. The other scenarios were analyzed with a range of forecast errors from +0.5 to +2.0 σ in 0.5 σ increments. Figure 9 shows the voltage profiles at hour 14 in the four cases and five scenarios. As shown in Figure 9a, it can be confirmed that overvoltage occurred in all scenarios except the base scenario. The most vulnerable node to overvoltage was node 18, and the voltages in each scenario were 1.050, 1.052, 1.054, 1.056, and 1.058, respectively. The MESS in Case 1 was located at station 2, which had low voltage sensitivity to node 18, and did not help mitigate overvoltage. Case 2 with day-ahead curtailment had a slightly lower overvoltage than Case 1, but it has been confirmed that there was still a overvoltage risk. The voltage of node 18 represents 1.050, 1.051, 1.052, 1.053, and 1.054 in the Case 2. On the other hand, in Case 3 and 4, MESS was located at station 4, which had high voltage sensitivity for nodes 18 and 33. As shown in Figure 9c, it can be seen that overvoltage occurred only at node 33. The voltage at node 33 represents 1.044, 1.046, 1.048, 1.050, and 1.052, respectively. This means that the travel schedule of the MESS can have a great impact on overvoltage relaxation. Finally, Figure 9d shows that overvoltage did not occur in all scenarios at hour 14. As a result, it means that appropriate day-ahead curtailment and MESS operation can help ensure stable network operation against renewable energy uncertainty.

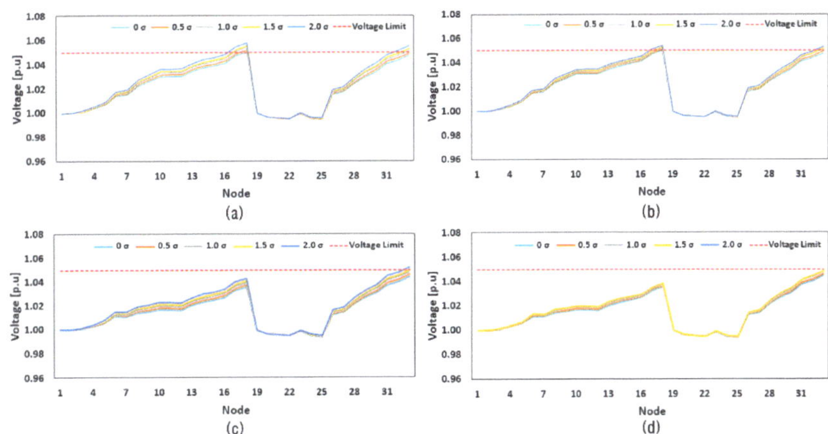

Figure 9. Results of the voltage profile at hour 14. (**a**) Case 1, (**b**) Case 2, (**c**) Case 3, and (**d**) Case 4.

Table 3 shows the operating costs of the distribution network in the four cases. It is based on forecast data, assuming that there was no forecast error. The operating costs in each case were KRW15,811,261, KRW15,815,593, KRW15,813,446, and KRW15,817,811, respectively. In Case 2, the solar output was lower than in Case 1 owing to the tight upper limit of PVs, and the operating cost was more than KRW4332. Case 3 had a KRW2185 higher operating cost than Case 1, which was mainly caused by the movement loss of the MESS. Finally, Case 4 had a KRW6550 higher operating cost compared to Case 1, which was related to the upper limit of the PVs and the movement loss of the MESS.

Table 3. The operating cost in each case.

Case	Operating Cost (KRW)
Case 1	15,811,261
Case 2	15,815,593
Case 3	15,813,446
Case 4	15,817,811

This paper also conducted simulations to quantify the curtailment by forecasting error. The curtailment was divided into power generation tripping to remove the overvoltage caused by the forecasting error and the output reduction caused by the upper limit of offline control PVs. Table 4 shows the results of the curtailment and operating costs in the four cases and five forecasting error scenarios. Case 1, which applied OPF without day-ahead curtailment, had no curtailment due to the upper limit. However, this case did not help resolve the overvoltage due to forecasting error as MESS was far from the bus, which was vulnerable to overvoltage. The total curtailments represent 0, 1186, 1236, 1262, and 2000 kW, respectively. Case 2 was eased more than Case 1 because the upper limit was applied to the offline control PVs. The total curtailments were 22, 828, 1290, 1406, and 2212 kW, respectively. The total curtailment in Case 3 only occurred for forecasting errors of +1.5 σ and +2.0 σ and the amount of curtailment was 1201 and 2000 kW, respectively. Finally, the total curtailments of Case 4 were 22, 51, 79, 149, and 1339 kW, respectively. The curtailment caused by PV tripping only occurred in the forecasting error +2.0 σ scenario, and the value was 1110 kW. In Cases 3 and 4 using CC-OPF, the MESS was located at station 4 during periods of high PV output to alleviate overvoltage, which greatly reduced curtailment due to PV tripping compared to Cases 1 and 2. In particular, Case 4, which is the proposed method, sets the upper limit of offline control PV output to cope with overvoltage due to renewable energy uncertainty and can significantly reduce the overvoltage risk. As shown in Table 4, it reduced the curtailment due to overvoltage to 0 kW in the forecasting error

+1.5 σ scenario, which was reduced by approximately 1262, 1258, and 1201 kW compared to Cases 1, 2, and 3, respectively. Therefore, the proposed method can greatly reduce the total curtailment in scenarios with forecasting error while the operating costs in the base scenario hardly increase compared to other methods.

Table 4. The curtailment in each case and scenario.

Case	Forecasting Error	Operating Cost (KRW)	Total Curtailment (kW)	Curtailment Due to Upper Limit (kW)	Curtailment Due to Trip (kW)
	0 σ	15,811,261	0	0	0
	+0.5 σ	16,044,332	1186	0	1186
Case 1	+1.0 σ	16,054,169	1236	0	1236
	+1.5 σ	16,059,278	1262	0	1262
	+2.0 σ	16,213,351	2000	0	2000
	0 σ	15,815,593	22	22	0
	+0.5 σ	15,975,341	828	51	778
Case 2	+1.0 σ	16,064,836	1290	79	1211
	+1.5 σ	16,087,617	1406	149	1258
	+2.0 σ	16,255,175	2212	229	1983
	0 σ	15,813,446	0	0	0
	+0.5 σ	15,813,446	0	0	0
Case 3	+1.0 σ	15,813,446	0	0	0
	+1.5 σ	16,049,431	1201	0	1201
	+2.0 σ	16,215,536	2000	0	2000
	0 σ	15,817,811	22	22	0
	+0.5 σ	15,823,394	51	51	0
Case 4	+1.0 σ	15,828,976	79	79	0
	+1.5 σ	15,842,709	149	149	0
	+2.0 σ	16,084,269	1339	229	1110

6. Conclusions

This paper presented an optimal scheduling strategy for distribution networks saturated with offline control units. Offline control PVs are not equipped with online communication and remote control systems, as they were mainly included into the grid before facility obligation regulation were enacted. This means that offline-controlled PVs can be easily tripped in distribution networks because it is difficult to respond in real time to overvoltage caused by renewable energy uncertainty. Therefore, DSOs must operate the distribution network considering renewable energy uncertainty to minimize PV power curtailment due to overvoltage. The proposed method uses CC-OPF with upper limits for offline control PVs to maintain the overvoltage risk below a certain probability. In addition, the proposed method can prevent PV tripping due to overvoltage by moving and charging the MESS to nodes with high voltage sensitivity. Several case studies were performed on IEEE 33-bus distribution test system and 15-bus transportation system to verify the performance of the proposed method. The results show that the proposed operation scheduling strategy can reduce curtailment for PVs due to renewable energy uncertainty.

In future work, we plan to verify the performance of the proposed operation scheduling strategy in distribution grids equipped with various DERs, such as wind farms, ESSs, and demand response resources. In addition, the proposed method can be expanded to operation strategies in distribution networks with system problems such as line congestion and overvoltage.

Author Contributions: Conceptualization, S.K. and J.L.; formal analysis, S.K.; funding acquisition, J.L.; methodology, S.K.; supervision, J.L.; visualization, S.K.; writing—original draft, S.K.; writing—review and editing, J.L. All authors have read and agreed to the published version of the manuscript.

Funding: This research received no external funding.

Data Availability Statement: The original contribution presented in the study are included in the article, further inquiries can be directed to the corresponding author.

Acknowledgments: This research was supported by Korea Electrotechnology Research Institute (KERI) Primary research program through the National Research Council of Science & Technology (NST) funded by the Ministry of Science and ICT (MSIT) (No. 24A01044).

Conflicts of Interest: The authors declare no conflicts of interest.

References

1. Sexauer, J.M.; Mohagheghi, S. Voltage quality assessment in a distribution system with distributed generation—A probabilistic load flow approach. *IEEE Trans. Power Deliv.* **2013**, *28*, 1652–1662. [CrossRef]
2. Kim, D.; Kim, J.-C.; Su, Q.; Joo, S.-K. Electricity blackout and its ripple effects: Examining liquidity and information asymmetry in U.S. financial markets. *Energies* **2023**, *16*, 4939. [CrossRef]
3. Lee, D.; Lee, D.; Jang, H.; Joo, S.-K. Backup capacity planning considering short-term variability of renewable energy resources in a power system. *Electronics* **2021**, *10*, 709. [CrossRef]
4. Kim, S.; Joo, S.K. Transmission pricing incorporating the impact of system fault and renewable energy uncertainty on the transmission margin. *IEEE Access* **2023**, *11*, 103779–103789. [CrossRef]
5. Chen, H.; Cong, T.N.; Yang, W.; Tan, C.; Li, Y.; Ding, Y. Progress in electrical energy storage system: A critical review. *Prog. Nat. Sci.* **2009**, *19*, 291–312. [CrossRef]
6. Cha, J.-W.; Joo, S.-K. Probabilistic short-term load forecasting incorporating behind-the-meter (BTM) photovoltaic (PV) generation and battery energy storage systems (BESSs). *Energies* **2021**, *14*, 7067. [CrossRef]
7. Sun, Y.; Li, Z.; Shahidehpour, M.; Ai, B. Battery-based Energy Storage Transportation for Enhancing Power System Economics and Security Country. *IEEE Trans. Smart Grid* **2009**, *19*, 291–312.
8. Divya, K.C.; Østergaard, J. Battery energy storage technology for power systems—An overview. *Electr. Power Syst. Res.* **2009**, *19*, 511–520. [CrossRef]
9. Zhao, H.; Wu, Q.; Hu, S.; Xu, H.; Rasmussen, C.N. Review of energy storage system for wind power integration support. *Appl. Energy* **2015**, *137*, 545–553. [CrossRef]
10. Ding, H.; Hu, Z.; Song, Y. Value of the energy storage system in an electric bus fast charging station. *Appl. Energy* **2015**, *157*, 630–639. [CrossRef]
11. Li, C.; Yang, H.; Shahidehpour, M.; Xu, Z.; Zhou, B.; Cao, Y.; Zeng, L. Optimal planning of islanded integrated energy system with solar-biogas energy supply. *IEEE Trans. Sustain. Energy* **2020**, *11*, 2437–2448. [CrossRef]
12. Jian, L.; Hu, Z.; David, B.; Zhao, Y.; Wang, Z. The future of energy storage shaped by electric vehicles: A perspective from China. *Energy* **2018**, *154*, 249–257. [CrossRef]
13. Mohammad, J.; Behnam, M.-I.; Mousa, M.; Pierluigi, S. Short-term self-scheduling of virtual energy hub plant within thermal energy market. *IEEE Trans. Ind. Electron.* **2021**, *68*, 3124–3136.
14. Xu, D.; Zhou, B.; Wu, Q.; Chung, C.Y.; Li, C.; Huang, S.; Chen, S. Integrated modelling and enhanced utilization of power-to-ammonia for high renewable penetrated multi-energy systems. *IEEE Trans. Power Syst.* **2020**, *35*, 4769–4780. [CrossRef]
15. Abdeltawab, H.H.; Mohamed, Y.A.R.I. Mobile energy storage scheduling and operation in active distribution systems. *IEEE Trans. Ind. Electron.* **2017**, *64*, 6828–6840. [CrossRef]
16. Lei, S.; Wang, J.; Chen, C.; Hou, Y. Mobile emergency generator prepositioning and real-time allocation for resilient response to natural disaster. *IEEE Trans. Smart Grid* **2018**, *9*, 2030–2041.
17. Prabawa, P.; Choi, D.H. Multi-agent framework for service restoration in distribution systems with distributed generators and static/mobile energy storage systems. *IEEE Access* **2020**, *8*, 51736–51752. [CrossRef]
18. Yao, S.; Wang, P.; Zhao, T. Transportable energy storage for more resilient distribution systems with multiple microgrids. *IEEE Trans. Smart Grid* **2019**, *10*, 3331–3341. [CrossRef]
19. Gao, H.; Chen, Y.; Mei, S.; Huang, S.; Xu, Y. Resilience oriented pre-hurricane resource allocation in distribution systems considering electric buses. *Proc. IEEE* **2017**, *105*, 1214–1233. [CrossRef]
20. Lei, S.; Chen, C.; Zhou, H.; Hou, Y. Routing and scheduling of mobile power sources for distribution system resilience enhancement. *IEEE Trans. Smart Grid* **2019**, *10*, 5650–5662. [CrossRef]
21. Yang, L.; Li, X.; Sun, M.; Sun, C. Hybrid policy-based reinforcement learning of adaptive energy management for the energy transmission-constrained island group. *IEEE Trans. Ind. Inform.* **2020**, *16*, 6423–6432. [CrossRef]
22. Pulazza, G.; Zhang, N.; Kang, C.; Nucci, C.A. Transmission planning with battery-based energy storage transportation for power systems with high penetration of renewable energy. *IEEE Trans. Power Syst.* **2021**, *36*, 4928–4940. [CrossRef]
23. Chen, Y.; Zheng, Y.; Luo, F.; Wen, J.; Xu, Z. Reliability evaluation of distribution systems with mobile energy storage systems. *IET Renew. Power Gener.* **2009**, *3*, 783–791. [CrossRef]
24. Kim, J.; Dvorkin, Y. Enhancing distribution resilience with mobile energy storage: A progressive hedging approach. In Proceedings of the Power & Energy Society General Meeting 2018 IEEE, Portland, OR, USA, 5–10 August 2018.

25. Jeon, S.; Choi, D.H. Optimal energy management framework for truck-mounted mobile charging stations considering power distribution system operating conditions. *Sensors* **2021**, *21*, 2798. [CrossRef] [PubMed]
26. Saboori, H.; Jadid, S.; Savaghebi, M. Spatio-temporal and power–energy scheduling of mobile battery storage for mitigating wind and solar energy curtailment in distribution networks. *Energies* **2021**, *14*, 4853. [CrossRef]
27. Yang, S.N.; Wang, H.W.; Gan, C.H.; Lin, Y.B. Mobile charging station service in smart grid networks. In Proceedings of the International Conference on Smart Grid Communications 2018 IEEE, Tainan, Taiwan, 5–8 November 2012.
28. Sun, Y.; Zhong, J.; Li, Z.; Tian, W.; Shahidehpour, M. Stochastic scheduling of battery-based energy storage transportation system with the penetration of wind power. *IEEE Trans. Sustain. Energy* **2017**, *8*, 135–144. [CrossRef]
29. Ogimoto, K.; Wani, H. Making renewables work: Operational practices and future challenges for renewable energy as a major power source in Japan. *IEEE Power Energy Mag.* **2020**, *18*, 47–63. [CrossRef]
30. Chen, T.; Xu, X.; Wang, H.; Yan, Z. Routing and scheduling of mobile energy storage system for electricity arbitrage based on two-layer deep reinforcement learning. *IEEE Trans. Transp. Electrif.* **2023**, *9*, 1087–1102. [CrossRef]
31. Tao, Y.; Qiu, J.; Lai, S. A learning and operation planning method for uber energy storage system: Order dispatch. *IEEE Trans. Intell. Transp. Syst.* **2022**, *12*, 23070–23083. [CrossRef]
32. Sun, W.; Qiao, Y.; Liu, W. Economic scheduling of mobile energy storage in distribution networks based on equivalent reconfiguration method. *Sustain. Energy Grids Netw.* **2022**, *32*, 100879. [CrossRef]
33. Samara, S.; Shaaban, M.F.; Osman, A.H. Optimal management of mobile energy generation and storage systems. *IEEE Access* **2020**, *8*, 203890–203900. [CrossRef]
34. Abdeltawab, H.; Mohamed, Y.A.R.I. Mobile energy storage sizing and allocation for multi-services in power distribution systems. *IEEE Access* **2019**, *7*, 176613–176623. [CrossRef]
35. Chen, X.; Wu, W.; Zhang, B. Robust restoration method for active distribution networks. *IEEE Trans. Power Syst.* **2016**, *31*, 4005–4015. [CrossRef]
36. Li, H.; Zhang, Z.; Yin, X.; Zhang, B. Preventive security-constrained optimal power flow with probabilistic guarantees. *Energies* **2020**, *13*, 2344. [CrossRef]
37. Schmidli, J.; Roald, L.; Chatzivasileiadis, S.; Andersson, G. Stochastic AC optimal power flow with approximate chance-constraints. In Proceedings of the Power & Energy Society General Meeting 2016 IEEE, Boston, MA, USA, 17–21 July 2016.
38. An, J.; Kim, D.-K.; Lee, J.; Joo, S.-K. Least squares monte carlo simulation-based decision-making method for photovoltaic investment in Korea. *Sustainability* **2021**, *13*, 10613. [CrossRef]
39. Lee, J.; Lee, J.; Wi, Y.-M.I. Impact of revised time of use tariff on variable renewable energy curtailment on Jeju island. *Electronics* **2021**, *10*, 135. [CrossRef]
40. Dolatabadi, S.H.; Ghorbanian, M.; Siano, P.; Hatziargyriou, N.D. An enhanced IEEE 33 bus benchmark test system for distribution system studies. *IEEE Trans. Power Syst.* **2021**, *36*, 2565–2572. [CrossRef]

 *energies*

Review

Review of Main Projects, Characteristics and Challenges in Flexibility Markets for Services Addressed to Electricity Distribution Network †

Giacomo Viganò, Giorgia Lattanzio and Marco Rossi *

Ricerca sul Sistema Energetico—RSE S.p.A., 20134 Milano, Italy; giacomo.vigano@rse-web.it (G.V.); giorgia.lattanzio@rse-web.it (G.L.)

* Correspondence: marco.rossi@rse-web.it

† This article is a revised and expanded version of a paper entitled Review on Local Market Flexibility Projects, Main Characteristics and Barriers, which was presented at the 8th International Conference on Clean Electrical Power (ICCEP), Terrasini, Italy, 2023, pp. 392–401.

Abstract: The expansion of distributed renewable resources, together with increased demand from the electrification of transport and heating sectors, impacts distribution networks significantly. Additionally, the emergence of non-programmable and intermittent generators is set to diminish the dominance of traditional rotating and programmable generation, thereby affecting the overall stability of the system. Nevertheless, the flexibility offered by distributed resources has the potential to alleviate the necessity for network reinforcement and contribute to system stability at competitive costs. Local flexibility procurement should be rooted in local markets, serving as mechanisms to address distribution congestion and coordinate the provision of flexibility for transmission network services. The multitude of existing systems and the interdependence of flexibility services have given rise to diverse solutions, still undergoing experimentation in various countries. This paper aims to scrutinize key projects that have established local flexibility markets, delineating their fundamental characteristics, the most common solutions, identifying prevalent barriers and suggesting potential future improvements. The investigation focuses on the most uncertain aspects of local markets: possible TSO-DSO coordination schemes, the time horizon for the acquisition of services and the baseline definition methodologies.

Keywords: distributed energy resources; local market models; TSO-DSO coordination schemes; flexibility; baseline

Citation: Viganò, G.; Lattanzio, G.; Rossi, M. Review of Main Projects, Characteristics and Challenges in Flexibility Markets for Services Addressed to Electricity Distribution Network. *Energies* **2024**, *17*, 2781. https://doi.org/10.3390/en17112781

Academic Editor: Hugo Morais

Received: 29 March 2024
Revised: 7 May 2024
Accepted: 4 June 2024
Published: 6 June 2024

1. Introduction

The European Green Deal, together with the Fit-for-55 package, aims to make Europe the first carbon-neutral continent. Energy uses are estimated to be responsible for 77% of greenhouse gas emissions among all other uses [1]. Therefore, to reach the final emissions targets, a fast increase in renewable energy production and electrification of transport and other consumptions is ongoing. It can be observed that small renewable plants placed near consumption points help reach the final target, and a large growth of distributed generation is foreseen. New non-programmable and intermittent generators will also impact the stability of the whole system, reducing the share of traditional rotating and programmable generation. Indeed, the opening to the services market at the national and European level launched in 2012 [2] requires a market reform able to manage the large increase in volatility and uncertainty in the system's stability. Two parallel strands overlap in this context: (1) Distribution System Operators (DSOs) must be revealed to be able to sustain the drastic changes that their grids are undergoing, and (2) DSOs must cooperate with Transmission System Operators (TSOs) to guarantee distribution generation availability and usability.

This article focuses on the first step, introducing the concept of flexibility of resources that can play a key role in overcoming difficulties and uncertainty caused by the increase in renewable resources and increase in consumption. Indeed, it is difficult to imagine that the current fit&forget approach, according to which every capacity problem is solved by reinforcing or building new lines, will be cost-efficient in a framework where production and consumption profiles are mostly volatile. Therefore, flexibility services at the local level are envisioned as a substitute for grid reinforcement; they are to be introduced both in long- and short-term operation procedures. Among the different procurement strategies aligned with recommendations of international/national regulators [1,3], the acquisition of local flexibility should be based on dedicated markets, which are designed to address local violations and, furthermore, to aggregate flexibility for transmission network services. A proficient local flexibility market should align with the current planning and operational procedures of the DSO, as delineated in [4]. This can be achieved via the adoption of a standardized approach that effectively aids the planning stage and enhances operations by facilitating the procurement of congestion management services. While regulatory guidelines [5] have outlined some fundamental aspects and major network associations [6] have proposed a technical strategy for acquiring flexibility services, the multitude of existing systems and the mutual dependency on flexibility services result in a wide variety of solutions that are yet to be fully determined.

On the basis of the highlighted necessities, many research and experimental projects, together with the scientific community, are currently investigating local market applications. All initiatives feature different characteristics, adapting the proposed implementation to the identified necessities. The aim of the paper consists of providing references to these projects and to extract and comment on the characteristics of the most relevant ones according to the authors' opinion.

This paper extends the work presented in [7] by considering and describing additional projects and review papers. This extension reports a more detailed and comprehensive discussion of identified barriers and proposes possible solutions. The analysis of some of the less-defined aspects of local markets is added, such as the time horizon for the acquisition of services and the baseline definition methodologies. The examination considers the main projects that have been established and/or implemented regarding local flexibility markets, particularly focusing on European initiatives addressing local congestion management at the distribution level. In Section 2, a literature analysis is carried out, many projects and platforms for local service procurement are studied and, on their basis, local market integration is analyzed. Specifically, TSO-DSO coordination schemes investigated in previous projects are reviewed (Section 2.1), and methodologies to define market timeframe and baseline profiles are confronted and critically observed (respectively Sections 2.2 and 2.3) without neglecting the main markets and bid characteristics (Section 2.4). In Section 3, given the acquired experience, possible alternatives to integrate local markets in the current regulatory framework are provided as a function of system needs and characteristics. In Section 4, barriers to the development of markets are analyzed and categorized according to three different typologies (technical, economic and regulatory), and, in the Conclusions Section, key aspects are considered for sustaining the future evolution of the power system.

2. Literature Analysis

The literature review analyzed more than seventy references, primarily concentrating on European projects featuring demonstration pilots on local markets (Table 1). Additionally, thirteen papers were identified that detail and compare a subset of projects [8–20]. However, in addition to considering a larger number of projects, the objective of this paper is not to undertake an exhaustive analysis and comparison of each project, which is deeply completed in [10–12], but rather to delineate the main characteristics of the adopted markets understanding that there is not a single best solution, but the best choice depends on the application context. From the literature review, several general traits of proposed local flexibility markets have been identified, also quantifying the number of projects that adopt a specific solution [13].

Table 1. List of considered projects on the exploitation of local flexibility for distribution services (square brackets indicate references, [-] indicates missing references and Others actors include balancing responsible parties and balancing service providers).

Project	Ref.	Actors			Active	Review											
		TSO	DSO	Others	[8]	[9]	[10]	[11]	[12]	[13]	[14]	[15]	[16]	[17]	[18]	[19]	[20]
ALPGRID	[21]		x	x				x									
ATLAS	[22]			x				x									
BeFlexible	[23]	x	x	x													
COORDINET	[24]	x	x	x		x	x	x	x	x			x	x	x	x	
Cornwall Local Energy Market	[25]	x	x						x								
CROSSBOW	[26]	x													x	x	
DA/RE	[27]	x	x		x			x							x		
De-Flex-Market	[28]	x		x				x		x							
DELTA	[29]	x	x	x													
DOMINOES	[30]	x	x										x				
DRES2Market	[31]	x	x						x								
EcoGrid 2.0	[32]	x	x	x				x		x							
EDGE	[33]		x	x	x												
E-LAND	[34]	x	x														
EMPOWER	[-]	x	x					x		x							
Enera	[35]	x	x	x		x		x	x	x	x		x	x	x	x	x
ENKO Flexibility Platform	[36]	x	x	x											x		
Equigy Crowd Balancing Platform	[37]	x		x	x										x		
ESIOS	[38]	x		x											x		
EUniversal	[39]	x	x	x											x		
EU-SysFlex	[40]	x	x					x	x	x	x		x			x	
EvolvDSO	[41]	x	x														
FEVER	[42]	x	x	x									x				
FLEXCoop	[43]										x						
Flex-DLM	[44]		x					x		x							
FLEXGRID	[45]	x	x														
Enedis	[46]		x	x													x
flexiblepower	[47]		x	x													
FLEXICIENCY	[48]		x					x		x		x					
FLEXITRANSTORE	[49]			x											x	x	
FlexMart	[-]		x					x		x							
FutureFlow	[50]	x											x				
GIFT	[51]		x														
GOFLEX	[52]		x						x				x	x			
GOPACS	[53]	x	x		x	x	x	x	x	x			x		x	x	x
IFLEX	[54]			x													
InteGridy	[55]		x	x													
INTERFLEX	[56]	x	x			x		x	x	x	x		x			x	
InteGrid	[-]	x	x					x	x	x						x	
Internet of Energy project (IO.E) in Belgium	[57]			x													
INTERPLAN	[58]	x	x														
INTERRFACE	[59]	x	x			x	x	x	x	x					x	x	
INVADE	[60]	x	x	x													
iPOWER	[61]		x	x													
IREMEL	[-]	x	x					x	x	x							

Table 1. *Cont.*

Project	Ref.	Actors			Active	Review												
		TSO	DSO	Others		[8]	[9]	[10]	[11]	[12]	[13]	[14]	[15]	[16]	[17]	[18]	[19]	[20]
Magnitude	[62]			x								x						
MERLON	[63]	x	x	x														
MiNDFlex	[64]		x	x	x													
MUSE GRIDS	[65]			x														
NODES-INTRAFLEX	[66]	x	x	x		x	x	x	x	x	x	x		x	x	x	x	
NORFLEX	[67]			x						x						x		
OneNet	[68]	x	x	x														
OSMOSE	[69]	x	x														x	
PICLO FLEX	[70]		x			x	x	x		x	x	x		x	x	x	x	x
PLATONE	[71]	x	x	x						x						x	x	
Power Potential	[72]	x	x															
Project LEO	[73]	x	x															
Redispatch 2.0	[74]	x	x		x													
REFLEX	[75]			x														
RomeFlex	[76]	x	x	x	x													
SENSIBLE	[-]	x	x							x		x		x				
SmarterEMC2	[77]	x	x															
SmartNet	[78]	x	x		x									x		x	x	
SMILE	[79]			x														
Soteria	[80]			x														
Store & Go	[81]			x														
TDX-ASSIST	[82]	x	x														x	
TenneT Cooperation Project Grid Stabilisation—Vehicle 2 Grid	[83]			x														
UPGRID	[84]	x	x															
USEF	[85]	x	x	x						x		x						
WiseGRID	[86]	x	x															

The main services procured via local flexibility markets include resolving current congestion and voltage violations on the distribution network managed by the DSO. Although a few projects consider additional services such as black-start and island operation, they are not considered in the present analysis as they are adopted by only a minority of projects.

Ref. [19] proposes the definition of harmonized products, considering the possibility of comparing relevant attributes and their respective numerical values. In this context, it is also observed how harmonized products could help in reducing effort in TSO DSO coordination. Indeed, in certain instances, the local platform is also utilized by the TSO to secure flexibility for balancing the system and contain current and voltage violations on the transmission network. In this regard, the interaction between the procurement of local and global services is one of the key aspects shaping the operational dynamics of the new local market in alignment with pre-existing ancillary services markets. This interaction is significant as the outcomes of local markets can impact dispatchment in the global system and vice versa.

Each of the following paragraphs focuses on a single market aspect, each of them selected according to the authors' opinions developed from the literature review. Different TSO-DSO coordination schemes are presented in Section 2.1, and implications for the choice of the scheme are then described in Section 2.2, where the importance of the timeframe for acquiring various services is discussed. For instance, if local flexibility is planned during the day-ahead market, local activation could be seamlessly integrated into the day-ahead bids of aggregators. Another crucial aspect explored in the literature is the definition of the baseline profiles of flexible resources (Section 2.3), which directly affects the quantification of activated flexibility. Finally, the main alternatives for market implementation are presented (Section 2.4), forming the groundwork for the classification described in Section 3.

2.1. TSO-DSO Coordination Schemes

In the review, seven distinct coordination schemes can be delineated, characterized by an increasing level of integration between local and global markets. These coordination schemes will be briefly outlined, with particular attention to the procurement priority of distributed energy resources (DER), following the terminology and the schemes of [13] and adding the new coordination scheme M0. In this coordination scheme (M0—Figure 1), local and global markets are entirely segregated, and the agents are responsible for proposing bids in separated markets. Participants are responsible for offering bids to one market without determining the imbalance in the other market. This coordination is recommended when there is a significant difference in the timeframe of services acquisition (e.g., local services are acquired during the day-ahead market, while global services are acquired near real-time) or when the expected activation of local power is small, thus exerting negligible impact on the overall system balance.

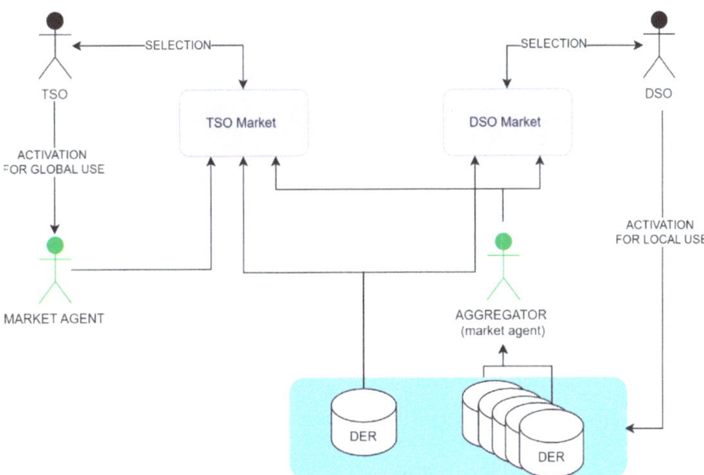

Figure 1. General representation of the coordination scheme M0.

The second coordination scheme (M1) does not foresee the local market: the sole responsibility of the DSO consists of pre-qualifying the resources for the TSO market (Figure 2). This market structure is implemented in scenarios where the is no requirement to procure local services, but the utilization of local resources for global needs could determine issues in the distribution network.

In the third market scheme (M2), precedence is given to the local market over the global one (Figure 3). The DSO is responsible for selecting resources necessary to address local congestion, with any remaining flexibility offers being directed toward the TSO. Furthermore, the DSO may validate the offers chosen by the TSO to ensure their activation does not impact the secure operation of the distribution network. Under this scheme, the local market may also aggregate local resources to offer them to the TSO. Consequently, bids submitted to the TSO would inherently adhere to the constraints of the distribution network. This arrangement implies that the local market can coordinate resources from various aggregators, a circumstance that warrants careful consideration from a regulatory standpoint to prevent discrimination.

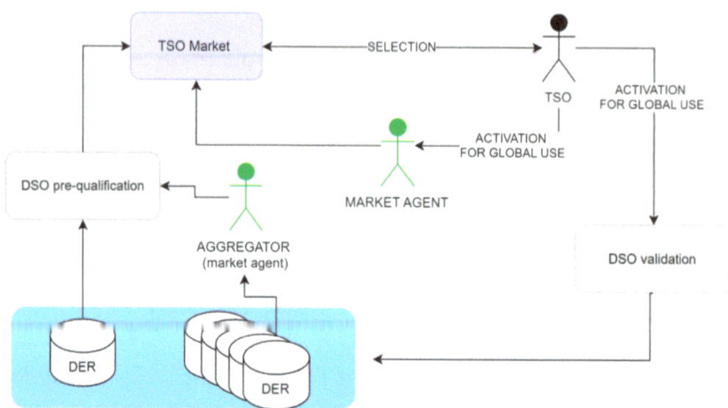

Figure 2. General representation of the coordination scheme M1.

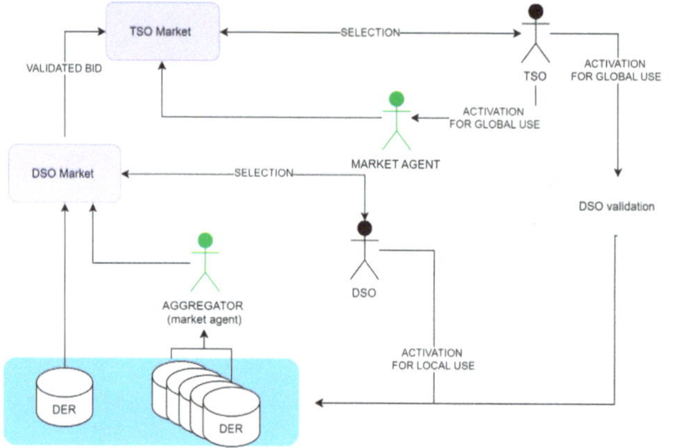

Figure 3. General representation of the coordination scheme M2.

In the fourth scheme (M3), the two markets establish predefined profiles that the DSO must adhere to. In this scenario, the DSO can use the local flexibility to address local congestion as well (Figure 4).

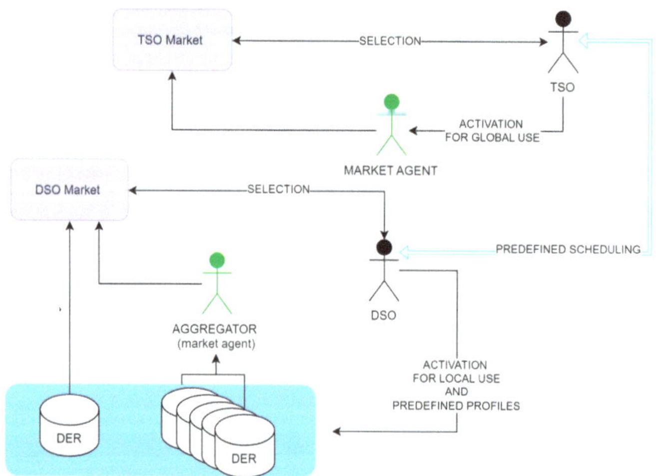

Figure 4. General representation of the coordination scheme M3.

In the fifth scheme (M2/3), only the DSO can use local resources. The overall imbalance resulting from resource activation is then communicated to the TSO (Figure 5), which manages any possible imbalances within its market.

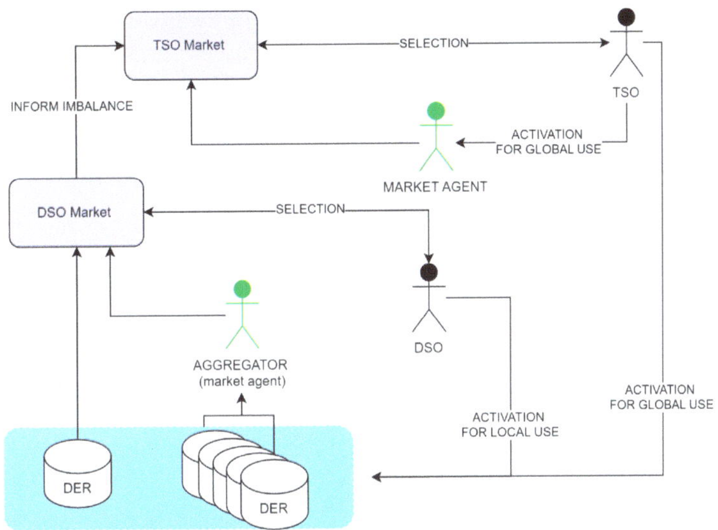

Figure 5. General representation of the coordination scheme M2/3.

In the sixth scheme (M4), a unified market exists for both DSO and TSO to acquire the needed services (Figure 6). In this case, both the DSO and TSO share equal priority in procuring local resources, and their selection process jointly takes into account constraints at both distribution and transmission network levels.

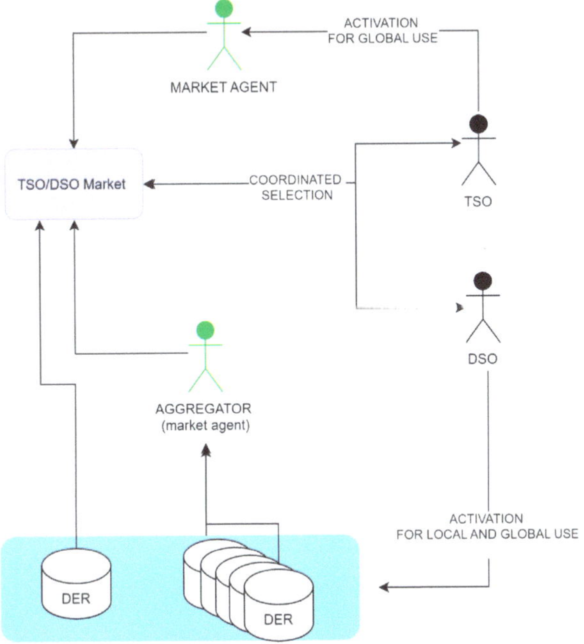

Figure 6. General representation of the coordination scheme M4.

The last scheme (M5) represents the most inclusive market environment, as it allows not only the DSO and TSO to procure services but also permits responsible parties to acquire flexibility to balance their own resource portfolios (Figure 7). While in other schemes, the balancing responsible parties can also procure balancing resources (e.g., in the global market), in M5, they directly compete with the network operators. Consequently, the DSO must offer higher prices relative to other market participants in order to prioritize the use of resources.

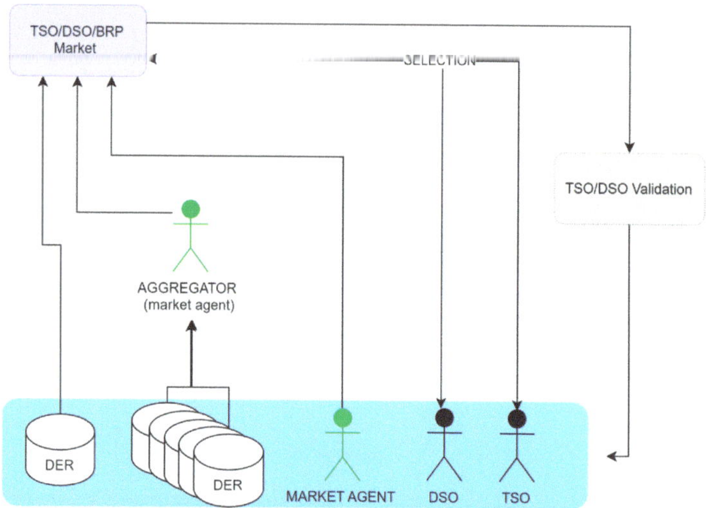

Figure 7. General representation of the coordination scheme M5.

The choice of coordination scheme relies on several factors. Typically, the necessity for coordination increases with the volume of local flexibility because of its impact on system balancing. Additionally, as local resources become more involved in supporting the transmission network operations and as the DSO requires resources to address local congestion, the need for coordination intensifies. During the initial experimental implementations, where the demand for local flexibility and the availability of resources were limited, the simplest coordination schemes were validated with success by testing their technical efficacy.

2.2. Service Acquisition Timeframe

Another important aspect of implementing local markets is establishing the timeframe for acquiring services. While not thoroughly investigated in all the analyzed projects, this factor significantly influences the selection of the coordination scheme. Typically, the timeframe is closely related to the ability of DSO and TSO to forecast violations, which depends on the variability of user exchange profiles and on the forecasting methodology employed. Usually, the DSO forecasts are the most critical since they consider smaller aggregation perimeters than TSO, which determines an intrinsic higher uncertainty due to the lower number of users.

The timeframe of the acquisition of local services contributes to defining how to integrate the acquisition of local services within the workflow of already existing markets. For instance, in scenarios where violations can be detected prior to the opening of the global market, they may be addressed independently without relying on the TSO market. This is the case of some currently active platforms (e.g., [47]), where the activation of resources for certain services is scheduled days in advance. However, in these cases, to consider possible errors in the forecast calculations, the DSO could acquire more resources than needed, increasing the cost of local flexibility. In the opposite case, when the congestion can be detected only close to real time, a strong integration with the global market is needed to compensate for unforeseen imbalances. For these reasons, the timeframe is defined by a trade-off between the over-acquisition of services from DSO and the imbalances in the TSO.

2.3. Baseline

The timeframe also influences the identification of the baseline profile of resources [12], which is essential for quantifying the delivered service. In the literature it is not found a general review of the baseline calculation methodologies and their impact on the local market design. For this reason, on the basis of the analyzed literature (Table 2), the main methods used to define the baseline are identified:

(a)   The DSO (or the local market operator) calculates the baseline on the basis of historical patterns, possibly considering meteorological forecast and real-time data [70,87–90];
(b)   The aggregator computes and provides the baseline profile to the DSO [32,45,91,92];
(c)   The baseline is established from the results of the previous (energy) market sessions [32,93];
(d)   The baseline profile is extrapolated from the measured power exchange before and after the service delivery [94].

The definition of the baseline is particularly relevant when the DSO and TSO share the same market: Since the remuneration of resources should consider the activation of the two markets, the baseline computation must be aligned. In this case, there is also the difficulty that the aggregation of perimeters for the global market can be different with respect to the aggregation perimeters for the local market. For example, a perimeter for global services could contain many perimeters for local services, and the baseline computation of the larger perimeter should consider the baseline calculation of all the smaller perimeters. A possible solution is to design the local and global markets in a way to that allows leaving the aggregators the responsibility of presenting coherent baselines. For example, if the aggregator knows that in the local market the baseline is computed by historical value, it can take this information into account when defining the baseline for participating in global markets. Instead, if the markets are separated, the baseline computation can be different.

The baseline calculation is particularly important for resources that cannot be programmed with high precision, like residential users or photovoltaic generation. The forecast and baseline calculation can be affected by a large error due to the high variability of such users. Given that the accuracy of power profiles impacts the uncertainty of flexibility margins and that non-programmable resources often have limited control margins, they reveal to be more suitable for providing services that do not require high precision, like in some local markets. For example, if the flexibility resources reduce the power more than the quantity required by the DSO, there are usually no issues for the operation of the network (the same does not apply in the case of balancing services). However, the uncertainties in determining and communicating baselines should be considered in the planning phase. In such cases, operational limits are kept conservative, and penalties apply for partial or missed activations with no remuneration for activations exceeding requests (Figure 8).

Furthermore, the challenges associated with the uncertainty of resource activation include those related to the definition of baseline. Since the compensation for resources depends on the difference between activation and the baseline profile, uncertainties in these two components combine, affecting the estimation of the service activation level. Therefore, the aggregator must account for these uncertainties in defining offers, and, subsequently, the market operator must ensure that the aggregates adhere to the baseline profile and activation, allowing for a margin of error on both. For all the methods to define the baseline profile, its precision decreases as the size of the aggregation perimeter decreases. For these reasons, the more granular the market, the more attention must be paid to baseline profile definition, and various methodologies can be employed, with varying effectiveness depending on the types of users considered. For instance, if a network node serves a controllable generator, the baseline profile may align with the production schedule defined by the aggregator within the day-ahead market. If a node predominantly serves residential users, statistical methods must be employed to predict its profile and to consider a defined confidence margin during the verification of the service delivery.

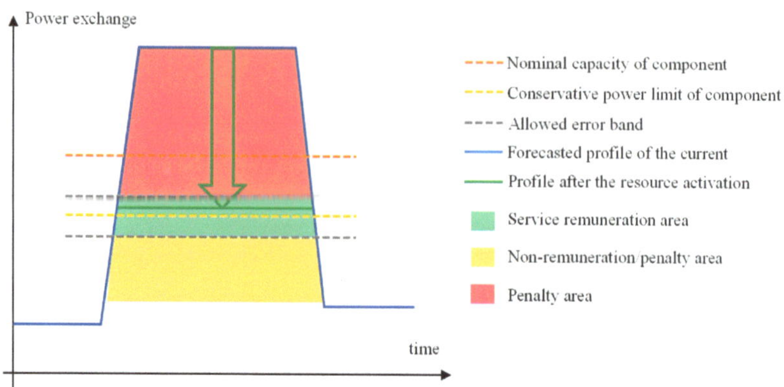

Figure 8. Simplified representation of service delivery remuneration and penalties application.

Table 2. Literature on baseline calculation methods.

Ref.	Baseline Calculation
[32]	The baseline profile is an estimate of the consumption of a set/group of loads, assuming that they are neither under external control nor show effects of past or future external control (external control refers to the actions that an aggregator would perform to activate flexibility). The load schedule is assumed to be equal to the expected consumption declared by the related aggregator on the day before. In cases where there are several metering points with the same aggregator, the aggregator will predict the entire consumption. If the aggregator does not respond to the day-before signals, the load schedule and baseline are the same.
[87]	The baseline profile is agreed upon by the parties on the basis of historical behavior of the involved energy resources, taking into account any changes in the year or weather forecasts.
[88]	The techniques for calculating the baseline profile vary depending on the availability of measurements. When a history of power exchange profiles is available, statistical regression methods use the collected data to calculate load forecasts. If the information is not available, the consumption of a few users of the same typology can be measured and then used as a reference for all other users of the same type.
[89]	The baseline profile is calculated using historical data by means of simple and well-defined methodologies. In particular, the methodology of similar days [95], which is widely accepted and used by the New England balancing operator, results in being the best among those analyzed [96].
[90]	The baseline profile is calculated as the average of the power exchange of the time frames similar to the reference time frame for the specific market session.
[91]	The actor in charge of calculating the basic profile may vary from case to case. It can be calculated by an independent external entity or agreed upon by the different market players. Since aggregators optimize the exchange profiles of all controlled resources to minimize the cost of procurement for each of them, this schedule can be used as a baseline profile. The DSO, also based on the basic profile communicated by the aggregators, predicts the evolution of the loads and generators that are used to determine the flexibility needs.
[92]	Aggregators optimize their portfolio of assets (e.g., in the day-ahead market) and communicate the resulting profile to the local flexibility market.
[93]	The results of previous markets (energy and/or ancillary services) can be used to determine the baseline profile with which flexibility needs can be defined.
[94]	There are two options for calculating the basic profile: • Market participants must declare the ex-ante baseline profile; • The market buyer (TSO, DSO, etc.) calculates the ex-post base profile using meter measurements.

Table 2. *Cont.*

Ref.	Baseline Calculation
[97]	The calculation of the baseline profile and the modulation of the load due to the activation of the service is carried out via the measurements of the meters of the resources participating in the market. The meter data must therefore be accessible to the market operator.
[98]	The day-ahead market results define the baseline profile.
[99]	The baseline can be calculated from historical data, and different methods of calculations are illustrated in [99]. The survey shows that the most promising methods are characterized by calculation procedures that are easy to implement, but at the same time accurate to avoid opportunistic behavior.
[100]	The baseline profile can consist of a forecast made by the aggregator. In other cases, the profile may be calculated by the market operator and communicated to the aggregator. In the first case, the profile takes on two roles: • Communicate the planned production/consumption profile so that the DSO can include it in a network security analysis; • Define the reference for flexibility exchanges between the aggregator and the DSO.

2.4. Market Schemes and Bid Characteristics

Two types of procurement schemes are proposed in the literature for acquiring local flexibility services. The first is similar to a capacity market, where the availability of resource modulation is acquired, and the related flexibility is activated on demand by the DSO based on their merit order list and the forecasted violations. This alternative is referred to as long-term procurement and, generally, concerns an availability price (EUR/kW) and an activation price (EUR/kWh). The other alternative, also called short-term procurement, consists of a spot market, which consists of immediately settling the full payment in exchange for the service delivery. The two schemes can be merged in a single procurement strategy: first, the DSO acquires the estimated necessary flexibility in advance, and then the activation is determined by running a spot market near real time where also additional resources can participate. Furthermore, there are two main acquisition mechanisms of the bids: continuous trading and periodic auctions. The latter is most used since it allows to consider in a single optimization process all the bids and constraints. Once the bids are collected, the optimum solution can be determined by two possible clearing mechanisms: pay-as-bid or pay-as-clear. Finally, the interaction schemes between the market participants can be one-sided if the only buyer of the resources is the DSO (or the TSO) or two-sided if also BRP can buy flexibility.

The characteristics of the market, the acquired services and the bid structure determine the model and the algorithm used for clearing the market [9,101]. Their complexity increases with the intricacy of the services and bids (Table 3, defined on the basis of [11,19,92,102–104]). For example, if only the congestion management on the distribution network is considered then the market could be cleared with a linear programming optimization (i.e., DC–Optimal Power Flow). Instead, if the market is also used to solve voltage violations and the bids are not divisible (i.e., a bid has to be completely accepted or not), it must be based on a non-linear and/or mixed-integer problems (i.e., AC model of the network, which is much more complex than a DC–Optimal Power Flow). A list of techniques for the optimization of local markets, together with the mathematical modeling and simulation tools, are reported in [105–107] with particular reference to the interaction between DSO and TSO.

Table 3. Main possible characteristics of flexibility bids [7].

Field	Description
Time window	The time window in which flexibility must/can be provided
Modulation power	Active or reactive power
Alert time	Time in advance communicated the request for actual activation
Congestion/location area	Geographic region within which resources must be located to provide the requested service
Quantity	Amount of flexibility required, typically a difference from an agreed base profile
Maximum duration of service	Maximum duration of service provision
Minimum duration of service	Minimum duration of service provision
Capacity remuneration	Price that the aggregator (or resource) offers for resource availability
Volume remuneration	Price for energy modulation
Maximum number of activations	Number of activations that can be requested in a given time interval
Recovery time	Minimum time interval between two successive activations
Penalties	Penalties in case of non-compliance with dispatching orders
Minimum quantity	Minimum quantity to offer
Maximum quantity	Maximum quantity offered
Ramp constraint	Maximum variation to rise or fall in power
Full activation time	Time needed for the resource to reach the proposed power delivery
Divisibility	Possibility of partial acceptance of the offer
Resource Type	Source type
Rebound effect	Describes the characteristics of the possible rebound effect due to resource activation

3. Characterization of Proposed Markets

Starting from the description of local markets, the number of proposed markets implementing each feature is evaluated to give an idea of which are the most implemented solutions (Table 4). It is important to notice that the characterization is merely indicative: each proposed market possesses unique attributes that may not align with any scheme. Furthermore, not all aspects of the market are consistently detailed.

One of the fundamental aspects is the coordination between DSO and TSO, as local services must be managed to avoid impacting the operation of the transmission network. While the number of projects does not necessarily indicate the effectiveness of a scheme, it does offer an indication of the most promising or readily implementable option. The M2 scheme emerges as the most frequently proposed, likely due to its balanced interaction between DSO and TSO: the DSO primarily validates bids for the TSO, maintaining a straightforward interaction. Conversely, the M2/3 scheme is the least used, possibly because it closely resembles M2 but lacks the capability to transfer unused local offers to the TSO. With the exception of the M2/3 scheme, all adoptions are investigated in various projects. However, the prevalent scheme in actual platforms utilized by network operators mainly falls within the first three categories: M0, M1 and M2. In these schemes, the role of each actor is well defined since each service is managed by a single operator in a specific market (e.g., congestion for distribution network in local markets). In this way it is also simpler to associate the power modulations of balance responsible parties to the specific network operator request.

The necessity for coordination with the TSO also influences the services considered by the local market. While resolving DSO congestion remains the most investigated service, as it aligns with the main objective of local markets, services directed to the TSO (balancing, congestion, etc.) are commonly integrated into the market. This integration is crucial because the activation of distribution resources has the potential to exert significant effects on the transmission network operation. However, as explained in the previous paragraph, the provision of local and global services is not usually considered in a single market, but the global services are negotiated only after the local market is cleared.

Table 4. Amount of investigated local markets for each particular feature [7]. The acronyms of coordination schemes are presented in Section 2.1.

Coordination Scheme between DSO and TSO						
M0	M1	M2	M2/3	M3	M4	M5
No coordination scheme	Centralized market	Separated local and global market	Coordinate balancing	Shared responsibility	Common local and global market	Integrated market
8	11	20	1	8	5	5
Proposed services						
DSO congestion	DSO voltage	Balancing	TSO congestion	TSO voltage	Island operation	
40	26	24	25	20	4	
Buyers/sellers coordination						
One-sided			Two-sided			
45			9			
Negotiations						
Periodic auctions			Continuous trading			
33			10			
Price formation						
Pay as bid			Pay as clear			
34			6			
Type of participants						
Any	Loads		Storages		Generation	
44	3		4		3	
Aggregators						
Allowed The DSO usually defines the perimeter of aggregation.		Nodal The aggregation is possible, but only at the nodal level		DSO The DSO is responsible for the aggregation		
37		7		2		
Time horizon						
Short period Market closure hours before real time			Long period Market closure Month or years before the real time			
35			10			

The majority of proposed markets follow a one-sided model, being characterized by a single buyer of the service (either DSO or TSO) and multiple sellers. However, in certain markets that incorporate balancing services, a two-sided mechanism is proposed, allowing balance responsible parties to procure flexibility for their own interest. The one-sided model has the advantage of associating each resource activation with the corresponding provided service. It allows for a more straightforward implementation of long-term contracts since each network operator can independently define and acquire its flexibility needs. The two-sided mechanism allows for clearing in a single market many flexibility needs, optimizing the network in a single run and avoiding the need for multiple markets. However, this requires a clear definition of the role of each actor in the common market and a more complex division of cost and benefit since a single power modulation can solve multiple services.

Periodic auctions are the favored method of negotiations, likely attributed to the necessity of incorporating network constraints into the market optimization processes. In particular, grid constraints introduce complexity and time constraints in the optimization algorithm of the market, making the implementation of continuous-trading markets less feasible. In addition, periodic auctions are more easily integrated into the already existing energy and flexibility markets, which have a well-defined time schedule.

The complexity of the market, network constraints, potential liquidity issues and the presence of only one buyer (the DSO) likely contribute to the preference for pay-as-bid price remuneration. Furthermore, in most cases, the market only determines the selection of resources rather than their activation. Activation signals are dispatched only upon identification of constraint violations based on the merit order list of received bids.

While most projects permit the participation of resources that comply with defined market rules, there are exceptions where implementations focus on specific resources (e.g., storage systems), although such cases are limited. This second choice, even if not technologically neutral, is driven by the issues of integrating a generic market resource that has specific characteristics (e.g., the capacity limitation of the storage system). In cases where there is an abundance of a specific asset, it is easier to implement a market that is designed accordingly to the resource specification. This issue is usually present when there are resources with a capacity constraint (e.g., storage system), and it is necessary to define if this constraint is integrated directly into the market algorithm that autonomously determines the best use of the resource or the responsibility of presenting offers that respect the capacity constraint is left to the balance responsible parties. Usually, when multiple resources are present, this second choice is the most common since it makes easier and more transparent the interpretation of the market results since all the resources, both with and without capacity constraints, are treated equally by the optimization algorithm.

Aggregation is typically permitted, with the DSO defining the perimeter of aggregation based on required services. In certain instances, aggregation is nodal (e.g., medium voltage node) to circumvent the need for defining perimeters, or it may be altogether unfeasible. In exceptional cases, the DSO is directly responsible for the aggregation of resources, in particular in the market scheme M2, where the DSO is responsible for submitting unused bids to the global market. The perimeter of aggregation strictly depends on the requested services: in case of congestion, only the resources that contribute to solving the issue can be aggregated (e.g., in case of overcurrent in an MV/LV transformer, only the underlying resources can be aggregated).

Most markets operate on a relatively short time horizon, with gate closures typically set at day-ahead or intraday intervals immediately preceding or following energy market closures. While some markets extend over longer time horizons (e.g., annually), various activation process alternatives exist beyond gate closure timing, such as sending activation signals immediately post-market closure or only in the event of a violation. The time horizon depends on the capacity of operators to foresee the service needs and on the time schedule of already existing markets. However, for brevity, these alternatives are not reported.

From Table 4, the characteristics of the most prevalent market can be outlined. The primary buyer of services is the DSO, tasked with defining service parameters (e.g., aggregation scope, time horizon) and forecasting required quantities. The local market operator collects flexibility requests from the DSO and offers from local resources, typically aggregated. This process considers periodic auctions with a daily or longer time horizon if flexibility procurement is needed further in advance. After the gate closure, a merit-order list of available resources is compiled based on bid prices. Subsequently, the local market operator, in collaboration with the DSO, reserves necessary bids for distribution operations and forward the remaining bids to the TSO. Additionally, the DSO communicates to the TSO that resources cannot be utilized for global services. As real time approaches (time horizon contingent on forecast quality and TSO market coordination), the DSO activates required resources based on anticipated violations and communicates activation levels to the TSO, possibly via the local market operator. Resources are then compensated based on their activation and bid prices, with additional compensation possibly allocated for resource availability in long-term procurement mechanisms.

The presented coordination schemes mainly consider how DSO and TSO can interact during the procurement of services in short-term markets without considering how this coordination can be affected by the presence of long-term contractualization. Indeed, in the literature [108], it is pointed out that voluntary markets (such as short-term procurement solutions) often do not represent the best alternative. In support of this statement, the INC-DEC strategy is explained: when service providers can anticipate market prices, they are inclined to bid strategically. Giving an example of bidding strategy: in a region characterized by low electricity supply, producers will bid at high prices to maximize their profits in the following re-dispatch market. Such behavior can aggravate network congestion, increase electricity prices and distort financial markets and price signals used to define where to build new plants. Along these lines, the same behavior could be expected during the procurement of local flexibility services. To prevent such phenomena, long-term contracts are a viable alternative. Indeed, when long-term procurement is deployed, availability and activation prices are established a priori; thus, the possibility of gaming is averted. However, when long- and short-term procurements are used together, the short-term market competitiveness can be compromised because of the effects of activation prices defined in the long-term procurement.

4. Identified Barriers

On the basis of the described review of established local markets for procuring ancillary services at the distribution level, numerous significant barriers have been identified during the testing phase. These barriers can be categorized into three distinct groups: technical, economic and regulatory.

Technical barriers consider all aspects related to available technologies and infrastructure, as well as constraints imposed by current market design and the need to define new technical parameters. Economic barriers, on the other hand, concern monetary considerations associated with the development and deployment of these technologies, the emergence of new costs linked to procuring such services and the necessity to establish new economic relationships among market participants. Lastly, regulatory barriers primarily relate to the lack of defined roles, coordination procedures and priorities, as well as the necessity to reform electricity markets to incorporate the new services offered by distributed flexibility resources. The analysis of these barriers is based on the interpretation of the authors, starting from the consultation of the literature reported in the following subsections. Particularly relevant for the scope of the analysis are observations collected in [94,109], from which most of the barriers have been extracted and integrated with the experience gathered from other projects.

4.1. Technical Barriers

Technical barriers pose a significant challenge to the implementation of local flexibility services, particularly concerning technology and telemetry prerequisites (Table 5). Strict market access requirements, established in previous years looking at traditional generation characteristics, introduce significant limitations to the integration of distributed resources. For instance, specific constraints for providing balancing services, such as minimum bid offers and deviation penalty designs [94], may not align with demand-side management participation, excluding them from the provision of balancing services. Therefore, bearing in mind that some services, such as balancing ones, are essential for ensuring system stability, access barriers for the provision of different services should be smoothened in order to foster the engagement of different resources. For example, looking at congestion management in the distribution network, fewer technical requirements exist. Resources can provide additional services beyond what is needed, and the unrequested modulation is typically not remunerated and does not determine issues. Given the varied technical characteristics of potentially beneficial activations (e.g., fewer resources and lower modulable power), it is important to establish dedicated requirements for services procured in local ancillary services markets to avoid the exclusion of significant resources. Basically, the idea is to go from technology-specific requirements to service-specific requirements, allowing a larger pool of resources to provide services in aggregated configuration or individually.

Table 5. Main technical barriers limiting the adoption of local markets.

The difficulty for some types of users, especially smaller user loads, to offer flexibility with sufficient performance is due, for example, to the following: • Impossibility of modulating the power exchanged in some types of converters (e.g., older photovoltaic systems); • The difficulty of calculating the baseline; • The scarcity of the quantities offered; • The uncertainty of correct activation.
The characteristics of bids for balancing markets (high minimum size…) exclude the participation of smaller resources and cannot therefore be replicated in local markets. It is then necessary the harmonization between offers presented in local markets and those presented in global markets.
Insufficient measurement and communication standards of smart meters (number of measurements, precision, update period, communication channels).
Poor experience of aggregators in providing services at the distribution level and DSO in quantifying and resolving violations via ancillary services.
Lack of standards for the following: • Data model; • Data exchange (protocols…); • Functionality of the markets; • Characteristics of the offers; • Methods of control of distributed resources; • Prequalification procedures.
Uncertainty that the local market can guarantee sufficient flexibility over a long time horizon.
Difficulty of DSO to forecast network violations both for long (years) and short (hours) time horizons due to the uncertainty of scenarios and the small aggregation perimeters of resources.
Definition of the baseline profiles of resources participating in the market and coordination with the baseline of other markets.
Training of employees, both DSOs and aggregators.
Difficulty in defining flexibility services for low voltage components, which are often poorly monitored, and also due to the low number of users involved, which makes efficient acquisition of regulation resources difficult.

Another technical barrier is identified in the determination of the baseline. The assessment of the baseline is a highly discussed topic when defining parameters, and ensuring a certain level of accuracy is mandatory as settlement, and consequently, service remuneration, depending on the baseline definition. In this regard, there are two potential approaches: service providers may declare the baseline ex-ante, or the market operator could calculate it using metering data and utilizing knowledge of reserved energy for the service. While both strategies are functional, neither guarantees a precise evaluation of the energy provided for the service. For example, oscillations in generation/consumption profiles due to external factors could be inaccurately assessed, being considered as a provided service even if not voluntarily activated. Moreover, it is essential for the baseline to be included in data exchange between the TSO and DSO. A dual and separate evaluation could lead to discrepancies and necessitate additional coordination efforts. This particular aspect should be incorporated into the interactions between system operators, which are mainly associated with regulatory barriers. Lastly, standardizing procurement timing and the amount of capacity allocated to a particular service is crucial to prevent distortions caused by local market activities on the global market.

The requirements must be taken into account in the establishment of the prequalification processes which are typically technology-specific rather than being service- or product-specific [109]. Furthermore, prequalification is presently defined at national level and separately for different technologies, which increases the necessary effort and introduces some discrepancies among national regulations.

Establishing a common set of attributes for new flexibility services and standardized prequalification procedures could facilitate the development of new local markets, introducing minimal standardization at the European level. This collection of parameters should also include requirements for verifying provided services, including measurement frequency, accuracy and communication channels. From the perspective of network operators, there are additional technical challenges that remain unresolved. Implementing a local market necessitates the capability of the network operator to identify and quantify forecasted violations. To achieve this objective, it is important for the DSO to develop both long- and short-term network observability tools for identifying congestion [110]: long-term tools assess the benefits of local markets compared to network reinforcements, while short-term tools are essential for accurately identifying flexibility needs and subsequently acquiring services within the market. A crucial component of the observability tool is the digitization of network assets, enabling the accurate modeling of violations and resources eligible to participate in the market.

Finally, another critical aspect consists of the necessity of training employees, both within DSOs and aggregators, to effectively manage and participate in the local market.

4.2. Economic Barriers

The establishment of local flexibility markets requires significant investment across the entire system, impacting the business cases of involved actors (Table 6). Indeed, the lack of expertise and historical data are always characteristics of first-of-a-kind mechanisms, such as the procurement of flexibility resources via local markets. The profitability of local markets is a subject of ongoing discussion and is not easily predictable due to various factors. These considerations pose challenges for balancing service providers in formulating business cases and for potential end users in adjusting their usual business practices.

Table 6. Main economic barriers limiting the adoption of local markets.

Lack of experience and data on the possible profitability of local markets that allows aggregators to develop business cases.
Difficulty for users, especially industrial ones, to change their business model.
Possible benefits for small and medium users are often limited.
Uncertainties about possible profits, but also about possible penalties in case of failure to provide the service.
High initial costs to enable resource flexibility.
Uncertainty for DSOs when comparing reinforcement and flexibility costs.
Lack of criteria to remunerate investments in information and telecommunications technologies by DSOs.

Ensuring the qualification of resources is essential to guarantee efficient service provision. However, market requirements must also be flexible enough to encourage an acceptable number of resources to participate. Achieving a satisfactory number of involved resources becomes crucial when the overall qualified capacity is significant, particularly to achieve a significant contribution from the demand-side response. This highlights the importance of industrial consumers consistently participating in local flexibility markets. Moreover, accurately evaluating and comparing the costs associated with grid investments versus those related to procuring flexibility services is complex for DSOs. The latter costs should encompass considerations such as payment for imbalances on the central market (resulting from local activations), investment in communication and information infrastructure, costs associated with unplanned unavailability of flexibility resources and others.

Furthermore, the potential for the exercise of market power should be considered, and mechanisms to counteract opportunistic behavior should be established. For example, while market-based procurements are typically preferred, in regions characterized by resource scarcity, must-run dispatches should be considered as a potential alternative.

4.3. Regulatory Barriers

Regulatory barriers primarily consist of gaps in defining the roles and responsibilities of market participants, such as aggregators and network or market operators, as well as the products and services for the local flexibility market (Table 7). Resource aggregation is fundamental for the management of very small resources; thus, it is essential to understand how aggregators should delineate the characteristics of a particular portfolio. In this context, challenges arise due to a lack of historical data, particularly when portfolios comprise different flexibility technologies with varying technical characteristics. Furthermore, network operators (i.e., DSO) must guarantee their independence from market participants, avoiding conflicts of interest during market procedures. This also requires the definition of open and transparent market-based procedures, highlighting technological neutrality.

Table 7. Main regulatory barriers limiting the adoption of local markets.

Lack of definition of possible local services (resolution of voltage violations via reactive power…) and mechanisms for providing local flexibility.
Absence of definition of roles, functions and responsibilities of the participating actors.
Lack of rules for the coordination between TSOs and DSOs and, more generally, between local services and system services.
Asymmetry of treatment between OPEX and CAPEX in the remuneration of DSOs: • Absence of local flexibility regulation. • DSOs are currently remunerated only on historical spending (especially for OPEX). • There is a lack of remuneration schemes based on long-term development plans. • Lack of incentives to extend the useful life of components.
Reduced possibility to implement sandboxes by DSO and TSO to implement local flexibility markets that extend the present regulation.
Lack of transparency on the residual capacity of the network to install new loads and generators and on the network reinforcement plans, which allows for identifying the areas where flexibility would be necessary and how much it could be remunerated.
Lack of guarantees of independence between the various market players (DSO, aggregators, market operators…).
Lack of connection costs and network tariff structure linked to connection capacity instead of energy consumption.
Lack of privacy protection of network data and users participating in the market.
Lack of regulation to avoid the risk of opportunistic behavior, for example, due to scarce liquidity in local markets.

Furthermore, a lack of definition regarding the services acquired in the local market remains. To provide support to regulatory development, network operators should conduct a comprehensive analysis focusing on system needs and what services could address them. This analysis should encompass issues experienced on their infrastructure, including the type of violation, frequency of occurrence and required power modulation, among others. Additionally, this analysis serves as the basis for defining products, which should also consider the technical input from aggregators and final users. For instance, exploring the feasibility of internal resource balancing within the same aggregator deserves further examination.

Furthermore in a system where flexibility services become a fundamental aspect, the potential and integration of these services should be fully exploited. Taking into account that both TSO and DSO will benefit from flexibility services offered by the same pool of resources, coordination becomes essential. Indeed, a well-designed regulatory framework should include a coordination scheme between TSO and DSO: this scheme should outline priorities between different network operators for reserving flexibility resources, services prioritization, data exchange protocols, common prequalification procedures and other relevant aspects.

Moreover, the regulatory framework can serve as a tool for overcoming economic barriers imposed by remuneration mechanisms currently applied. In this regard, incentivizing mechanisms for TSO and DSO primarily focus on capital expenditure (CAPEX), considering the required investment. While remuneration mechanisms targeting capital investments facilitate the deployment of flexibility resources and support infrastructure and telemetry installation, incentives should also consider operational expenditure (OPEX) associated with developing an additional market. Furthermore, DSOs are currently remunerated based on OPEX evaluated from historical expenditures. Evaluating these costs should be reviewed to introduce incentives that support DSOs in managing and operating local flexibility markets, which could help remove economic barriers. To accurately determine when the local market is more effective compared to traditional network reinforcement, these incentives should be incorporated into network planning procedures, which ought to consider both OPEX and CAPEX costs [111]. Additionally, this integration enables DSOs to account for the management of their assets, such as network reconfiguration and the use of voltage-regulating transformers.

For the same reason, the potential contribution of distributed resources to flexibility provision should be considered in the TSO planning process, necessitating better coordination with the DSO planning procedures. An illustrative example of this coordination phase can be found in the FlexPlan project, where a grid expansion planning tool considers flexibility assets alongside conventional grid expansion. This facilitates correlation between TSO and DSO network expansion via a method known as T&D decomposition [112]. For instance, it would be appropriate to include considerations for costs arising from the need to balance positions on the global market when a resource is activated by local ancillary services. Moreover, there is a lack of transparency in grid expansion planning, with data usage and sharing restrictions imposed by privacy policies, which is an obstacle to the development of flexibility resources in critical and strategic locations. Recently, steps forward have been taken by the European Commission, with DSOs now required to share their grid expansion planning [1]. Additionally, new guidelines for the utilization of demand response in local markets have been issued by ACER [113].

Additional barriers are identified in the definition of regulated tariffs. The current design and definition of tariffs are not always favorable for potential flexibility providers and often fail to support the deployment of distributed flexibility resources. For example, existing tariffs for on-site exchange, which remunerate the injection of internally produced energy in terms of EUR/MWh, typically feature fixed prices that may not align with market trends. This reduces the benefits of resources being able to adjust their power output on the basis of network requirements.

Overall, following EU regulatory guidelines, flexibility resources should compete in local markets for the provision of services, similarly to what is developed in global markets. However, considering a transition period where there is no assurance that enough flexibility will be available, permissions are granted when the amount of flexible capacity is not sufficient to reach a satisfactory level of liquidity. In case of scarce liquidity, different strategies can be deployed, such as the use of long-term availability/options contracts, mitigating the risk of market power exertion from strategic, flexible entities.

5. Conclusions and Proposals

The manuscript aims to report an exhaustive list of references to the current and recently concluded initiatives on local flexibility markets. From the analysis of their documentation, relevant projects have been identified, also considering a wide spectrum of accessible and promising implementations. Indeed, experiences with various alternatives allowed us to investigate distinct market models and TSO-DSO coordination schemes. Finding the optimal architecture has proven to be challenging as each approach carries its own set of advantages and disadvantages, depending on existing markets, such as the global ancillary service market. From the analysis of the investigated projects, the authors summarize a list of the most compelling barriers and briefly report them in the following paragraphs.

Common barriers emerge from current regulations, encompassing technical requirements and economic parameters. One significant challenge lies in ensuring that flexibility resources can participate in local markets at a level playing field. Existing qualification procedures and technical requirements must be tailored to suit distribution system needs, which differ from those of transmission networks. Consequently, parameters like the minimum required bid should be adjusted to account for the typically smaller sizes of demand-side management and other distributed resources. Moreover, there is a need to reevaluate prequalification procedures, shifting from technology-specific assessments to service-specific evaluations. This change could facilitate access to both local and global flexibility markets, encompassing resources capable of providing a limited range of services. Additionally, adopting a broader perspective beyond national boundaries may facilitate the establishment of standards for distributed resource participation in local markets. Given the similarities in distribution network needs across different countries, a common definition of services and resource requirements could be advantageous.

To date, DSOs have predominantly applied the Fit&Forget approach. However, with the definition of local markets, this grid expansion strategy becomes inadequate. It is necessary for DSOs to acquire expertise in long-term planning procedures to identify and assess potential grid violations. In this context, coordination between TSO and DSO can be proven to be beneficial. DSOs could leverage the TSO experience to implement long-term network observability tools. Additionally, transparency should be prioritized, especially for nodes where system flexibility is crucial, to guide service providers in investing in strategic locations.

Regarding TSO-DSO coordination, efforts should be directed toward defining interactions between local and global markets. Ideally, resources should be permitted to offer services in both markets, with final selection and activation aimed at minimizing overall system costs. However, such a solution would drive significant computational effort and complexity. Therefore, coordination for selecting shared resources should be identified to facilitate the development of local flexibility markets. Furthermore, common requirements for shared services should be defined to eliminate the need for additional testing procedures once a resource is enabled in the local market, thereby simplifying the process.

There are numerous evident gaps in current regulations, particularly in the absence of clear definitions of roles and responsibilities. Regulatory updates are necessary to deepen the concept of resource aggregation and to promote a transparent market that integrates distributed resources on a level playing field, whether aggregated or not. Criteria for prioritizing DSO and TSO needs should be established to encourage the development of impartial processes for resource selection. These criteria should aim to minimize overall system costs while considering social welfare aspects such as health and established common objectives.

Furthermore, incentivization mechanisms should be formulated to encourage the development and deployment of flexibility resources. These mechanisms should also address the additional costs incurred by system operators for managing local markets and the required infrastructure. A market reform could facilitate the development of strategies for procuring flexibility services: while local markets should always be the preferred solution, in cases where liquidity is limited due to resource scarcity, flexibility contracts should be initiated to leverage the potential of strategic resources.

Author Contributions: Investigation, G.V., G.L. and M.R.; writing—original draft preparation, G.V., G.L. and M.R.; writing—review and editing, G.V., G.L. and M.R.; visualization, G.V., G.L. and M.R. All authors have read and agreed to the published version of the manuscript.

Funding: This work has been financed by the Research Fund for the Italian Electrical System under the Three-Year Research Plan 2022–2024 (DM MITE n. 337, 15 September 2022), in compliance with the Decree of 16 April 2018.

Conflicts of Interest: The authors declare no conflicts of interest. The funders had no role in the design of the study; in the collection, analyses, or interpretation of data; in the writing of the manuscript; or in the decision to publish the results.

References

1. European Parliament, Greenhouse Gas Emissions by Country and Sector (Infographic), March 2023. Available online: https://www.europarl.europa.eu/topics/en/article/20180301STO98928/greenhouse-gas-emissions-by-country-and-sector-infographic (accessed on 1 March 2024).
2. Directive (EU) 2012/27 of the European Parliament and of the Council, October 2012. Available online: https://eur-lex.europa.eu/eli/dir/2012/27/oj (accessed on 1 March 2024).
3. Directive (EU) 2019/944 of the European Parliament and of the Council, June 2019. Available online: https://eur-lex.europa.eu/legal-content/EN/TXT/?uri=celex:32019L0944 (accessed on 1 March 2024).
4. Council of European Energy Regulators, CEER Views on Electricity Distribution Network Development Plans. 2021. Available online: https://www.ceer.eu/documents/104400/-/-/2da60a45-6262-c6bc-080a-4f24b4c542cd (accessed on 1 March 2024).
5. Council of European Energy Regulators, Distribution System Working Group, CEER Paper on DSO Procedures of Procurement of Flexibility, July 2020. Available online: https://www.ceer.eu/documents/104400/-/-/f65ef568-dd7b-4f8c-d182-b04fc1656e58 (accessed on 1 March 2024).
6. CEDEC; EDSO; ENTSO-e; EURELECTRIC; GEODE. An Integrated Approach to Active System Management, with the Focus on TSO–DSO Coordination in Congestion Management and Balancing. 2019. Available online: https://docstore.entsoe.eu/Documents/Publications/Position%20papers%20and%20reports/TSO-DSO_ASM_2019_190416.pdf (accessed on 1 March 2024).
7. Viganò, G.; Lattanzio, G.; Rossi, M. Review on Local Market Flexibility Projects, Main Characteristics and Barriers. In Proceedings of the 2023 International Conference on Clean Electrical Power (ICCEP), Terrasini, Italy, 27–29 June 2023.
8. Mendicino, L.; Menniti, D.; Pinnarelli, A.; Sorrentino, N.; Vizza, P.; Alberti, C.; Dura, F. DSO Flexibility Market Framework for Renewable Energy Community of Nanogrids. *Energies* **2021**, *14*, 3460. [CrossRef]
9. Badanjak, D.; Pandžić, H. Distribution-Level Flexibility Markets—A Review of Trends, Research Projects, Key Stakeholders and Open Questions. *Energies* **2021**, *14*, 6622. [CrossRef]
10. FINGRID; ENERGINET; Statnett; KRAFTNAT. Local Flexibility Projects in the Nordica. 2020. Available online: https://www.fingrid.fi/globalassets/dokumentit/fi/sahkomarkkinat/kehityshankkeet/local-flexibility-nordics-june2020.pdf (accessed on 1 March 2024).
11. Silva, R.; Alves, E.; Ferreira, R.; Villar, J.; Gouveia, C. Characterization of TSO and DSO Grid System Services and TSO-DSO Basic Coordination Mechanisms in the Current Decarbonization Context. *Energies* **2021**, *14*, 4451. [CrossRef]
12. Valarezo, O.; Gómez, T.; Chaves-Avila, J.P.; Lind, L.; Correa, M.; Ulrich Ziegler, D.; Escobar, R. Analysis of New Flexibility Market Models in Europe. *Energies* **2021**, *14*, 3521. [CrossRef]
13. Euniversal. Observatory of Research and Demonstration Initiatives on Future Electricity Grids and Markets. 2019. Available online: https://euniversal.eu/wp-content/uploads/2021/02/EUniversal_D1.2.pdf (accessed on 1 March 2024).
14. Scottish & Southern Electricity Networks; Origami. Analysis of Relevant International Experience of DSO Flexibility Markets. 2019. Available online: https://ssen-transition.com/wp-content/uploads/2019/08/TRANSITION-Analysis-of-relevant-international-experience-of-DSO-flexibility-markets.pdf (accessed on 1 March 2024).
15. Li, F.; Kong, W.; Yang, X.; Zhang, C.; Li, R. *Mapping of DSO Projects*; Report for the Customer-Led Distribution System Project; Northern Powergrid: Newcastle, UK, 2019.
16. Schwidtal, J.M.; Agostini, M.; Bignucolo, F.; Coppo, M.; Garengo, P.; Lorenzoni, A. Integration of Flexibility from Distributed Energy Resources: Mapping the Innovative Italian Pilot Project UVAM. *Energies* **2021**, *14*, 1910. [CrossRef]
17. N-SIDE. Market Approaches for TSO-DSO Coordination in Norway. 2021. Available online: https://www.statnett.no/contentassets/525e71910628494db2e4c627eb00dddc/market-approaches-for-tso-dso-coordination-in-norway.pdf (accessed on 1 March 2024).
18. Frontier Economics; ENTSO-e. Review of Flexibility Platforms. 2021. Available online: https://www.entsoe.eu/news/2021/11/10/entso-e-publishes-new-report-on-flexibility-platforms/ (accessed on 1 March 2024).

19. ONENET. Set of Standardised Products for System Services in the TSO-DSO-Consumer Value Chain. 2021. Available online: https://www.onenet-project.eu//wp-content/uploads/2022/10/D22-A-set-of-standardised-products-for-system-services-in-the-TSO-DSO-consumer-value-chain.pdf (accessed on 1 March 2024).
20. Dronne, T.; Roques, F.; Saguan, M. Local Flexibility Markets for Distribution Network Congestion-Management in Center-Western Europe: Which Design for Which Needs? *Energies* **2021**, *14*, 4113. [CrossRef]
21. Alpine Space. Available online: https://www.alpine-space.eu/project/alpgrids/ (accessed on 1 March 2024).
22. Architecture of Tools for Load Scenarios (ATLAS). Available online: https://www.enwl.co.uk/future-energy/innovation/smaller-projects/network-innovation-allowance/enwl008---architecture-of-tools-for-load-scenarios-atlas/ (accessed on 1 March 2024).
23. Project BeFlexible. Available online: https://beflexible.eu/ (accessed on 1 March 2024).
24. Project COORDINET. Available online: https://coordinet.netlify.app/projects/coordinet (accessed on 1 March 2024).
25. Cornwall Local Energy Market Research Reports and Papers. Available online: https://www.centrica.com/stories/2018/cornwall-local-energy-market/ (accessed on 1 March 2024).
26. Project BROSSBOW. Available online: https://cordis.europa.eu/project/id/773430/it (accessed on 1 March 2024).
27. Platform DARE. Available online: https://www.dare-plattform.de/ (accessed on 1 March 2024).
28. Project DEMAND FLEX MARKET. Available online: https://www.demandflexmarket.com/ (accessed on 1 March 2024).
29. Project DELTA. Available online: https://www.delta-h2020.eu/ (accessed on 1 March 2024).
30. Project DOMINO. Available online: https://cordis.europa.eu/project/id/645760 (accessed on 1 March 2024).
31. Project DRES2MARKET. Available online: https://cordis.europa.eu/project/id/952851/it (accessed on 1 March 2024).
32. Project ECOGRID. Available online: http://www.ecogrid.net/ (accessed on 1 March 2024).
33. Project EDGE. Available online: https://www.e-distribuzione.it/progetti-e-innovazioni/il-progetto-edge.html (accessed on 1 March 2024).
34. Project E-LAND. Available online: https://elandh2020.eu/ (accessed on 1 March 2024).
35. Project ENERA. Available online: https://projekt-enera.de/ (accessed on 1 March 2024).
36. Platform ENKO. Available online: https://www.enko.energy/ (accessed on 1 March 2024).
37. EQUIGY. Crowd Balancing Platform. Available online: https://equigy.com/ (accessed on 1 March 2024).
38. Red Electrica, System Operator's Information System. Available online: https://www.ree.es/en/activities/operation-of-the-electricity-system/e-sios (accessed on 1 March 2024).
39. Project EUNIVERSAL. Available online: https://euniversal.eu/ (accessed on 1 March 2024).
40. Project EU-SYSFLEX. Available online: https://eu-sysflex.com/ (accessed on 1 March 2024).
41. Project EVOLVDSO. Available online: https://cordis.europa.eu/project/id/608732/it (accessed on 1 March 2024).
42. Project FEVER. Available online: https://fever-h2020.eu/ (accessed on 1 March 2024).
43. Project FLEXCOOP. Available online: http://www.flexcoop.net/ (accessed on 1 March 2024).
44. EUniversal. Deliverable D3.3, System-Level Assessment Framework for the Quantification of Available Flexibility for Enabling New Grid Services. Available online: https://euniversal.eu/wp-content/uploads/2022/03/EUniversal_D3.3_System-level-assessment-framework-for-the-quantification-of-available-flexibility-for-enabling-new-grid-services.pdf (accessed on 1 March 2024).
45. Project FLEXGRID. Available online: https://flexgrid-project.eu/ (accessed on 1 March 2024).
46. ENEDIS. Flexibilities to Enhance the Energy Transition and the Performance of the Distribution Network. 2019. Available online: https://www.enedis.fr/sites/default/files/documents/pdf/flexibilities-enhance-energy-transition-performance-distribution-network.pdf (accessed on 1 March 2024).
47. Platform FLEXIBLEPOWER. Available online: https://www.flexiblepower.co.uk/ (accessed on 1 March 2024).
48. Project FLEXICIENCY. Available online: https://cordis.europa.eu/project/id/646482/results (accessed on 1 March 2024).
49. Project FLEXITRANSTORE. Available online: https://cinea.ec.europa.eu/featured-projects/flexitranstore_en (accessed on 1 March 2024).
50. Project FUTUREFLOW. Available online: https://www.futureflow.eu/ (accessed on 1 March 2024).
51. Project GIFT. Available online: https://www.gift-h2020.eu/ (accessed on 1 March 2024).
52. Project GOFLEX. Available online: https://goflex-project.eu/ (accessed on 1 March 2024).
53. Project GOPAX. Available online: https://en.gopacs.eu/ (accessed on 1 March 2024).
54. Project IFLEX. Available online: https://www.iflex-project.eu/ (accessed on 1 March 2024).
55. Project INTEGRIDY. Available online: http://www.integridy.eu/ (accessed on 1 March 2024).
56. Project INTERFLEX. Available online: https://interflex-h2020.com/ (accessed on 1 March 2024).
57. Project INTERNET OF ENERGY. Available online: http://www.artemis-ioe.eu/ioe_project.htm (accessed on 1 March 2024).
58. Project INTERPLAN. Available online: https://ses.jrc.ec.europa.eu/eirie/en/project/integrated-operation-planning-tool-towards-pan-european-network (accessed on 1 March 2024).
59. Project INTERRFACE. Available online: http://www.interrface.eu/ (accessed on 1 March 2024).
60. Project INVADE. Available online: https://h2020invade.eu/ (accessed on 1 March 2024).
61. Platform IPOWER. Available online: https://ipower-net.weebly.com/ (accessed on 1 March 2024).
62. Project MAGNITUDE. Available online: https://www.magnitude-project.eu/ (accessed on 1 March 2024).
63. Project MERLON. Available online: https://www.merlon-project.eu/ (accessed on 1 March 2024).

64. Project MiNDFlex. Available online: https://www.unareti.it/it/media/progetti/sperimentazione-mindflex-rete-elettrica-milano (accessed on 1 March 2024).
65. Project MUSE GRID. Available online: https://cordis.europa.eu/project/id/824441 (accessed on 1 March 2024).
66. Platform NODES. Available online: https://nodesmarket.com/ (accessed on 1 March 2024).
67. Project NORFLEX. Available online: https://flextools.com/reference-project/norflex/ (accessed on 1 March 2024).
68. Project ONENET. Available online: https://onenet-project.eu/ (accessed on 1 March 2024).
69. Project OSMOSE. Available online: https://www.osmose-h2020.eu/ (accessed on 1 March 2024).
70. Platform PICLOFLEX. Available online: https://picloflex.com/ (accessed on 1 March 2024).
71. Project PLATONE. Available online: https://www.platone-h2020.eu/ (accessed on 1 March 2024).
72. National Grid, Power Potential. Available online: https://www.nationalgrideso.com/future-energy/projects/power-potential (accessed on 1 March 2024).
73. Project LEO. Available online: https://project-leo.co.uk/ (accessed on 1 March 2024).
74. Project REDISPATCH. Available online: https://www.bayernwerk-netz.de/de/energie-einspeisen/redispatch-2-0/fuer-anlagenbetreiber.html (accessed on 1 March 2024).
75. Project REFLEX. Available online: https://reflex-project.eu/ (accessed on 1 March 2024).
76. Project RomeFlex. Available online: https://www.areti.it/conoscere-areti/innovazione/progetto-romeflex (accessed on 1 March 2024).
77. Project SMARTEREMC. Available online: http://www.smarteremc2.eu/ (accessed on 1 March 2024).
78. Project SMARTNET. Available online: http://smartnet-project.eu/ (accessed on 1 March 2024).
79. Project SMILE. Available online: https://cordis.europa.eu/project/id/731249 (accessed on 1 March 2024).
80. Project SOTERIA. Available online: https://www.ioenergy.eu/soteria/ (accessed on 1 March 2024).
81. Project STORE&GO. Available online: https://www.storeandgo.info/ (accessed on 1 March 2024).
82. Project TDX-ASSIST. Available online: http://www.tdx-assist.eu/ (accessed on 1 March 2024).
83. TENNET, Vehicle2Grid. Available online: https://www.tennet.eu/our-key-tasks/innovations (accessed on 1 March 2024).
84. Project UPGRID. Available online: https://bridge-smart-grid-storage-systems-digital-projects.ec.europa.eu/node/57 (accessed on 1 March 2024).
85. Project USEF. Available online: https://www.usef.energy/ (accessed on 1 March 2024).
86. Project WISE GRID. Available online: https://www.wisegrid.eu/ (accessed on 1 March 2024).
87. MUSE GRIDS. Demosite D1.2—DSM Scheme Assessment and Users' Engagement Strategies. 2019. Available online: https://muse-grids.eu/wp-content/uploads/2020/07/D1.2-DSM-schemes.pdf (accessed on 1 March 2024).
88. WISE GRID. WiseGRID Requirements, Use Cases and Pilot Sites Analysis. 2017. Available online: https://cdn.nimbu.io/s/76bdjzc/channelentries/oan2oj6/files/D2.1_WiseGRID_requirements_Use_cases_and_pilot_sites_analysis.pdf?gej0qha (accessed on 1 March 2024).
89. CENTRICA. LEM Flexibility Market Platform Design and Trials Report. 2020. Available online: https://www.centrica.com/media/4614/lem-flexibility-market-platform-design-and-trials-report.pdf (accessed on 1 March 2024).
90. MAGNITUDE. Benchmark of Markets and Regulations for Electricity, Gas and Heat and Overview of Flexibility Services to the Electricity Grid. 2019. Available online: https://www.magnitude-project.eu/wp-content/uploads/2019/07/MAGNITUDE_D3.1_EDF_R1_Final_Submitted.pdf (accessed on 1 March 2024).
91. INVADE. Overall INVADE Architecture. 2018. Available online: https://h2020invade.eu/wp-content/uploads/2017/05/D4.1-Concept-design.pdf (accessed on 1 March 2024).
92. FLEXGRID. The Overall FLEXGRID Architecture Design, High Level Model and System Specifications. 2020. Available online: https://flexgrid-project.eu/assets/deliverables/FLEXGRID_D2.2_final_31032020.pdf (accessed on 1 March 2024).
93. ONENET. Overview of Market Designs for the Procurement of System Services by DSOs and TSOs. 2021. Available online: https://www.onenet-project.eu//wp-content/uploads/2022/10/D31-Overview-of-market-designs-for-the-procurement-of-system-services-by-DSOs-and-TSOs.pdf (accessed on 1 March 2024).
94. ONENET. Review on Markets and Platforms in Related Activities. 2021. Available online: https://www.onenet-project.eu//wp-content/uploads/2022/10/D2.1-Review-on-markets-and-platforms-in-related-activities.pdf (accessed on 1 March 2024).
95. ENERNOC. The Demand Response Baseline. 2011. Available online: https://www.naesb.org/pdf4/dsmee_group3_100809w3.pdf (accessed on 1 March 2024).
96. ISO New England. Measurement and Verification of Demand Reduction Value from Demand Resources. 2014. Available online: https://www.iso-ne.com/static-assets/documents/2017/02/mmvdr_measurement-and-verification-demand-reduction_rev6_20140601.pdf (accessed on 1 March 2024).
97. SmarterEMC2. Analysis of Reference Usage Scenarios for Market/Retail and Distribution System Operation Services. 2015. Available online: https://cordis.europa.eu/project/id/646470/results/es (accessed on 1 March 2024).
98. DOMINOES. Scalable Local Energy Market Architecture (Second Release). 2019. Available online: https://ec.europa.eu/research/participants/documents/downloadPublic?documentIds=080166e5cad9d2d1&appId=PPGMS (accessed on 1 March 2024).
99. FEVER. Flexibility Related European Electricity Markets: Modus Operandi, Proposed Adaptations and Extension and Metrics Definition. 2020. Available online: https://www.fever-h2020.eu/data/deliverables/FEVER_D4.1_-_Flexibility_related_European_electricity_markets.pdf (accessed on 1 March 2024).

100. USEF. USEF: The Framework Explained. 2021. Available online: https://www.usef.energy/app/uploads/2021/05/USEF-The-Framework-Explained-update-2021.pdf (accessed on 1 March 2024).
101. Jin, X.; Wu, Q.; Jia, H. Local flexibility markets: Literature review on concepts, models and clearing methods. *Appl. Energy* **2020**, *261*, 114387. [CrossRef]
102. CEDEV; EDSO; EURELECTRIC; GEODE. Flexibility in the Energy Transition—A Toolbox for Electricity DSOs. 2018. Available online: https://cdn.eurelectric.org/media/2395/flexibility_in_the_energy_transition_-_a_tool_for_electricity_dsos-2018-2018-oth-0002-01-e-h-F857DD9F.pdf (accessed on 1 March 2024).
103. Armenteros, A.S.; Heer, H.D.; Van Der Laan, M. Flexibility Deployment in Europe, USEF. 2021. Available online: https://www.usef.energy/app/uploads/2021/03/08032021-White-paper-Flexibility-Deployment-in-Europe-version-1.0-3.pdf (accessed on 1 March 2024).
104. SmartNet. Network and Market Models. 2019. Available online: https://smartnet-project.eu/wp-content/uploads/2019/02/20190215113154_D2.2_20190215_V1.0.pdf (accessed on 1 March 2024).
105. Dukovska, I.; Slootweg, H.; Paterakis, N.G. A Review of Network Modeling and Services Integration in Peer-to-Peer Electricity Markets. In Proceedings of the 2021 IEEE Power & Energy Society General Meeting (PESGM), Washington, DC, USA, 26–29 July 2021.
106. Bahramara, S.; Mazza, A.; Chicco, G.; Shafie-khah, M.; Catalão, J.P.S. Comprehensive review on the decision-making frameworks referring to the distribution network operation problem in the presence of distributed energy resources and microgrids. *Int. J. Electr. Power Energy Syst.* **2020**, *115*, 105466. [CrossRef]
107. Kazmi, S.A.A.; Shahzad, M.K.; Khan, A.Z.; Shin, D.R. Smart Distribution Networks: A Review of Modern Distribution Concepts from a Planning Perspective. *Energies* **2017**, *10*, 501. [CrossRef]
108. Hirth, L.; Schlecht, I. Market-Based Redispatch in Zonal Electricity Markets: Inc-Dec Gaming as a Consequence of Inconsistent Power Market Design (Not Market Power). 2019. Available online: https://www.econstor.eu/handle/10419/194292 (accessed on 1 March 2024).
109. ENTSO-E; CEDEC; EDSO; Eurelectric; GEODE. Roadmap on the Evolution of the Regulatory Framework for Distributed Flexibility. 2021. Available online: https://www.geode-eu.org/wp-content/uploads/2021/07/210728_TSO-DSO-Roadmap-on-Distributed-Flexibility.pdf (accessed on 1 March 2024).
110. EDSO. Grid Observability for Flexibility. 2022. Available online: https://www.edsoforsmartgrids.eu/edso-publications/grid-observability-for-flexibility-report (accessed on 1 March 2024).
111. ARERA. Consultazione 12 Luglio 2022 317/2022/R/com—Ambito di Applicazione dell'Approccio ROSS e Criteri di Determinazione del Costo Riconosciuto Secondo l'Approccio ROSS BASE-Orientamenti. 2022. Available online: https://www.arera.it/atti-e-provvedimenti/dettaglio/22/317-22 (accessed on 1 March 2024).
112. Rossini, M.; Ergun, H.; Rossi, M. An open-source optimization toolkit for the smart scheduling of DERs in distribution grids. In Proceedings of the 2023 Open Source Modelling and Simulation of Energy Systems (OSMSES), Aachen, Germany, 4–5 April 2022.
113. ACER. Framework Guideline on Demand Response (Draft for Public Consultation). 2022. Available online: https://documents.acer.europa.eu/Official_documents/Public_consultations/Pages/PC_2022_E_05.aspx (accessed on 1 March 2024).

Article

Parameter Estimation Method for Virtual Power Plant Frequency Response Model Based on SLP

Zheng Shi [1,*], Haixiao Zhu [2], Haibo Zhao [1], Peng Wang [1], Yan Liang [1,3], Kaikai Wang [1], Jie Chen [1], Xiaoming Zheng [1] and Hongli Liu [1]

[1] Economic and Technical Research Institute of State Grid Shanxi Electric Power Company Taiyuan 030000, China; zhb8711@163.com (H.Z.); wangpeng@163.com (P.W.); liangyan_lyly@163.com (Y.L.); wangkaikai510@163.com (K.W.); serenachen@126.com (J.C.); ysuzhxm@126.com (X.Z.); liuhonglifriend@126.com (H.L.)

[2] School of Electrical and Power Engineering, Taiyuan University of Technology, Taiyuan 030024, China; 2023510359@link.tyut.edu.cn

[3] School of Electrical and Electronic Engineering, North China Electric Power University, Beijing 100000, China

* Correspondence: shizheng_sgcc@163.com; Tel.: +86-1326-161-1011

Abstract: In adapting to the double-high development trend of high-voltage direct current (HVDC) receiving-end power systems and solving optimization problems in emergency frequency control (EFC) supporting virtual power plants (VPPs) in large-scale power systems, a parameter estimation method for a VPP frequency response model based on a successive linear programming (SLP) method is proposed. First, a "centralized/decentralized" hierarchical control architecture for VPP participation in EFC is designed. Second, the frequency response characteristics of multiple flexible resources are scientifically analyzed, and the system frequency response (SFR) model and equivalent model of VPP are constructed. Subsequently, parameter estimation of the VPP frequency response model is carried out based on the SLP method, aiming to balance the accuracy and computational efficiency of the model. Finally, the effectiveness of the proposed methodology is verified by using PSD-BPA to simulate and test the three-zone HVDC recipient area grid.

Keywords: virtual power plant; emergency frequency control; system frequency response model; parameter estimation

Citation: Shi, Z.; Zhu, H.; Zhao, H.; Wang, P.; Liang, Y.; Wang, K.; Chen, J.; Zheng, X.; Liu, H. Parameter Estimation Method for Virtual Power Plant Frequency Response Model Based on SLP. *Energies* **2024**, *17*, 3124. https://doi.org/10.3390/en17133124

Academic Editor: Ahmed Abu-Siada

Received: 18 May 2024
Revised: 15 June 2024
Accepted: 21 June 2024
Published: 25 June 2024
Corrected: 23 September 2024

1. Introduction

In the process of building a new power system in China, the rapid growth of installed capacity and the power generation of inverter interface power sources, such as wind power, energy storage, HVDC, etc., has brought to the fore the issue of grid frequency security [1,2]. As a potential frequency modulation resource, the virtual power plant (VPP) [3–5] can aggregate a wide range of types of flexible resources to holistically participate in grid regulation and assist the large grid in emergency frequency control [6,7]. Therefore, it is extremely important to construct a reasonable VPP frequency response model to accurately and quickly reflect its frequency response characteristics and regulation capability.

Accurate modeling of the VPP is the basis and prerequisite for its participation in the scheduling and control of large power grids. Reference [8] proposes a multi-timescale responsiveness assessment model for VPP with EVs as the main resource based on the responsiveness model of a single EV. Reference [9] considers the uncertainty impact of a wind power grid connection and V2G, and establishes a double-layer inverse robust optimization scheduling model with VPP. Reference [10] establishes an extended SFR model that takes into account the transient frequency modulation process of the VPP, which leads to a new thought for the overall participation of the VPP in frequency control but does not consider the way of controlling each resource within the VPP. Reference [11] treats wind turbines, energy storage devices, and motor loads as frequency modulation resources in the

VPP and constructs a primary frequency modulation model for the VPP to participate in the frequency control of the power system, but the study lacks a description of the overall frequency control characteristics of the VPP. Reference [12] proposes an SFR model based on inverter droop control for various types of distributed resources, but the model has a uniform and simple model structure, which makes it difficult to accurately reflect the control characteristics of each flexible resource.

Existing research related to VPP is mostly related to the power market, economic dispatch, and other fields [13,14], which are slightly insufficient in the optimization of parameter estimation for VPP frequency response models. Reference [15] presents a VPP model for transient stability analysis with VPP resources for current source and voltage source inverters. Reference [16] develops an equivalent aggregation model based on VPP frequency effects, which mainly considers different response delays among different distributed resources. In terms of parameter estimation, Reference [17] comes up with a way to estimate the equivalent inertia of a system based on polynomial fitting of frequency curves. Reference [18] raises a method for estimating the equivalent inertia and damping coefficients of synchronous and nonsynchronous power supplies based on closed-loop control. Reference [19] utilizes section information to estimate the equivalent frequency response model parameters of interconnected multi-area systems.

In summary, the following difficulties arise in modeling the frequency response characteristics of VPP. (1) VPP brings together many heterogeneous and flexible resources, and each resource participates in frequency modulation with different control modes and response characteristics. Therefore, how to collaborate various resources and construct a VPP frequency response model accurately reflecting the characteristics of the multi-flexible resources to support the emergency frequency control of the power system needs to be studied. (2) The VPP frequency response model is complex and of high-order and has many control parameters; the detailed model can ensure the calculation accuracy but also limit the decision-making efficiency of emergency frequency control; and the VPP model parameters are time-varying, which are different under different operating conditions of the grid, which exacerbates the complexity of the decision making of emergency frequency control taking into account the VPP. How to obtain model parameter estimations that take into account both accuracy and computational efficiency is still an open problem.

To overcome the problems mentioned above, this paper first comes up with a VPP hierarchical control architecture suitable for emergency frequency control. Secondly, the detailed frequency response model of VPP is constructed based on the response characteristics of multiple flexible resources. Then, the complex feedback control branch is aggregated to obtain its equivalent model, and the parameter estimation model is constructed and solved based on successive linear programming (SLP). Finally, the scientific validity of the proposed method is confirmed by multi-fault scenario testing based on a three-region HVDC receiving area grid.

2. Hierarchical Control Architecture of VPPs for Emergency Frequency Control

In this paper, we present a centralized/decentralized hierarchical architecture for emergency frequency control of VPPs, as shown in Figure 1. The architecture consists of three control layers: the centralized control layer, the VPP aggregation layer, and the local control layer.

The main body of the central control layer includes the grid control core and the VPP control master. The control core issues specific power control command values $\sum_i^{N_V} P_{eov\_q}$, a system minimum frequency threshold f^{eo}_{nadir}, and a quasi-steady-state frequency threshold f^{eo}_{qss} according to the power deficit and frequency response capability of VPP, and monitors the grid frequency deviation in real time. The VPP control master reports in real-time the maximum support power $\sum_i^{N_V} P_{arv\_m}$ that all VPPs in the network can provide, the steady-state support power $\sum_i^{N_V} P_{arv\_q}$, the minimum value of the system frequency f^{ar}_{nadir}, and the quasi-steady-state frequency(QSSF) f^{ar}_{qss} after the VPPs join the process of emergency frequency control according to the given power command in case of a foreseen failure.

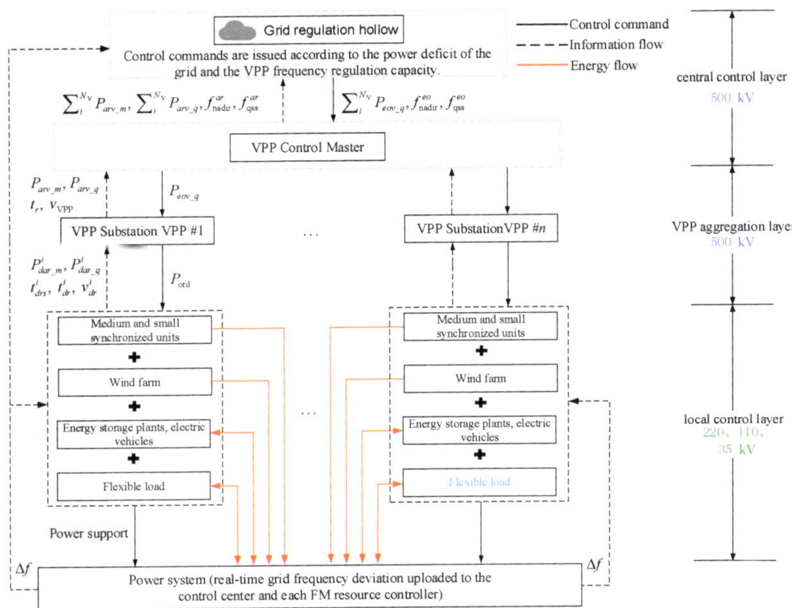

Figure 1. "Centralized/decentralized" VPP hierarchical control framework.

The main body of the VPP aggregation layer is the VPP sub-station under each 500 kV bus. The VPP sub-station receives, on the one hand, the maximum support power $P^i_{dar\_m}$, the steady state power $P^i_{dar\_q}$, the control response time t^i_{drs}, the power climbing time t^i_{dr}, and the climbing rate v^i_{dr} that each control resource can provide and, accordingly, reports the maximum regulation power P_{arv_m}, the steady-state regulation power P_{arv_q}, the climbing time t_r, and the climbing rate v_{VPP} that this sub-station can provide. On the other hand, according to the power support quantity P_{eov_q} issued by the VPP master station and the operation status of each control resource in the sub-station, the VPP sub-station co-ordinates and allocates the power command value P_{ord} to the multiple types of flexible control resources in the station.

The subjects in the local control layer include small- and medium-sized turbine units, distributed wind turbines, energy storage power stations, cluster electric vehicles, and flexible loads. When an active deficit fault occurs in the HVDC recipient area grid, as control terminals, each control resource will participate in emergency frequency control, provide power support, and safeguard grid frequency security. It should be noted that there are differences in the control characteristics of the resources, which have different forms of power support.

This architecture mainly includes three control levels: central control layer, VPP aggregation layer, and local control layer. In terms of communication, the communication between the power grid control center and the control master station, the control master station and sub-stations, and the control sub-stations and various distributed resources is bidirectional. In terms of data, the power grid control center and the control master station implement the issuance of specific control instructions and monitoring of system frequency. On the one hand, the control sub-station can receive data provided by various control resources; on the other hand, it allocates instructions to flexible control resources within the station based on the power support issued by the control master station and the status of the control sub-station. In terms of control strategy, the control strategy is issued from the control center to the control master station, from the control master station to the control sub-station, and from the control sub-station to the local control layer. When there is an active power shortage fault in the HVDC receiving-end power system, as the control terminal, each control resource will participate in emergency frequency control, provide power support, and ensure system frequency security.

3. Modeling the Frequency Response Characteristics of a VPP

3.1. Frequency Characteristic Modeling of Multiple Flexible Resources

Units with voltage levels of 220 kV and below include coal-fired units, gas-fired units, and mixed-use power plants (e.g., waste power plants, steel power plants, and paper power plants). The medium-sized unit speed control system connected to 220 kV can be modeled by the GS-TA model, and the small-sized unit speed control system connected to 110 kV and 35 kV can be modeled by the IEEEG2 model [20–22]. Distributed wind turbines can be involved in frequency modulation based on rotor kinetic control, the control architecture diagram of which is shown in Figure 2, where H_W is the virtual inertia time constant provided by the turbine, K_W is the tuning coefficient of the turbine, and T_w is the turbine control response time constant.

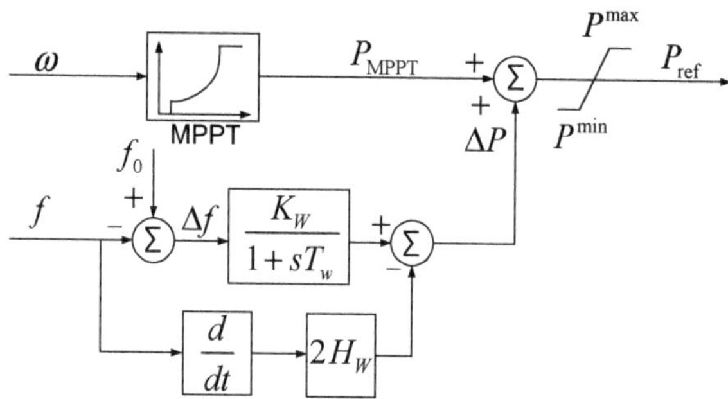

Figure 2. Diagram of WF active power reference control loop with the frequency regulation ability.

The energy storage plant can participate in frequency modulation based on fast power control with the power control function shown in Equation (1):

$$P_{ES} = P_{ES}^{ord}[1 - (1 + \frac{T_{ES}}{T_e})e^{-\frac{t}{T_e}}] \tag{1}$$

In the above equation, P_{ES}^{ord} and P_{ES} are the power command value and the actual output power value of the ESS issued by the VPP, and T_{ES} and T_e are the communication delay time and the control response time constant of the ESS from the occurrence of the fault to the control response.

The EV cluster can be equated to an energy storage power station, and the reverse charging is realized by the EV–grid interaction technology (vehicle to grid, V2G), whose power control function is shown in Equation (2):

$$P_{EV} = P_{EV}^{ord}[1 - (1 + \frac{T_{EV}}{T_{ev}})e^{-\frac{t}{T_{ev}}}] \tag{2}$$

In the above equation, P_{EV}^{ord} and P_{EV} are the power command value and the actual output power value of the EV issued by the VPP. T_{EV} and T_{ev} are the communication delay time and the control response time constant of the EV from the occurrence of the fault to the control response.

The way flexible loads (FLs) participate in frequency control is mainly divided into open-disconnected mode and regulation mode. Among them, the model of an open-break FL participating in frequency control is shown in Equation (3):

$$P_{FL,S} = P_{FL,S}^{ord}[(t - T_{FL,S})] \tag{3}$$

In the above equation, $P_{FL,S}^{ord}$ and $P_{FL,S}$ are the control command value and the actual load-shedding amount of the open-break FL. $T_{FL,S}$ is the time interval between the occurrence of the fault and the load-shedding action of the open-break FL.

Based on Reference [23], the frequency response model of the modulated FL in one frequency modulation time scale is shown in Figure 3.

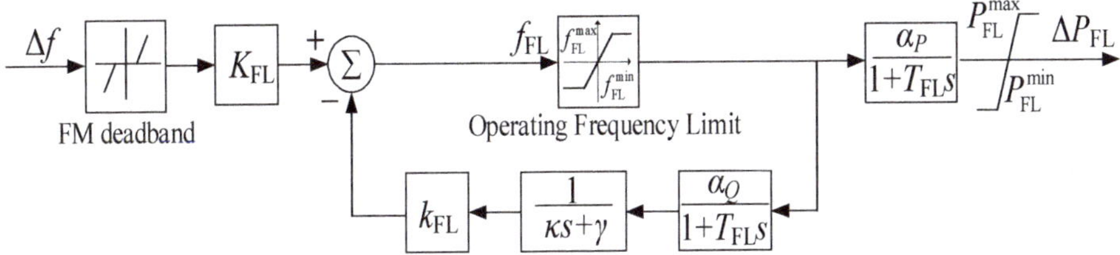

Figure 3. Frequency response model of adjustable FL.

In this model, K_{FL} is the tuning coefficient, Δf is the system frequency deviation, k_{FL} and T_{FL} denote the tuning coefficient and inertia time constant of the compressor controller, κ and γ are the equivalent thermodynamic parameters, α_P is the constant coefficient of electric power for normal operation of the FL, α_Q is the constant coefficient of refrigeration/heat of the FL, f_{FL} is the operating frequency of the FL, and f_{FL}^{max} and f_{FL}^{min} denote the upper and lower bounds of the operating frequency of the FL, respectively. P_{FL}^{max} and P_{FL}^{min} are the maximum and minimum limits of FL power. ΔP_{FL} is the amount of the power change of the FL.

3.2. VPP Frequency Response Modeling

Considering the frequency control characteristics of each flexible resource, the VPP frequency response model is shown in Figure 4:

$$\begin{cases} H_{VPP} = \sum_{i=1}^{N_G} H_{Gi}(S_{Gi}/S_{sys}) + \sum_{i=1}^{N_{WF}} H_{Wi}(S_{Wi}/S_{sys}) \\ D_{VPP} = \sum_{i=1}^{N_G} D_{Gi}(S_{Gi}/S_{sys}) + \sum_{i=1}^{N_{WF}} D_{Wi}(S_{Wi}/S_{sys}) \end{cases} \quad (4)$$

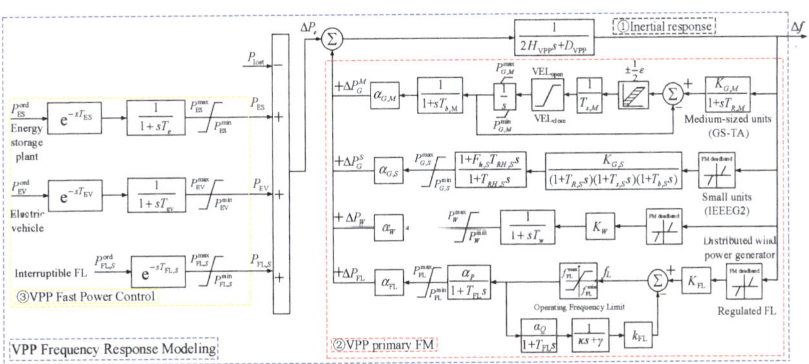

Figure 4. Frequency response model of VPP.

In the above equation, H_{VPP} is the equivalent inertia time constant provided by the VPP, H_{Gi} is the inertia time constant of the ith turbine unit, and NG is the number of units in the VPP in grid-connected operation, including small- and medium-sized units. H_{Wi} is the virtual inertia time constant provided by the ith wind turbine, and NWF is the number of wind turbines. S_{Gi}, and S_{Wi} are the rated powers of the ith turbine unit and wind turbine, respectively, and S_{sys} is the total installed system capacity. DVPP is the equivalent damping coefficient of the system, and D_{Gi} is the damping factor of the ith turbine unit. D_{Wi} is the virtual damping factor provided by the ith wind turbine.

(1) Primary frequency modulation. The primary frequency modulation parameters for turbine units, distributed fans, and regulated FLs are as follows:

$$
\begin{cases}
K_{G,M} = \sum_{i=1}^{N_{MG}} \alpha_{G,Mi} K_{G,Mi}, \ K_{G,S} = \sum_{i=1}^{N_{SG}} \alpha_{G,Si} K_{G,Si} \\
K_W = \sum_{i=1}^{N_{WF}} \alpha_{Wi} K_{Wi}, \ K_{FL} = \sum_{i=1}^{N_{FL}} \alpha_{FLi} K_{FLi}
\end{cases}
\tag{5}
$$

In the above equation, $K_{G,M}$, $K_{G,S}$, K_W, and K_{FL} are the modulation coefficients for medium and small units, turbines, and regulated FLs. α is a constant coefficient representing the ratio of the rated power of the ith feedback control branch resource to the total capacity of that resource, and N_{MG}, N_{SG}, N_{WF}, and N_{FL} stand for the number of medium and small turbine units, turbines, and regulated FLs.

$$
\begin{cases}
X_{MG} = \sum_{i=1}^{N_{MG}} \lambda_i X_{MGi}, X_{MG} = \left\{ T_{b,M}, T_{s,M}, T_{R,M} \right\} \\
X_{SG} = \sum_{i=1}^{N_{SG}} \lambda_i X_{SGi}, X_{SG} = \left\{ T_{b,S}, T_{s,S}, T_{R,S}, T_{RH,S}, F_{h,S} \right\} \\
T_w = \sum_{i=1}^{N_{WF}} \lambda_i T_{wi} \\
X_{FL} = \sum_{i=1}^{N_{FL}} \lambda_i X_{FLi}, X_{FL} = \left\{ T_{FL}, \alpha_P, \alpha_Q, k_{SL} \right\}
\end{cases}
\tag{6}
$$

In the above equation, X_{MG}, X_{SG}, and X_{FL} represent the feedback control branch parameter sets of medium- and small-sized units and regulated FLs. $T_{b,M}$, $T_{s,M}$, and $T_{R,M}$ are the steam volume time constant, servo time constant, and governor response time constant for medium-sized units. $T_{b,S}$, $T_{s,S}$, $T_{R,S}$, $T_{RH,S}$, and $F_{h,S}$ are the steam volume time constant, servo time constant, governor response time constant, reheater time constant, and high-pressure cylinder power share for small-sized units, respectively. T_w is the response time constant for the turbine control and λ_i is the weighting coefficient of the ith feedback control branch.

(2) Fast power control. The control parameters of ESS, EV, and on-off FL are as follows:

$$
\begin{cases}
T_{ES} = \sum_{i=1}^{N_{ES}} \frac{P_{ESi}}{P_{ES}} T_{ESi} , T_e = \sum_{i=1}^{N_{ES}} \frac{P_{ESi}}{P_{ES}} T_{ei} \\
T_{EV} = \sum_{i=1}^{N_{EV}} \frac{P_{EVi}}{P_{EV}} T_{EVi} , T_{ev} = \sum_{i=1}^{N_{EV}} \frac{P_{EVi}}{P_{EV}} T_{evi} \\
T_{FL,S} = \sum_{i=1}^{N_L} \frac{P_{FL,Si}}{P_{FL,S}} T_{FL,Si}
\end{cases}
\tag{7}
$$

In the above equation, P_{ESi}, P_{Evi}, and $P_{FL,Si}$ represent the actual output power of the ith branch of ESS, EV, and open FL. N_{ES}, N_{EV}, and N_L are the number of ESS, EV, and open FL.

4. SLP-Based Parameter Estimation Method for Frequency Response Model of VPP

4.1. Equivalent Methods for Frequency Response Modeling of VPPs

Since the VPP aggregates multiple types of heterogeneous resources, its feedback control branch is characterized by complex high-order and many parameters. The detailed model of frequency response taking into account multiple types of flexible resources has high accuracy, but it puts high requirements on simulation efficiency and model parameter estimation; at the same time, it is difficult to integrate in emergency frequency control optimization decisions. Aiming at the above difficulties, this section aggregates its complex feedback control branches to obtain the VPP frequency response equivalent model.

In the VPP frequency response model, the sum of each feedback control branch is a higher order transfer function. Therefore, the generalized transfer function $G_{\mathrm{VPP}}(s)$ is used to describe the VPP equivalent feedback control branch model in the general form:

$$G_{\mathrm{VPP}}(s) = \frac{b_m s^m + b_{m-1} s^{m-1} + \cdots + b_0}{a_n s^n + a_{n-1} s^{n-1} + \cdots + a_0}, \quad n \geq m \tag{8}$$

In the above equation, a_i, and b_j are the coefficients of the denominator and numerator polynomials, respectively. When the model order is one, its transfer function has an equivalent relationship with the simplified turbine governor model.

4.2. Parameter Estimation Optimization Model

For the SFR model after equating the VPP feedback control branch, the optimization model for VPP parameter estimation is established to minimize the frequency minimum point difference $\Delta f_{\mathrm{nadir},k}$ and QSSF difference $\Delta f_{\mathrm{qss},k}$ calculated from the detailed SFR model and the equating model under multiple types of foreseen fault scenarios, where Equation (9) is the objective function, and Equations (10) and (11) represent the system frequency nadir difference constraint and QSSF difference constraint, respectively. To avoid the frequency estimation value being higher than the true value of frequency, which causes the system leakage in the emergency frequency control optimization decision, the difference between the true value of the frequency index and the estimated value is set to be greater than zero. Equations (12) and (13) represent the limit constraints of each parameter in the estimation model.

$$\min \sum_{k=1}^{N_K} \left(\Delta f_{\mathrm{nadir},k} + \Delta f_{\mathrm{qss},k} \right) \tag{9}$$

$$f_{\mathrm{nadir},k} - \hat{f}_{\mathrm{nadir},k}(a_i, b_j) \geq 0, \quad \forall i \in N_a, \quad \forall j \in N_b \tag{10}$$

$$f_{\mathrm{qss},k} - \hat{f}_{\mathrm{qss},k}(a_i, b_j) \geq 0, \quad \forall i \in N_a, \quad \forall j \in N_b \tag{11}$$

$$a_i^{\min} \leq a_i \leq a_i^{\max}, \forall i \in N_a \tag{12}$$

$$b_j^{\min} \leq b_j \leq b_j^{\max}, \forall j \in N_b \tag{13}$$

In the above equation, (9) is the objective function of parameter estimation, and NK represents the number of expected fault scenarios. $f_{\mathrm{nadir},k}$ and $\hat{f}_{\mathrm{nadir},k}$ are the minimum values of system frequency calculated by the detailed frequency response model and the equivalent model, respectively. N_a and N_b are the numbers of parameters in the denominator and numerator of the equivalent model, respectively. $f_{\mathrm{qss},k}$ and $\hat{f}_{\mathrm{qss},k}$ are the quasi-steady-state frequencies obtained by the detailed frequency response model and the equivalent model, respectively. $a_i^{\max}$, $a_i^{\min}$, $b_i^{\max}$, and $b_i^{\min}$ are the upper and lower bounds of the parameters of the equivalent model, respectively.

The SFR is affected by various factors such as grid operation state, feedback control branch parameters, power deficit, etc., and its frequency dynamics are complex and nonlinear, which leads to the parameter estimation problem that VPP is not suitable to be solved directly. Therefore, the frequency difference constraints in Equations (10) and (11) are linearized based on the trajectory sensitivity method to establish the coupling relationship between the model parameters and the minimum and quasi-steady-state frequencies. The above optimization model can be rewritten as the following linear programming model:

$$\min \sum_{k=1}^{N_K} \left\{ \left[f_{\mathrm{nadir},k} - \hat{f}_{\mathrm{nadir},k}(a_i, b_j) \right] + \left[f_{\mathrm{qss},k} - \hat{f}_{\mathrm{qss},k}(a_i, b_j) \right] \right\} \tag{14}$$

$$\sum_{i=0}^{N_a} (\beta^k_{\text{nadir},i} \Delta a_i) + \sum_{j=0}^{N_b} (\beta^k_{\text{nadir},j} \Delta b_j)$$
$$\leq f_{\text{nadir},k} - \hat{f}_{\text{nadir},k}(a_i^0, b_j^0),$$
$$\forall i \in N_a, \forall j \in N_b, \forall k \in N_K \tag{15}$$

$$\sum_{i=0}^{N_a} (\beta^k_{\text{qss},i} \Delta a_i) + \sum_{j=0}^{N_b} (\beta^k_{\text{qss},j} \Delta b_j)$$
$$\leq f_{\text{qss},k} - \hat{f}_{\text{qss},k}(a_i^0, b_j^0),$$
$$\forall i \in N_a, \forall j \in N_b, \forall k \in N_K \tag{16}$$

$$-a_i^0 \leq \Delta a_i \leq a_i^{\max} - a_i^0, \forall i \in N_a \tag{17}$$

$$-b_j^0 \leq \Delta b_j \leq b_j^{\max} - b_j^0, \forall j \in N_b \tag{18}$$

In the above equation, $\hat{f}_{\text{nadir},k}(a_i^0, b_j^0)$ and $\hat{f}_{\text{qss},k}(a_i^0, b_j^0)$ are the lowest values of the system frequency and the quasi-steady-state frequency of the equivalent model at the initial parameter settings. Δa_i, Δb_j is the variation of the model parameters. $\beta^k_{\text{nadir},i}$, $\beta^k_{\text{nadir},j}$ and $\beta^k_{\text{qss},i}$, $\beta^k_{\text{qss},j}$ are the frequency nadir sensitivity coefficients and QSSF sensitivity coefficients of the model parameters, respectively, as shown in Equation (19):

$$\begin{cases} \beta^k_{\text{nadir}/\text{qss},i} = \dfrac{\hat{f}_{\text{nadir}/\text{qss},k}(a_i,b_j,\kappa_{ai}) - \hat{f}_{\text{nadir}/\text{qss},k}(a_i,b_j)}{\kappa_{ai}} \\ \beta^k_{\text{nadir}/\text{qss},j} = \dfrac{\hat{f}_{\text{nadir}/\text{qss},k}(a_i,b_j,\kappa_{bj}) - \hat{f}_{\text{nadir}/\text{qss},k}(a_i,b_j)}{\kappa_{bj}} \end{cases} \tag{19}$$

In the above equation, κ_{ai} and κ_{bj} are the uptakes of the model parameters a_i, b_j.

4.3. SLP-Based Solution Algorithm

SLP is a mathematical optimization method used to solve nonlinear programming problems. It gradually approaches the optimal solution by decomposing nonlinear problems into a series of linear problems. The advantage of the SLP algorithm is that it decomposes complex nonlinear problems into a series of relatively simple linear problems, making the algorithm easier to implement and solve. The above parameter estimation optimization model is solved based on the SLP algorithm. The computational flow is shown in Figure 5.

Step 1: Initialization parameters. Extract the operating state data of multiple types of frequency modulation resources in the power grid and set the initial values of the equivalence model parameters a_i, b_j, and the model order. Set the number of expected fault scenarios N_K, calculate the minimum frequency $f_{\text{nadir},k}$, quasi-steady-state frequency $f_{\text{qss},k}$, and the minimum frequency $\hat{f}_{\text{nadir},k}(a_i^0, b_j^0)$ and quasi-steady-state frequency $\hat{f}_{\text{qss},k}(a_i^0, b_j^0)$ of the system under each fault scenario from the detailed frequency response model, and the minimum frequency and quasi-steady-state frequency of the system from the equivalence model at the given initial values. Set the time-domain simulation time and step size, and the number of iterations $n = 0$.

Step 2: Optimize the solution. Firstly, the frequency key index difference constraint is linearized, and the VPP parameter estimation problem is transformed into a linear programming model. Set the regimes of a_i and b_j as 0.1, and calculate the minimum frequency of the system at the nth iteration and the QSSF sensitivity coefficient a_i^n and b_j^n according to Equation (19). Then, solve the linear programming model (Equations (14)–(18)) to obtain the model parameter variations Δa_i^n, Δb_j^n, and update the estimates at the nth iteration $a_i^{n+1} = a_i^n + \Delta a_i^n$, $b_j^{n+1} = b_j^n + \Delta b_j^n$.

Step 3: Termination judgment and parameter output. Judge whether the change of model parameters under the nth iteration is less than the cutoff error ε_a, ε_b or whether the parameter values a_i^n, b_j^n are out of the parameter design range. If satisfied, the calculation is terminated and the parameter estimation results are output; otherwise, make $n = n + 1$ and return to Step 2.

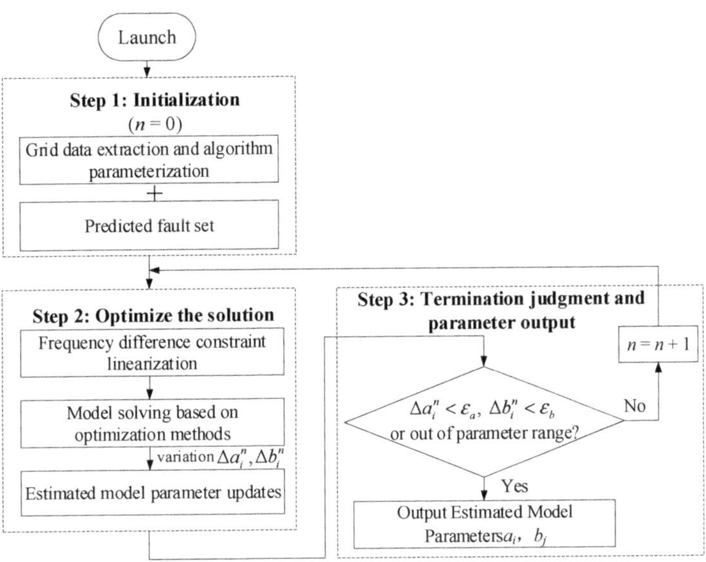

Figure 5. Parameter estimation process based on the SLP algorithm.

5. Simulation Analysis

5.1. Test Systems

This section is based on the electromechanical transient simulation software PSD-BPA (5.1.2) for verification. The test system is a three-region grid model (see Figure 6), and the VPP is configured in the HVDC receiving-end regional grid system C (see Figure 7); the main parameters are shown in Reference [24]. The test system is set up to access four large thermal power units at 500 kV, totaling 3.85 GW, and 18 VPPs, totaling 3.79 GW. In addition to the power fed by HVDC, the VPP capacity in the local grid region accounts for 49.6% of the generation resources. Taking VPP-1 as an example, the information on the types of resources and equipment capacities involved in frequency control within it is shown in Table 1.

Table 1. Frequency regulation resources in VPP-1.

Frequency Modulation Resources	Device Capacity/MW	Operating Power/MW	Control Method	Adjusted Power/MW	Trigger Method
gas turbine	54.0	42.0	inertia response and primary frequency modulation	12.0	response-driven
coal-fired unit	50.0	40.0	inertia response and primary frequency modulation	10.0	response-driven
wind power	22.5	22.5	rotor kinetic energy control	3.0	response-driven
energy storage plant	30.0	25.0	fast power control	5.0	event-triggering
electric vehicle	5.0	3.0	fast power control (V2G)	6.0	event-triggering
flexible load	5.0 (regulated)	3.0	one-time frequency modulation	3.0	response-driven
	3.0 (on-off)	2.0	load-shedding	2.0	event-triggering

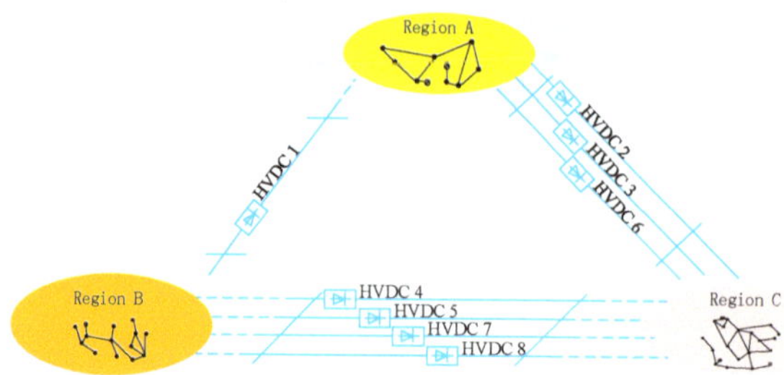

Figure 6. Schematic diagram of the HVDC interconnected power grid model with three regions.

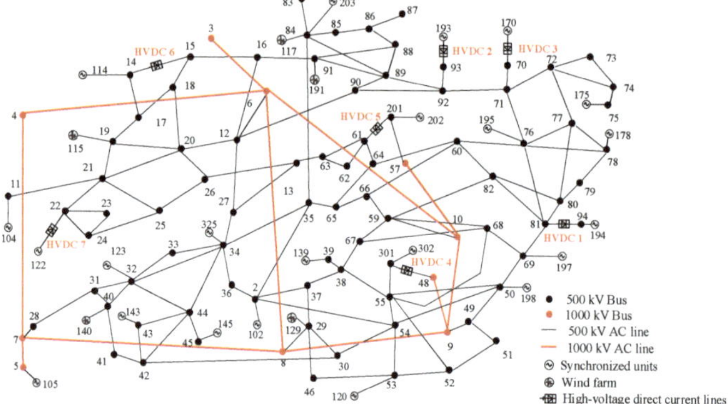

Figure 7. Topology of the receiving-end power system at Region C.

5.2. Verification of the Accuracy of the SFR Model of the VPP

Considering the randomness and diversity of faults, the fault scenarios are set as the operating conditions when a unipolar/bipolar blocking fault occurs in HVDC1-HVDC7 at 1 s, or when the load of the grid is perturbed according to the load increase of 0.1%, 0.2%, ..., 9.9%, or 10.0% of the total load, or when the two types of faults are randomly combined.

(1) Different model orders

Set the simulation time to 20 s, the simulation step size to 0.01 s, and the number of iterations $n = 0$. The three models—the first-order model, the second-order model, and the third-order model—are used to estimate the parameters of the VPP feedback control branch (the cutoff errors ε_a and ε_b of the model parameters a_i and b_j are 0.01, respectively) and set the number of fault scenarios to 1000. In each fault scenario, the frequency minimum and QSSF obtained from the SFR modeling of the detailed feedback control branch counting the VPP are used as the real values. The initial values of the first-order model parameters are set as $a_{0,\text{est}(0)} = 0.5$, $a_{1,\text{est}(0)} = 7.0$, $b_{0,\text{est}(0)} = 15.0$, and $b_{1,\text{est}(0)} = 15.0$.

As shown in Table 2, after the first iteration, the parameter estimates increase significantly, realizing one approximation to the optimal solution. After the second iteration, the parameter estimates continue to increase, but the increment decreases. Although the change of estimated parameters $a_{1,\text{est}}$ $\Delta a_{1,\text{est}}$ is smaller than the cutoff error ε_a at this time, the change of other parameters still does not satisfy the termination condition, so it is necessary to continue the calculation. After the third iteration, the change of each estimated parameter satisfies the cutoff error, and the optimization calculation is finished, and the results under the third iteration are recorded as the estimated parameters of the first-order model. Similarly, the solution process of second-order and third-order models conditional on increasing numbers of iterations are shown in Tables 3 and 4.

Table 2. Parameter changes of iterations under the first-order model.

Number of Iterations	$a_{0,est}$	$a_{1,est}$	$b_{0,est}$	$b_{1,est}$
0	0.500	7.000	15.000	15.000
1	0.863	8.995	15.827	32.320
2	0.991	9.002	16.136	38.881
3	0.998	9.007	16.142	38.889

Table 3. Parameter changes of iterations under the second-order model.

Number of Iterations	$a_{0,est}$	$a_{1,est}$	$a_{2,est}$	$b_{0,est}$	$b_{1,est}$
0	0.500	7.000	1.000	15.000	15.000
1	0.859	8.376	2.018	15.715	32.002
2	0.995	8.949	2.612	16.134	38.878
3	0.997	9.291	2.695	16.140	38.885
4	0.998	9.293	2.697	16.141	38.884

Table 4. Parameter changes of iterations under third-order model.

Number of Iterations	$a_{0,est}$	$a_{1,est}$	$a_{2,est}$	$a_{3,est}$	$b_{0,est}$	$b_{1,est}$
0	0.500	7.000	1.000	0.5	15.000	15.000
1	0.857	8.816	3.761	0.712	15.795	32.202
2	0.995	9.336	4.680	0.777	16.134	38.876
3	0.996	9.591	5.594	0.805	16.138	38.884
4	0.998	9.594	5.597	0.807	16.141	38.885
5	0.999	9.595	5.601	0.808	16.142	38.885

Equivalent feedback control branch model order affects the frequency response model accuracy. For the three model orders, the error between the estimated and true values of the lowest frequency and quasi-steady-state frequency is shown in Figure 8.

The absolute errors between the true and estimated values of the minimum frequency and QSSF are greater than zero in each of the predicted fault scenarios, which reflects conservativeness of the model. As the order of the model increases, the errors of the lowest value of frequency and the quasi-steady-state frequency become smaller. When the third-order model is used, the absolute errors of the frequency minimum and the QSSF are about less than 0.01 Hz and 0.005 Hz. Even if the first-order model is used, the error of the frequency minimum is less than 0.05 Hz, i.e., 0.001 p.u., and that of the QSSF is less than 0.015 Hz, i.e., 0.0003 p.u., which meets the requirements of engineering applications.

(2) The influence of different initial values

To analyze the impact of different initial values on parameter estimation, based on the first-order model, three sets of different initial values were set. The first set was consistent with the initial values in Table 2. The second set of initial values approached the lower limit of the parameters, $a_{0,est(0)} = 0.1$, $a_{1,est(0)} = 4.0$, $b_{0,est(0)} = 12.5$, and $b_{1,est(0)} = 5.0$. The initial value of the third group was set close to the upper limit of the parameter, $a_{0,est(0)} = 2.0$, $a_{1,est(0)} = 11.0$, $b_{0,est(0)} = 33.0$, and $b_{1,est(0)} = 100.0$. The remaining settings of the algorithm remain unchanged. The parameter estimation process under the second and third initial values is shown in Figure 9.

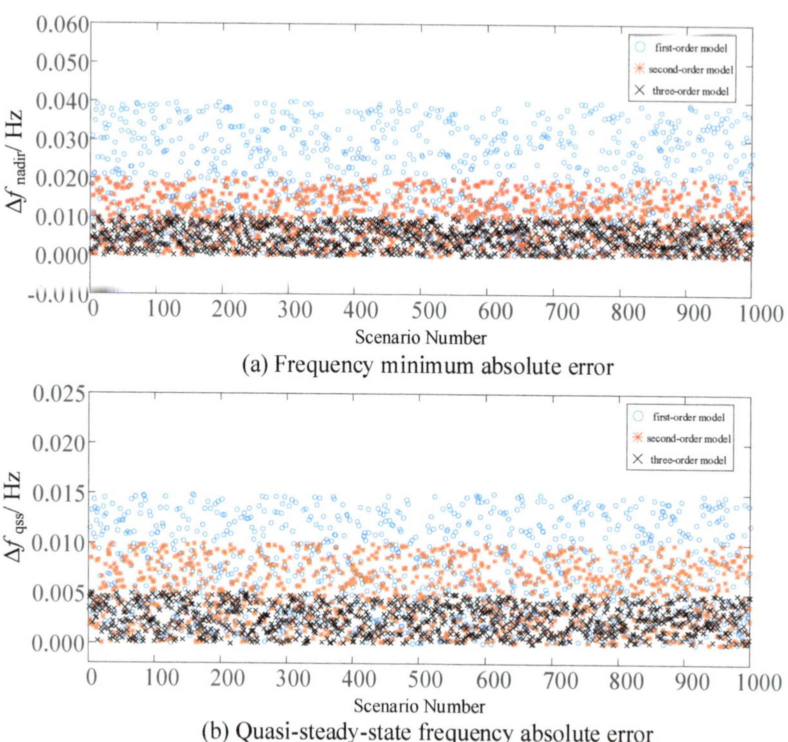

(a) Frequency minimum absolute error

(b) Quasi-steady-state frequency absolute error

Figure 8. Absolute error of frequency nadir and quasi-steady-state frequency.

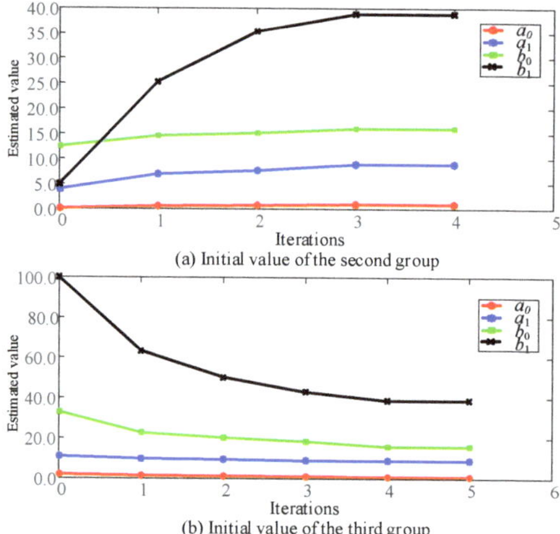

(a) Initial value of the second group

(b) Initial value of the third group

Figure 9. Parameter changes of iterations under different initial values.

According to the results in Figure 9, when setting different initial values for parameter estimation, the estimated values obtained through algorithm iteration are close. Based on the estimated parameters under the first set of initial values, the absolute error of the estimated values for the other two sets of parameters is in the order of 10^{-3}. Regardless of whether the initial value is set high or low, it can still approach the optimal solution after solving, indicating that the parameter estimation optimization model proposed in this paper has good adaptability under different initial parameters.

(3)   Comparative analysis

In this chapter, the proposed parameter estimation method is compared with the parameter estimation methods in time-domain simulation, least squares method, and Reference [19]. Among them, the third-order model with higher accuracy is used in the proposed method in this paper, and the initial values are set to be consistent with those in Table 4. The fitting function in the nonlinear least squares method adopts Equation (10) in Reference [25], which is an expression describing the relationship between the SFR and each control parameter. In Reference [19], the typical initial values are set to be $T_R = 0.2$ s, $T_b = 0.3$ s, $T_{RH} = 7$ s, and $F_h = 0.3$. The fault scenario is set as a precipitous increase in grid load of 700 MW at 1 s. The SFR curves under different methods are shown in Figure 10, and the frequency key indexes and errors are shown in Table 5.

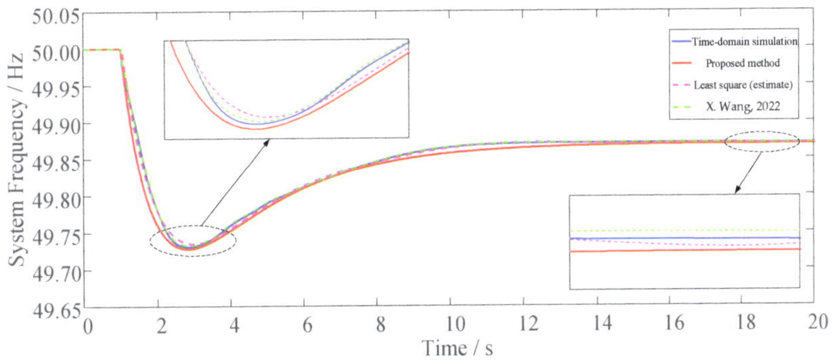

Figure 10. System frequency response curves under different methods [19].

Table 5. Frequency indices and relative errors under different methods.

Methodology	Minimum Frequency/Hz	Quasi-Stationary Frequency/Hz	Relative Error/%	
			Minimum Frequency	Quasi-Stationary Frequency
Time-domain simulation	49.7308	49.8704	-	-
Least square (estimate)	49.7347	49.8699	0.0078	0.0010
Reference [19]	49.7323	49.8712	0.0030	0.0016
Methodology of this paper	49.7278	49.8692	0.0060	0.0024

Due to the high computational accuracy of time-domain simulation, other methods often use it as a reference value for comparison. According to the results in Figure 10 and Table 5, the lowest point of the system frequency obtained using the nonlinear least squares method is higher than the calculated value of the time-domain simulation, with a relative error of 0.0078%, which is somewhat optimistic. Reference [19] constructed an optimization model with the objective of minimizing the sum of squares of the difference between the true and estimated frequency values, and solved it using optimization methods such as steepest descent and Newton's method to obtain the estimated parameters. This method has high computational accuracy and is suitable for analyzing the dynamic characteristics of system frequency response. However, it cannot guarantee the conservatism of key frequency indicators, that is, the lowest frequency value and quasi-steady-state frequency value of the system may be higher than the results of time-domain simulation, making it difficult to use for emergency frequency control optimization decision making. In addition, both of the above methods require a large amount of sampling of the system frequency after the fault to obtain the computational data required by the algorithm. However, the method proposed in this paper can achieve good computational accuracy with simpler inputs, simpler implementation, and good accuracy. In this scenario, the relative error of the frequency index is in the order of 10^{-3}. And the method proposed in this article only needs to take the frequency key indicator values (the minimum frequency and quasi-steady-state frequency) as inputs. The frequency key indicator constraint of the optimization model stipulates that the frequency value obtained by the method in this article is lower than the time-domain simulation. This can reflect the conservatism of the method in this article and avoid missed judgments in emergency frequency control decisions.

6. Conclusions

VPP participates in emergency frequency control by aggregating multiple types of flexible resources, but it is difficult to construct the frequency response model of VPP because the SFR characteristics of each resource are different and the VPP model is characterized by complexity, high order, and unknown parameters. In this paper, the SLP algorithm is used to perform parameter estimation on VPP frequency response model considering multiple flexible resources, and the methodological innovations are as follows:

(1) The VPP frequency response model proposed in this paper takes into account the frequency response characteristics of various types of resources and analyzes the role of each control mode on frequency support based on the inertia response, primary frequency modulation, and fast power control in multiple dimensions.
(2) The SLP-based parameter estimation method for the VPP SFR model only needs to take the value of frequency key index as input, which is simple to implement and has high accuracy. In addition, the parameter estimation results are conservative, which can prevent the system from misjudging in the emergency frequency control decision.
(3) The scientific validity and soundness of the proposed methodology in this paper is verified by simulation through a multi-fault test scenario based on a three-region HVDC receiving area grid C. Even if a first-order model is used for parameter estimation, the error of its frequency key index is less than 0.001 p.u., which meets the requirements of engineering applications.

The method in this article is to aggregate distributed resources at a voltage level of 500 kV and construct a system frequency response model considering VPP, which can support emergency frequency control. This article analyzes and considers the frequency response model construction of VPP from the system level, so there is no more detailed modeling of various distributed resources at the device level, which is also the direction for our future improvement.

Author Contributions: All authors contributed to the research in the paper. Conceptualization, Z.S. and H.Z. (Haixiao Zhu); data curation, H.Z. (Haibo Zhao); formal analysis, P.W. and K.W.; funding acquisition, Y.L.; investigation, Z.S. and K.W.; methodology, H.Z. (Haibo Zhao) and J.C.; resources, X.Z.; software, Z.S. and H.L.; supervision, H.Z. (Haibo Zhao), P.W. and Y.L.; validation, Z.S.; visualization, J.C.; Writing—original draft, H.L.; writing—review and editing, Z.S. and H.Z. (Haixiao Zhu). All authors have read and agreed to the published version of the manuscript.

Funding: This research received no external funding.

Data Availability Statement: The original contributions presented in the study are included in the article, further inquiries can be directed to the corresponding author.

References

1. Cui, R.Y.; Hultman, N.; Cui, D.; McJeon, H.; Yu, S.; Edwards, M.R.; Sen, A.; Song, K.; Bowman, C.; Clarke, L.; et al. A plant-by-plant strategy for high-ambition coal power phaseout in China. *Nat. Commun.* **2021**, *12*, 1468. [CrossRef] [PubMed]
2. Cao, Y.; Zhang, H.; Zhang, Y.; Li, C.G. Event-driven Fast Frequency Response Control Method for Generator Unit. *Autom. Electr. Power Syst.* **2021**, *45*, 148–154.
3. Chen, C.; Li, N.; Zhong, P.; Zeng, M. Review of Virtual Power Plant Technology Abroad and Enlightenment to China. *Power Syst. Technol.* **2013**, *37*, 2258–2263.
4. Hernandez, L.; Baladron, C.; Aguiar, J.M.; Carro, B.; Sanchez-Esguevillas, A.; Lloret, J.; Chinarro, D.; Gomez-Sanz, J.J.; Cook, D. A multi-agent system architecture for smart grid management and forecasting of energy demand in virtual power plants. *IEEE Commun. Mgz.* **2013**, *51*, 106–113. [CrossRef]
5. Wei, Z.; Yu, S.; Sun, G.; Sun, Y.; Yuan, Y.; Wang, D. Concept and Development of Virtual Power Plant. *Autom. Electr. Power Syst.* **2013**, *37*, 1–9.
6. Zhou, B.; Zhang, K.; Chan, K.W.; Li, C.; Lu, X.; Bu, S.; Gao, X. Optimal coordination of electric vehicles for virtual power plants with dynamic communication spectrum allocation. *IEEE Trans. Ind. Inform.* **2020**, *17*, 450–462. [CrossRef]
7. Chen, W.; Ai, X.; Fan, Y. Replicator Dynamics Algorithm Based Equilibrium Dispatching Strategy for Distributed Energy Resources in Virtual Power Plant. *Power Syst. Technol.* **2014**, *38*, 589–595.
8. Zhang, Y.; Mu, Y.; Jia, H.; Wang, M.; Zhou, Y.; Xu, Z. Response Capability Evaluation Model with Multiple Time Scales for Electric Vehicle Virtual Power Plant. *Autom. Electr. Power Syst.* **2019**, *43*, 94–103.
9. Lu, Z.; Wang, H.; Zhao, H.; Feng, H. Double layer inverse robust optimization scheduling strategy for virtual power plants containing V2G. *Power Syst. Technol.* **2017**, *41*, 1245–1252.
10. Xu, Q.; Cao, Y.; Zhang, H. Preventative Control Method for Virtual Power Plant Considering Transient Frequency Regulation Capability. *Autom. Electr. Power Syst.* **2022**, *46*, 83–89.
11. Li, R.; Xu, T.; Li, Y.; Zhang, E.; Cui, L. Study on primary frequency modulation characteristics of virtual power plant using different control strategies. *Mod. Electron. Tech.* **2021**, *44*, 95–99.
12. Wang, J.; Liao, S.; Yao, L.; Pu, T.; Xu, J.; Liu, Y.; Cheng, F. A Coordinated Frequency Control for HVDC Receiving-End Power Grid with Distributed Frequency Regulation Resources Based on Consensus Algorithm. *Power Syst. Technol.* **2022**, *46*, 888–900.
13. Wang, S.; Wu, W. Aggregate flexibility of virtual power plants with temporal coupling constraints. *IEEE Trans. Smart Grid* **2021**, *12*, 5043–5051. [CrossRef]
14. Jiang, Z.; Zhang, F.; Hu, F.; Sun, Y.; Jiang, W.; Deng, Y. Aggregation response capability evaluation method of distributed resources in virtual power plant. *Electr. Power Eng. Technol.* **2022**, *41*, 39–49.
15. Chen, J.; Liu, M.; Milano, F. Aggregated model of virtual power plants for transient frequency and voltage stability analysis. *IEEE Trans. Power Syst.* **2021**, *36*, 4366–4375. [CrossRef]
16. Feng, C.; Chen, Q.; Wang, Y.; Kong, P.-Y.; Gao, H.; Chen, S. Provision of Contingency Frequency Services for Virtual Power Plants with Aggregated Models. *IEEE Trans. Smart Grid* **2022**, *14*, 2798–2811. [CrossRef]
17. Inoue, T.; Taniguchi, H.; Ikeguchi, Y.; Yoshida, K. Estimation of power system inertia constant and capacity of spinning-reserve support generators using measured frequency transients. *IEEE Trans. Power Syst.* **1997**, *12*, 136–143. [CrossRef]
18. Liu, M.; Chen, J.; Milano, F. On-line inertia estimation for synchronous and non-synchronous devices. *IEEE Trans. Power Syst.* **2020**, *36*, 2693–2701. [CrossRef]
19. Wang, X.; Li, X.; Guo, L.; Zhao, Z.; Wang, H.; Wang, C. Parameter Estimation Method of a Frequency Response Model of Multi-Area Interconnected Power System Based on Cross-Section Information. *Power Syst. Technol.* **2022**, *46*, 4277–4291.
20. Han, S.; Xu, Z.; Wu, X.; Jin, X. Comparison of Mathematical Models for Transient Stability between PSD-BPA and PSS/E. *South Power Syst. Technol.* **2010**, *4*, 67–71.
21. Song, X.; Wu, X.C.; Liu, W.Z.; Xu, A.D.; Bu, G.Q.; Jing, X.M. New Quasi-Steady-State HVDC Models for PSD-BPA Power System Transient Stability Simulation Program. *Power Syst. Technol.* **2010**, *34*, 62–67.
22. Mauricio, J.M.; Marano, A.; Gomez-Exposito, A.; Ramos, J.L.M. Frequency regulation contribution through variable-speed wind energy conversion systems. *IEEE Trans. Power Syst.* **2009**, *24*, 173–180. [CrossRef]
23. Hui, H.; Ding, Y.; Chen, T.; Rahman, S.; Song, Y. Dynamic and Stability Analysis of the Power System With the Control Loop of Inverter Air Conditioners. *IEEE Trans. Ind. Electron.* **2021**, *68*, 2725–2736. [CrossRef]
24. Shi, Z.; Xu, Y.; Wang, Y.; He, J.; Li, G.; Liu, Z. Coordinating multiple resources for emergency frequency control in the energy receiving-end power system with HVDCs. *IEEE Trans. Power Syst.* **2023**, *38*, 4708–4723. [CrossRef]
25. Shi, Q.; Li, F.; Cui, H. Analytical method to aggregate multi-machine SFR model with applications in power system dynamic studies. *IEEE Trans. Power Syst.* **2018**, *33*, 6355–6367. [CrossRef]

Article

Decarbonization through Active Participation of the Demand Side in Relatively Isolated Power Systems

Sophie Chlela *, Sandrine Selosse * and Nadia Maïzi

Centre for Applied Mathematics, Mines Paris-PSL, 06560 Valbonne, France
* Correspondence: sophie.chlela@minesparis.psl.eu (S.C.); sandrine.selosse@minesparis.psl.eu (S.S.)

Abstract: In the context of power system decarbonization, the demand-side strategy for increasing the share of renewable energy is studied for two constrained energy systems. This strategy, which is currently widely suggested in policies on the energy transition, would impact consumer behavior. Despite the importance of studying the latter, the focus here is on decisions regarding the type, location, and timeframe of implementing the related measures. As such, solutions must be assessed in terms of cost and feasibility, technological learning, and by considering geographical and environmental constraints. Based on techno-economic optimization, in this paper we analyze the evolution of the power system and elaborate plausible long-term trajectories in the energy systems of two European islands. The case studies, Procida in Italy and Hinnøya in Norway, are both electrically connected to the mainland by submarine cables and present issues in their power systems, which are here understood as relatively isolated power systems. Renewable energy integration is encouraged by legislative measures in Italy. Although not modeled here, they serve as a backbone for the assumptions of increasing these investments. For Procida, rooftop photovoltaics (PV) coupled with energy storage are integrated in the residential, public, and tertiary sectors. A price-based strategy is also applied reflecting the Italian electricity tariff structure. At a certain price difference between peak and off-peak, the electricity supply mix changes, favoring storage technologies and hence decreasing imports by up to 10% during peak times in the year 2050. In Norway, renewable energy resources are abundant. The analysis for Hinnøya showcases possible cross-sectoral flexibilities through electrification, leading to decarbonization. By fine-tuning electric vehicle charging tactics and leveraging Norway's electricity pricing model, excess electricity demand peaks can be averted. The conclusions of this double-prospective study provide a comparative analysis that presents the lessons learnt and makes replicability recommendations for other territories.

Keywords: decarbonization; power systems; European islands; optimization; energy planning; rooftop PV; storage; electric vehicles; flexibility; renewable energy; demand response

Citation: Chlela, S.; Selosse, S.; Maïzi, N. Decarbonization through Active Participation of the Demand Side in Relatively Isolated Power Systems. *Energies* **2024**, *17*, 3328. https://doi.org/10.3390/en17133328

Academic Editor: Wen-Hsien Tsai

Received: 28 March 2024
Revised: 28 June 2024
Accepted: 3 July 2024
Published: 7 July 2024

1. Introduction

1.1. Background of the Study

Many countries are counting on the massive development of renewable energy to meet their commitments to the CO_2 emission reduction targets in the framework of the Paris Agreement and more broadly in the fight against global warming. As part of its overall energy policy, particularly the European Green Deal, the European Union is aiming, for example, for a minimum of 42.5% renewable energy participation in its energy mix by 2030. This target has been chosen to reach the goal of reducing greenhouse gas emissions by 55% in 2030, with respect to 1990 levels [1]. Among these renewable energies, solar and wind power are leading figures, with an increase of 309 GW and 389 GW in their respective capacities planned between 2015 and 2050 [2]. However, the integration of producing electrical energy using these variable renewable sources involves rethinking the electrical system to ensure its stability and reliability. Grid stability is crucial for power system operation, with a balance needed between production and consumption. The variation between these two factors impacts the system's inertia, which affects its ability to maintain voltage and frequency stability after a disturbance. Lower inertia leads to faster frequency changes and challenges concerning system reliability [3].

Thus, to adapt to the increasing variability of electricity production from renewable energy, that is, to guarantee supply and take advantage of the energy produced at all times, the electricity system must be more flexible. This encourages setting up strategies that necessitate the participation of the demand side to ensure robust, decarbonized energy systems.

Many European islands possess abundant renewable energy resources, such as wind, solar, geothermal, and wave power, offering them opportunities to transition toward self-sufficient, sustainable energy generation [4]. The favorable regulatory framework of the Clean Energy for all Europeans Package (CEP) [5] indicates that the European Commission is dedicating a Clean Energy for EU Islands Initiative that "provides a long-term framework to help islands generate their own sustainable, low-cost energy" [6]. Studies related to the energy or electricity systems of islands have until now focused on renewable energy integration for energy autonomy, such as the cases of Reunion Island [7] and Madeira Island [8]. Some have also focused on the reduction of greenhouse gas emissions through the inclusion of climate mitigation targets, such as Nationally Determined Contributions (NDC). The International Renewable Energy Agency (IRENA) dedicates specific studies to the energy system of Small Island Developing States (SIDS) [9]. One example is the study of Antigua and Barbuda, where pathways for renewable energy integration, electric vehicles, and green hydrogen were elaborated [10]. Importance is assigned to assessing the potential of RE on islands in [11] and analyzing the energy system through indicators related to energy resources, energy profiles, and energy management in [4]. Through technical scenarios and future planning, the authors of [12] were able to determine the alignment between national and successful policies, while focusing on five critical policy areas. As can be seen, some islands aim at achieving complete energy autonomy, while other goals include reducing emissions by having low percentages of diesel in their future energy supply mix. In other words, they seek diversification of their energy resources through harnessing the full potential on their territories. In some cases, interest is directed to policy evaluation. The literature highlights the benefits and importance of involving the demand side yet fails to provide guidance on its long-term facilitation in the energy systems of islands, especially considering their restricted interconnections with the mainland.

1.2. Motivation and Contribution

In this study, the subjects delve deeper into decarbonization and renewable energy integration in insular energy systems, focusing on strategies that would enable flexibility. First, it relies on the applied methods that would be employed in the energy systems. Hence, this research isolates and evaluates the impacts of these methods within the energy systems themselves. It provides a unique lens on the feasibility of integrating renewable energy and flexibility into the system without altering demand.

Second, the study highlights the need to assess the energy transition and action plans implemented by islands, which are often not adequately addressed in national policies. For this reason, islands are taken as case studies: Procida in Italy, and Hinnøya in Norway. As a novelty, the paper emphasizes island-specific energy transitions and underlines the importance of localized energy solutions that cater to the unique needs and potentials of island communities.

Third, utilizing the TIMES model generator from the IEA Energy Technology Systems Analysis Program (ETSAP), the study models the future evolution of power systems for these islands, aiming to accommodate higher shares of renewable energy for their decarbonization. This long-term, high-resolution modeling provides a detailed framework for understanding how renewable energy and flexibility solutions can be optimally integrated over extended timeframes, considering local resources and data [13,14], and demand participation strategies. Furthermore, it provides an assessment of different scenarios or hypotheses that are crucial for making decisions now on investments for the sustainability of decarbonization pathways.

Fourth, owing to the time structure of the two energy systems, the load curve variations are obtained across different timescales (seasons, days, and hours of the day). This method is particular to this long-term study and provides detailed analysis of the optimal use of the potential of renewable energy. It enhances the understanding of how flexibility solutions and strategies can be tailored to account for the grids' needs (peak demands, imports, etc.) in different timeslots.

Fifth, the decarbonization of end-use sectors is also part of the study. The analysis for each island offers a comprehensive view of how different elements can synergistically contribute to decarbonization.

Sixth, the selected case studies are representative of two contrasting EU island territories. Thus, this study offers broader policy recommendations and insights for replicability in other territories. Indeed, it includes a comparative summary in the discussion (Section 5) on implementation challenges, success factors, and cross-regional lessons. It aims to enhance the generalizability and applicability of its findings.

1.3. The Different Roles and Strategies of Active Demand Participation for Increasing Renewable Energy

Power system flexibility encompasses maintaining the supply–demand balance, based on [15], the existence of storage capacity for balancing [16], the adjustment of demand (i.e., demand-side management (DSM)), and the operation in the context of a market that allows flexibility [17]. Active participation of the demand side, as explained by the authors of [18], involves activities such as load management and self-production, which are integral components of DSM. In addition, the International Energy Agency analysis of the demand response (described in the following section) notes that "this flexibility will become increasingly important as grids become progressively dominated by variable power generation, such as wind and solar PV". These mechanisms, including load levelling, valley filling, and load shifting, enable consumers to actively engage in the energy system. To facilitate DSM, enabling technologies, such as IoT-driven automation, connect devices such as PV systems, battery storage, and smart meters to exchange information on energy consumption and consumer responsiveness to price signals [19]. This discussion also delves into the adoption of load management and self-consumption solutions in the studied territory. Three types of DSM activities in relation to the analysis are non-exhaustively described below.

(a) Demand response

Historically, electricity production and transmission have evolved to meet rising consumption demands. However, the demand response is now shifting this dynamic by encouraging consumption to adapt to electricity production and transportation conditions [20]. The demand response involves temporarily reducing a site's consumption in response to an external request. This reduction modifies the initially planned consumption, possibly involving a time shift [21]. The advantages of the demand response include reduced network connection investments and power system decarbonization. Flexible consumption can lower emissions from conventional peak-hour power plants. The demand response comes in two forms: explicit and implicit [22]. The first involves consumers facing time-variable electricity prices and participating by responding to price signals that change over time [23]. In Europe, time-of-use (TOU) pricing, which divides the day or year based on peak/off-peak hours or seasons, prevails for the energy component, but real-time pricing (RTP), with hourly rate adjustments based on market price adoption, is on the rise [24]. The second form is explicit, based on incentive programs, whereby consumers who shift demand receive payments through the demand response. Moreover, according to the IEA Net-Zero scenario, 500 GW of demand response will be deployed by 2030, which corresponds to 10 times the value of the year 2020 [22].

(b) Self-consumption

The concept of self-consumption of electricity is widespread across Europe [25]. The interpretation of this concept is found in the EU Directive 2019/944 (article 2 (8), related to "active customers") and the REDD II, which defines the "renewable self-consumer" in EU Directive 2018/2001 (article 2 (14)). However, the former definition offers more opportunities to customers, as it states that they can participate in flexibility and energy efficiency schemes.

Nowadays, the most competitive renewable energy sources are solar photovoltaics (PV) and wind [26]. According to [25], solar PV in Europe has approached "grid-parity costs for the kWh produced". A common solution in territories constrained by surface area, such as islands, is PV electricity self-generation with solar panels integrated on building rooftops. Residents become "prosumers" by self-consuming generated electricity, selling excess power to the grid, or storing it for later use, fostering energy transition, raising awareness, and addressing environmental concerns. These systems are usually connected to storage technologies for optimal use. Nevertheless, the focus is on reducing the carbon footprint of these solutions. As the industry grows, more transparency is needed in the value chain of these technologies, while the options of second-life use and recycling enhance its sustainability. In addition, policy plays a significant role in regulating the battery market [27].

(c) Electric vehicle charging

Creating synergies across the energy system can also identify possibilities for enhancing the flexibility and economic efficiency of the energy system [28], as well as making electricity prices more elastic for residential users [18]. This can be applicable with the electrification of the transportation sector. By controlling the charging of electric vehicles (EVs), the operator can avoid creating additional demand during peak hours and employ EVs to perform peak shaving or valley filling [18]. Two strategies exist in electric vehicle (EV) charging control: passive control, which encourages charging during low-electricity tariff periods, and active control (smart charging), which not only targets off-peak or low-tariff periods, but also involves modulation of charging power, allowing for better electricity demand management through input current adjustments. Smart charging offers two further options: unidirectional (V1G) and bidirectional (V2G), enabling power injection back into the grid [29].

1.4. Organization of the Paper

This paper is organized as follows: it begins by describing the modeling methodology in Section 2, including a presentation of the energy system modeling, followed by a focus on the tool used. Section 3 introduces the main assumptions integrated into the two models to understand the energy systems of each island and their evolution. Section 4 details the case tests implemented to form scenarios, showing renewable energy and storage investments and load curve variation. Section 5 elaborates on the impacts of the pricing structure with the integration of renewables and offers recommendations for future pathways, before concluding with final remarks in Section 6.

2. Method

To analyze the future development of territories' power systems considering the different challenges facing them (e.g., decarbonization, integration of high amounts of renewable energy, choice of investments, etc.), it is crucial to consider the plausible long-term trajectories of their evolution. Energy system models are frequently used in this context, bearing in mind that investments in the energy sector are often expensive and their success mainly depends on planned, well-defined scenarios covering a certain time horizon [30]. In this section, a review of energy system analysis and TIMES modeling is presented. Nevertheless, short-term energy planning is also employed for scheduling the operation of electricity grids and evaluating the integration of intermittent renewable energy, storage technologies, and pricing structures. This was the case of [31], which used battery storage as a flexible demand response framework and TOU to make a financial and economic analysis.

2.1. Background on Energy System Modeling

Generally, in energy system modeling, the scientific community distinguishes two approaches to integrating and representing the energy sector: bottom-up and top-down approaches [32]. Bottom-up optimization is known to provide a detailed representation of the energy system (technologies, processes, emissions, economic and financial properties, etc.) and the energy demand is satisfied by a certain market of technology option [33]. In contrast, top-down models provide a representation of the interaction of the whole energy sector with other economic sectors, for example, by identifying relationships between the economy and energy through macroeconomic indicators, or even by adding socio-technical aspects to the energy transition [32]. However, these models are not appropriate to assess the evolution of the energy system with all available low-carbon technologies. For instance, in [34], it was shown that to model carbon capture and utilization (CCU), bottom-up models still dominate due to the lack of information on this technology's socioeconomic impact. Top-down and bottom-up models can be linked together, where the output of the former is used as input for the latter. These models are known as hybrid models combining techno-economic details and constraints to assess economic (or socioeconomic) impacts, such as the case of the POLES-JRC model [35], which integrates the energy sector with a macro-economic evaluation and climate policy assessment [32].

The modeling tool employed in this study adopts a bottom-up optimization featuring two approaches for long-term planning: perfect foresight and myopic foresight, referring to the expectations of the economic actors [30,32]. In the first case, the development of energy prices, future improvements in certain technologies, and future decommissioning of power plants are known a priori. As for the myopic model, decisions are made based on the status quo of these factors, which requires that the optimization problem be divided into recursive subproblems, yielding suboptimal solutions. Babrowski et al. [30] switched from perfect foresight to a myopic approach in PERSEUS-NET, a bottom-up, linear optimizing energy system model used for studying the evolution of the German electricity generation system until 2030. As a consequence, the computation time was reduced ten-fold. In case of sudden events, such as the development of a new technology, perfect foresight provides an upper limit for an optimal performance of the system, but the myopic approach yields a more realistic output. However, for foreseen events, such as an expected rise in CO_2 prices, myopic approach solutions are suboptimal. This highlights the importance of exploiting the complementarity of both approaches.

A set of characteristics helps energy modelers to compare these tools, and they mainly utilize the techno-economic detail (scores can be allocated to the inclusion of flexibility options, especially when addressing future low-carbon energy systems), time resolution, geographical representation, sectoral coupling, and its representation [32,33]. Some of the well-known energy system models dealing with long-term planning, employing a bottom-up analytical approach and perfect foresight, are TIMES/MARKAL (version 4.8.1) from the IEA-ETSAP [36], LEAP (https://leap.sei.org/ (accessed on 19 January 2022) [37] from the Stockholm Energy Institute (SEI) [38], MESSAGE (https://docs.messageix.org/en/lates (accessed on 1 March 2024) [39] from the International Institute for Applied Systems Analysis (IIASA) [40], and OSeMOSYS (http://www.osemosys.org/ (accessed on 1 March 2024) [41] by the Royal Institute of Technology (KTH) [42]. The solution algorithm for LEAP is based on a heuristic algorithm, in contrast to the others, which use linear programming. This method provides a suboptimal solution in a rapid computational time. In addition, emerging energy system modeling tools are built with the objective of addressing the high integration of renewable energy in electric systems, thus handling operation and investments. This is the case of PyPSA, (v0.28.0) which addresses linking "power system analysis software and general energy system modeling tools" [43], but has other features such as sector coupling with other energy carriers, such as synthetic fuels and carbon removal with direct air capture (DAC). PyPSA is open source and it has been used to study various subjects, including the European transmission network for the ENTSO-E area and the optimization of power systems with renewable energy integration [44]. Another model aimed at exploring carbon-neutral energy systems is the Lappeenranta University of Technology (LUT) model. Its key features include representing the energy system in detail and finding the least-cost solution considering sets of constraints. It has been used for various energy transition scenarios, such as variable renewable energy integration in the Indian power system [33].

Observing that each model possesses distinct advantages over the other, the choice of tool often relies on the modelers' expertise, experience, and primary research objectives. For instance, TIMES/MARKAL is known for its high granularity and capability of showcasing the sector coupling, which avoids losing the coupling's impact. OSeMOSYS (http://www.osemosys.org/ (accessed on 1 March 2024), on the other hand, is open source and thus does not require upfront financial investment. Furthermore, tools such as the TIMES/MARKAL family are being constantly developed and extended, such as stochastic programming for optimization with imperfect foresight for assessing the robustness of policies under uncertainties [32]. TIMES modeling is further elaborated anddescribed in the following section.

2.2. TIMES Modeling

In this section, an overview describing the TIMES paradigm is provided, with a focus on the mathematical formulation of its objective function and characteristics relevant to this study. TIMES is a model generator developed by the IEA-ETSAP [45], whose code is available online (https://github.com/etsap-TIMES/TIMES_model (accessed on 1 February 2024)).

2.2.1. Description of a TIMES Model

A TIMES model describes the energy sector of a chosen region by sets of processes connecting primary energy sources and final energy demand through explicit and implicit input and output technologies [46] or commodity flows (commodities can be energy carriers, materials, emissions, etc.) [47]. Hence, present and future sources of primary energy supply, their potentials, and available and future technologies can be integrated using their characteristics. This allows adding new technologies that are currently too expensive but may be competitive in the future [47]. Another important capability of this energy system model is that it catches dynamic aspects of the power sector, such as flow equilibrium conditions, to follow load curves [46]. TIMES allows the user to divide the time horizon into years and subdivide it into four sub-annual segments: annual, seasonal, weekly, and daily [47].

The quantities and prices of the different commodities are in a state of equilibrium, meaning that their prices and quantities at each point in time are aligned in a way that ensures that suppliers supply precisely the amounts demanded by consumers. This state of balance possesses the characteristic of maximizing the overall economic surplus. Figure 1 conceptualizes the mathematical description of TIMES.

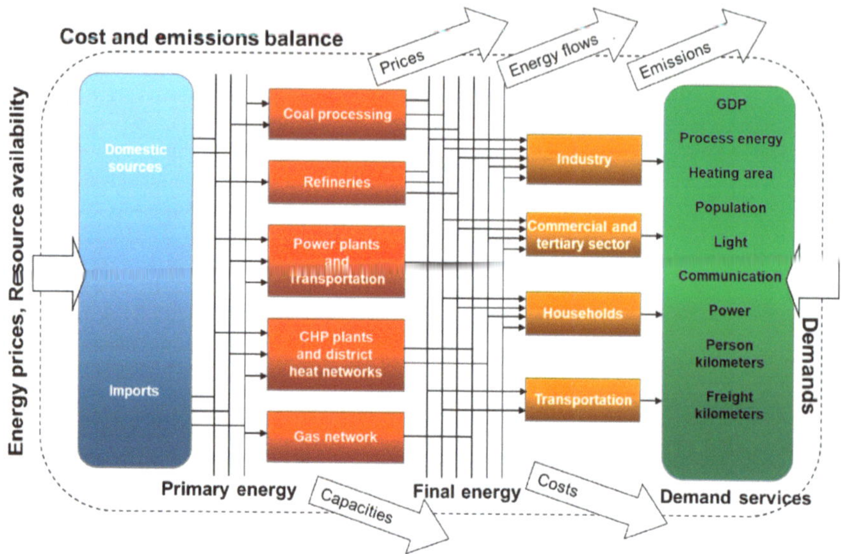

Figure 1. Representation of a reference energy system (RES) in TIMES, showing inputs and outputs (Reprinted from with permission from [47]).

2.2.2. The Objective Function and Scenarios for Possible Energy Futures

The objective function is considered as a linear mathematical expression of decision variables subject to linear constraints. It represents a minimization of the total discount costs of a power system over a long time horizon, as per the equation below. The decision variables are determined by the optimization (i.e., endogenously), whereas the constraints are equations or inequalities involving these variables and must be satisfied by the optimal solution. They could represent a number of environmental, technical, and demand constraints [48]. A TIMES model is vertically and horizontally integrated. Processes or technologies can be linked through inputs and outputs, where the outputs are expressed as linear functions of the inputs.

All these properties imply that TIMES is a partial equilibrium model that uses a linear programming (LP) approach. The objective function is, therefore, expressed as [45]:

$$min(NPV) = min(\sum_{r \in R} \sum_{y \in Y} (1 + d_{r,y})^{T_0 - y} \cdot ANNCOST(r,y))$$

where:

- NPV is the net-present value of the total cost for all regions, r, and years, y,
- R is the set of all the regions,
- Y is the set of years with costs,
- T_0 is the reference year,
- $d_{r,y}$ is the general discount rate in region, r, and year, y,
- $ANNCOST(r,y)$ is the total annual cost in region, r, and year, y. It includes capital costs from investing or dismantling processes, maintenance and operation, and trade costs.

Long-term simulations with TIMES make the scenario approach the best option, while short-term simulations may use econometric methods. Scenarios are based on coherent assumptions about future trajectories, organizing the system under study. Scenario builders need to test assumptions for internal coherence through a credible storyline. The demand component is essential when building a TIMES scenario. Demand drivers can be exogenous, such as using population, GDP, households, etc. National and sectoral output growth rates can be obtained from general equilibrium models, such as GEM-E3 [45,49].

2.2.3. Time Granularity

The time horizon in TIMES can be divided into several time periods, where each period contains a number of years that are possibly different. In a period, each year is considered identical, except for the cost objective function. It includes differentiation between payments in each year of a period. Input and output values, related to a period, t, are applied for each year in t. This concerns capacities, commodity flows, operating levels, etc., except investment variables, which are implemented once in a period (unless the period exceeds the technical life of the investments). There is a possibility to create time divisions using time slices (seasons, portions of the day/night, etc.). This is useful with production technologies that have different characteristics depending on the time of the year, such as wind turbines. Technologies storing commodities for later discharge can be defined and modeled. Different time slices may require different production technology deployment and investment decisions for peak reserve capacity. The initial period and its quantities of interest are fixed by the user as part of the calibration, an important step in creating a TIMES model.

3. Case Studies and Model Description

3.1. Case Studies: Italian and Norwegian Islands

The islands studied here are each situated in different geographical locations of Europe, and thus subject to different weather conditions and natural resources' availability. Their size and population density are other parameters that constitute a contrasting setting. In addition, the challenges they face, especially from an energy system perspective, are common to electricity systems on relatively isolated islands (i.e., existence of electricity connection to mainland), but the consequent issues remain diverse for these islands. A description of each territory is presented in Table 1, followed by an overview of the policy setting related to renewable energy in the respective countries.

Table 1. Territory characteristics [3].

	Procida, Italy	Hinnøya, Norway
Territory characteristics	Smallest island in the Gulf of Naples. Area = 4.26 km^2 Inhabitants = 10,428 [50] Density = 2449.1 inhabitants/km^2	Fourth largest island in Norway, Harstad City is the center of economic activities. It forms an island cluster with the smaller islands Grytøya, Bjarkøya, and Sandsøya [51]. Area = 4533 km^2 Inhabitants = 66,690 Density = 16.4 population/land km^2 [52]
Electric system	Dependent on electricity imports.	Situated in Nordpool region 4 (NO4) and imports its electricity from this market, which is based on hydroelectric power generation [53,54].
Challenges	Limited by space and surrounded with a protected marine area. Grid congestion and seasonality of demand.	Limited possibilities for new grid connections.

For the case of Italy, the legal framework surrounding climate change policies has been developed in the context of European recommendations, while the government is responsible for implementing various measures. Also, within the framework of the Covenant of Mayors, local administrations have implemented climate plans and measures. Italy has a goal of integrating the use of renewables at nearly 37% of the gross final energy consumption in 2030 to be in line with the Fit for 55 goals. The main types of support mechanisms to promote renewables are the feed-in tariff (FiT) and the feed-in premium (FiP). Other mechanisms include energy efficiency certificates and net billing ("scambio sul posto") preceded by the "conto energia" feed-in tariffs and green certificates that ended in 2013. To further promote the growth of renewable electricity generation, the Italian government encourages self-consumption and energy communities, including on small islands. By the Ministerial decree RES1 [55], FiT/FiP schemes became the main support, where FiT targets small plants and FiP targets large ones. The support scheme for islands' "RES auctions" was established by the FER1 Decree and provides a premium in addition to a base incentive paid for the PV electricity produced, whether it is fed into the grid or self-consumed [56]. Such mechanisms are related to energy and capacity markets and may impact investors and prosumers' profitability. In [57], it was found that the demand response and FiP drive an increase in renewable energy investments (wind energy in the case of the study) and a decrease in baseload generation and, hence, emissions. FiP increases the costs for consumers; thus, regulations, such as premiums, are established to protect their participation and reduce trade-offs. Currently, Italy's support for the IEA Digital Demand-Driven Electricity Networks initiative underscores the purpose of the demand response by modernizing power systems and establishing the appropriate regulatory framework [55].

Norway's climate policies target a 90–95% reduction in greenhouse gas emissions compared to 1990 levels by 2050, excluding carbon sinks. Its carbon pricing mechanism (included in the modeling, see Section 3.3) and abundant renewable resources in hydro and wind provide a strong basis for achieving this goal [58]. Nevertheless, the country still faces challenges to decarbonize its industry and transport sectors. In terms of legislative measures for increasing the integration of RES, the country only implemented an electricity certificates system in 2015, following Sweden, which introduced it in 2012. All producers and some consumers are obliged to buy certificates for a certain percentage of their production (or consumption). This percentage was gradually increased until 2020 before decreasing until 2035 when the system will be withdrawn. Italy considers a demand response mechanism as a way to ensure the stability and flexibility of its grid, whereas Norway adopts the mechanism to deal with high electricity prices, especially in the southern part of the country, and to delay investments in new grid capacities.

These island case studies present a similar goal of decarbonizing the energy system with the integration of high shares of renewable energy in the power mix. The solutions to achieve the targets are analyzed in this study according to the specificity of each island. This offers decision-makers and local authorities a broader view of the types of solution needed, but also opens the discussion on the replicability of solutions on other European islands or territories.

3.2. Procida Energy Model: TIMES-Procida

We opted to model Procida's electricity system (Figure 2), as it is the energy commodity that is mostly used in this territory. From the supply side, the island acquires 99% of its electricity through the marine cable connecting it to Ischia. The electricity demand of the island in 2018 was 19,915,668 kWh and, considering the limited data refinement, the sector distribution estimation is based on information from Procida's 2015 sustainable energy action plan (PAES), submitted as part of the European Covenant of Mayors program [13]. The rooftop solar photovoltaics (PV) represent a small percentage (1%), making the island extremely energy-dependent and suitable for improvements by increasing the share of renewable energy with self-consumption PV investments. The residential sector stands as the largest consumer, accounting for 62.7% of demand, trailed by the tertiary (service) sector at 28.4%. The public sector constitutes 5.4% of demand, while agriculture, industry, and transport represent 2.1%, 1.3%, and 0.1%, respectively.

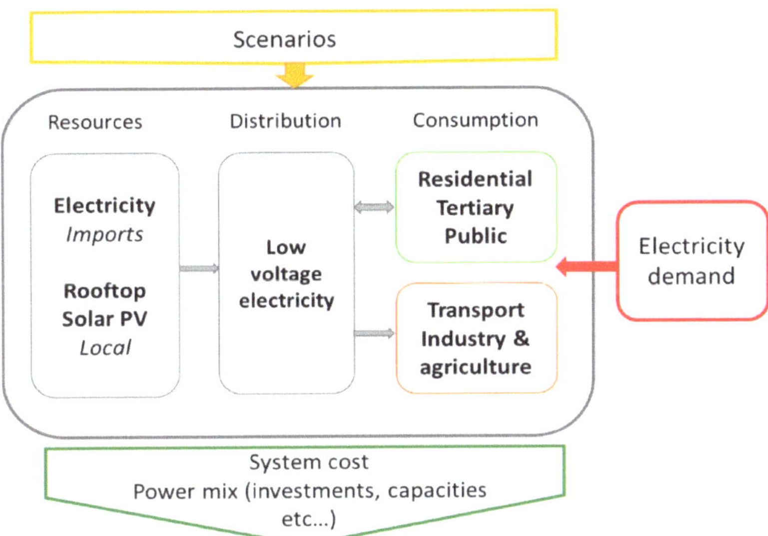

Figure 2. Reference energy system of the TIMES-Procida model.

To increase the renewable energy share, new photovoltaics have been considered for installation on building rooftops from 2019. The aim is to install PVs on residential, public, and tertiary buildings to analyze their contributions. The PV Watts calculator by the National Renewable Energy Laboratory (NREL) calculates the maximum PV capacity for the island's buildings. Lithium-ion batteries are considered for storing electricity daily in residential, tertiary, and public buildings to enhance flexibility and manage electricity generation. Battery characteristics vary for each sector where PVs are installed. Different batteries are chosen for each sector based on their capacity and C-rate. Constraints are set to determine the upper limit of battery capacity for each sector. The maximum allowable capacity for each sector is calculated using linear equations, multiplying the average battery capacity in a sector by the number of buildings. Only distributed storage systems that can be coupled with PVs are implemented, meaning storage does not charge from the grid but can inject electricity in case of excess production (more details are found in [13]).

Three scenarios were chosen to evaluate the results from the implementation of self-consumption on the island. These scenarios are based on the deployment of PV at low and high investment capacities, and the third one considers high deployment of PV with storage technologies (HIGH_STG), as storage integration becomes relevant with considerable amounts of solar electricity production. TIMES-Procida includes the future technologies of hydrogen storage and electric transportation, and the choice was made to avoid concentrating on these two aspects within this analysis, since hydrogen storage is costly, and it is only chosen by the techno-economic optimization when enforced, as investment and transportation demand for electricity is negligible on this island compared to the remaining sectors. Time periods were divided into 15 time slices: 3 seasons (winter, summer, and intermediate) and 5 daytime intervals. The dates separating the seasons were set by the Italian decree (DPR n. 412 del 26 Agosto 1993), which outlines heating periods for different areas based on climatic zones (Procida falls under zone C). Winter spans from 15 November to 31 March, while the summer season lasts from 15 May to 15 September, chosen due to higher average irradiation values observed in Procida from April to September (Figure 3). The intermediate season covers 1 April–14 May and 16 September–14 November. The defined time slices of TIMES-Procida for the summer and winter seasons are presented in Table 2 and can be read by aggregating the season with the time slice name; for example, for the period between 6:00 am and 10:00 am in the summer, the time slice would be denoted SMOR, while for the period between 7:00 pm and 12:00 am in the intermediate season, the time slice is IEVE. The horizon of the study was 32 years (from 2018 to 2050), where the sub-periods of one year were defined until 2035 and then became wider sub-periods (seasons). To determine the daytime level, the annual load curve was compared to the global clear-sky irradiance [13].

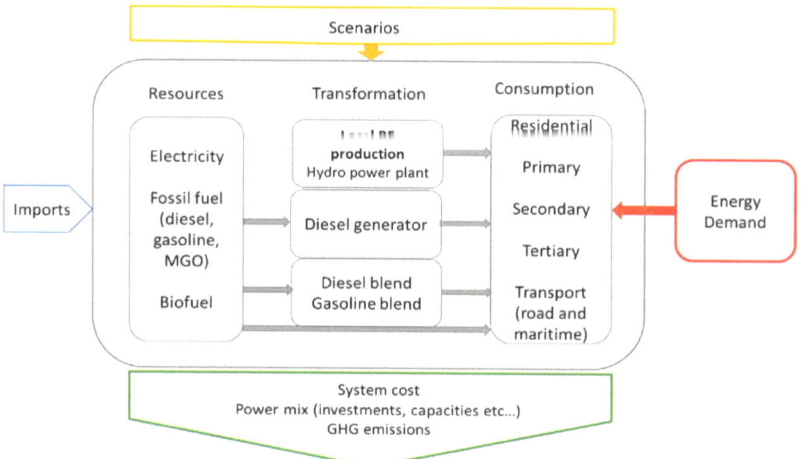

Figure 3. Reference energy system of the TIMES-Hinnøya model.

Table 2. Seasons and time slices of TIMES-Procida.

Season	Time	Timeslice Name	Hours
Intermediate "I"	Night	NGT	0–6:00
	Morning	MOR	6–10:00
Winter "W"	Midday	MID	10:00–15:00
	Afternoon	AFT	15:00–19:00
Summer "S"	Evening	EVE	19:00–24:00

3.3. Hinnøya Energy Model: TIMES-Hinnøya

A first approximation of the RES is shown in Figure 3. The reference year for the model was 2015. The energy system of the island relies on imports of electricity from the main grid connection of the country of Norway, which is the Nordpool electricity market [54]. Fuel imports are mainly used for the transportation sector, with only diesel used for producing electricity to power the fish farming sector [51] and included in the primary sector. The residential sector consumes electricity, which pertains to households' use of electricity for heating, lighting, appliances, and other domestic activities. Economic activities are also present on the island, such as farming, wholesale, tourism, and kindergarten, and they mainly use electricity. Thus, the demand side covers the primary, secondary, and tertiary sectors. The electricity in Norway is hydro-based, and thus these sectors are already considered decarbonized. The remaining sector is transportation, and demand for mobility is disaggregated in the model into private, public, transport for merchandise, and marine transport, with a distinction between short and long distances.

According to Harstad municipality in their Climate Budget of 2021 [59], the transport sector makes up the majority of emissions (69% in 2018), which includes maritime transport (i.e., shipping; Figure 4). A possible measure to decrease these emissions is the deployment of low-emission transportation, hence contributing to the decarbonization of the island.

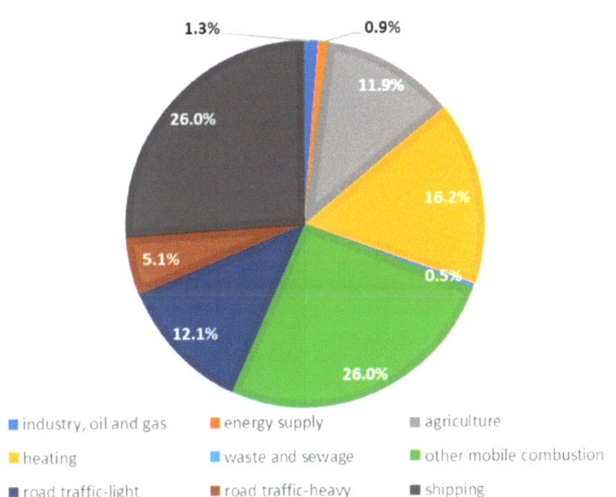

Figure 4. Sources of greenhouse gas emissions in Harstad in 2018 [59].

The evolution of the electricity supply with renewable energy, namely, hydroelectricity and wind, was based on the study of the Norwegian Water Resources and Energy Directorate (NVE), which shows that there is a potential for investing in new hydroelectricity power plants. The choice of technologies took into consideration on-shore and off-shore wind potential [14].

Moreover, the Norwegian government adopts several policies for the deployment of low-emission transportation. For instance, in 2025, all new purchases of passenger cars or light vans should be zero emission. Also, a taxation on fossil fuel in the transport sector is a central policy [60] and is implemented in TIMES-Hinnøya [61].

This taxation is applied in the following form:

$$\text{TAX} = C_{emission_i} \times \sum Emission_c$$

where $C_{emissionsCO2}$ = 50 Euros/tCO$_2$ and $C_{emissionsNOx}$ = 2280 Euros/tNOx [60], and c is the emitting commodity whose emission is defined in terms of kt/TJ, from [62].

This deployment also includes proper incentives and taxation rules. The new low-emission vehicles are exempted from tax on registration, traffic insurance, and road use tax and tolls ([3], found in [63]). Hence, one solution is the electrification of this sector, which is accelerated in the whole of Norway to reduce emissions and to benefit from the demand response mechanism. This is reflected on the local level, where numbers of battery electric vehicles (BEVs; in this case, private cars) are noticeably increasing in Harstad [51]. Consequently, the hypothesis of investing in battery electric vehicles for passenger car transportation (Table 2) is upheld, along with the examination of user flexibility through demand reduction or the shift to electricity. However, the model considers several modes of transportation (transport passenger cars, light and heavy duty, maritime transport, etc.). Transport passenger cars were chosen due to their potential to provide flexibility, with battery capacities of 15 kWh, 30 kWh, and 60 kWh. Based on the information provided in [51], the average commuting distance is 30 km and, therefore, the power consumed by these vehicles is sufficient for consideration. The BEVs are modeled in TIMES-Hinnøya as investment capacities in storage batteries, so their technical parameters are included in the declaration of technologies. The TIMES model invests in these technologies in a cost-efficient manner while optimizing the battery charging in such a way as to meet demand by the transport sector, which is provided as exogenous input. This allowed us to define the place of BEVs in the energy transition of the island. In addition, BEV chargers were included in the new technologies, including fixed and O&M costs. Storage technologies in the model consist of a hydrogen bromide flow battery [64]. However, this was not included in demand participation since it is likely to be piloted by the DSO of the island.

For the tariff structure, Norway adopts the RTP pricing scheme, which is widely accepted in this country, where 71% of households and 88% of small and medium enterprises and small industries use this scheme [24]. Also, it is adequate for customers to reduce their bills, as the heating sector is highly electrified in Norway, for instance, in residential homes, and the deployment of electric vehicles is on the rise, as mentioned earlier, so this pricing structure was implemented in the model. Electricity trade was calibrated for 2015 (the base year, Figure 5). Prices were obtained from the Nordpool database. NVE forecasts were used for future electricity prices ([65], found in [3]). The temporal structure in Figure 5 shows three levels, distinguishing between the months forming a year, weekdays and weekends, and finally, hours. In total, there were 576 time-steps that could be used for studying supply and demand in the long term [3].

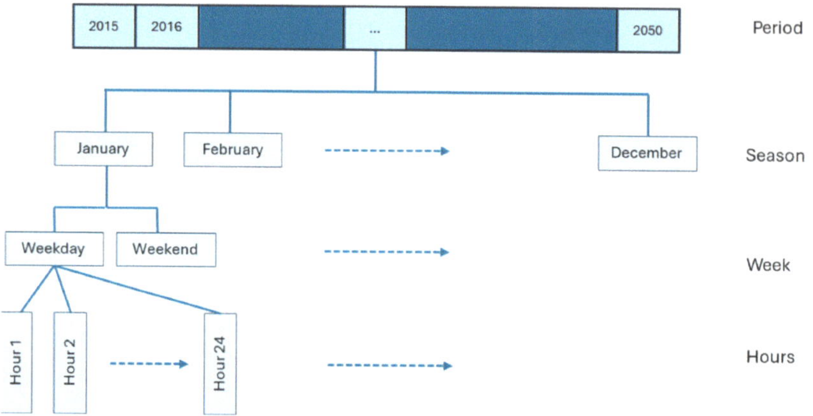

Figure 5. Seasons and time slices of TIMES-Hinnøya.

Table 3 summarizes the main assumptions included in the two models.

Table 3. Summary of the main assumptions of the models.

Assumptions	Procida	Hinnøya
Demand evolution	Italian GDP growth	Population growth for residential sector GDP growth for the remaining sectors Transport service demand
Price evolution	Constant (EUR 184.7/MWh) [66]	Variable on different timescales (hour, month, and year)
Horizon	2018–2050	2015–2050
Time slices	15 (3 seasons, 5 daytimes)	576 (12 months, weekday, weekend day, 24 h)
Discount rate	6%	6.50%
Currency	Euros (€)	Million euros (M€)
Regions	1 region (Procida)	3 regions (Harstad, Grytoya, and the rest of Hinnøya)

Table 3. *Cont.*

Assumptions	Procida	Hinnøya
Future technologies	Solar PV Storage: Li-ion *Smart Energy Hub* [67] (hydrogen electrolyzer, tank and fuel cell + Li-ion battery) Long-term hydrogen storage Electric vehicles and bikes	Hydro power Wind power Storage: Li-ion and flow battery (Hydrogen Bromide *Elestor* [64]) Electric transport
Import/export	Electricity imports Electricity exports not allowed	Electricity, fossil fuel (diesel and gasoline), biofuel, MGO Electricity exports allowed
Trade	N/A	Between the regions
Emissions	N/A	Emissions from fossil fuels

4. Results

In this section, the outcomes obtained from the TIMES optimization with the relevant scenarios are examined. For the case of Procida, solar PV and storage investments as part of the "self-consumption" measures were analyzed for the three scenarios. Then, a pricing structure based on the time of use was applied. For the case of Hinnøya, the transport sector was scrutinized, highlighting its electrification with renewable energy integration and demand-side load-shifting measures. Table 4 shows the establishment of testing used for validating these decarbonization approaches.

Table 4. Different cases set in the hypothesis for the two islands and the main results.

	Procida	Hinnøya
Objective: renewable integration and decarbonization	Solar PV integration. Reduced imports from mainland.	Decarbonization of the transport sector with RE increase. Reduced grid tension.
Implementation	Rooftop PV and storage investments. Electricity tariffs (TOU).	Electric vehicle charging structure. Electricity tariffs (RTP).
Hypothesis or test cases	Test case 1: Modest to high shares of PV with constant electricity prices: $Capacity\,PV_{High} \geq CapacityPV_t \geq Capacity\,PV_{low}$ Test case 2: Allowing investments in batteries + high PV shares + constant prices of electricity: $C_{max,BAT,s} = \overline{C_{BAT,s}} \cdot N_{B,s}$ Test case 3: Time-of-use structure + high PV + storage: $P_{peak} = (1 + 20\%) \times P_{off-peak}$ where P is the price of electricity	Test case 1: Taxation's impact on future passenger car deployment: $TAX = C_{emission_i} \times \sum Emission_c$ Test case 2: The introduction of EV and constant prices. ➤ Load curve variation analysis: $P = constant\ \forall\ i$ where P is the price of electricity and i is the hourly time slice. Test case 3: Introduction of EV with RTP (charging adaptation V1G). ➤ Load curve variation analysis

Table 4. *Cont.*

	Procida	**Hinnøya**
Results	Shift in the electricity supply with up to 10% decrease in imports at peak times in 2050.	Management of the additional electricity demand.
	Favoring storage technologies, with TOU, optimizes the use of RE and improves system flexibility (Section 4.1).	Preventing peak loads by load shifting to low-demand hours (1:00 am to 6:00 am and 7:00 pm to 12:00 am) (Section 4.2).

4.1. Procida Island

First, three scenarios were chosen to represent the integration of solar PV on building rooftops in Procida. The trend of these investments is shown in Table 5, where buildings' geographical constraints on PV installation were taken into account, as per [13]. The LOW scenario represents a case of modest PV, which follows the trend of current installations. Figure 6 shows the results obtained from the optimization of the model under the HIGH scenario, which was allowed to reach 300 kW of PV installed by mid-century. In terms of the optimal solution, the investments reached the allowable capacity in all the studied sectors.

Table 5. PV and storage investments according to three scenarios.

LOW Scenario			**HIGH and HIGH_STG Scenarios**		
PV investments:			PV investments:		
2018–2020	50	kW/year	2018–2020	50	kW/year
2020–2025	80	kW/year	2020–2025	150	kW/year
2020–2030	80	kW/year	2020–2030	200	kW/year
2030–2040	100	kW/year	2030–2040	250	kW/year
2040–2050	120	kW/year	2040–2050	300	kW/year
Storage investments: not included as possible technology.			Storage investments: allowed only for the HIGH_STG scenario.		

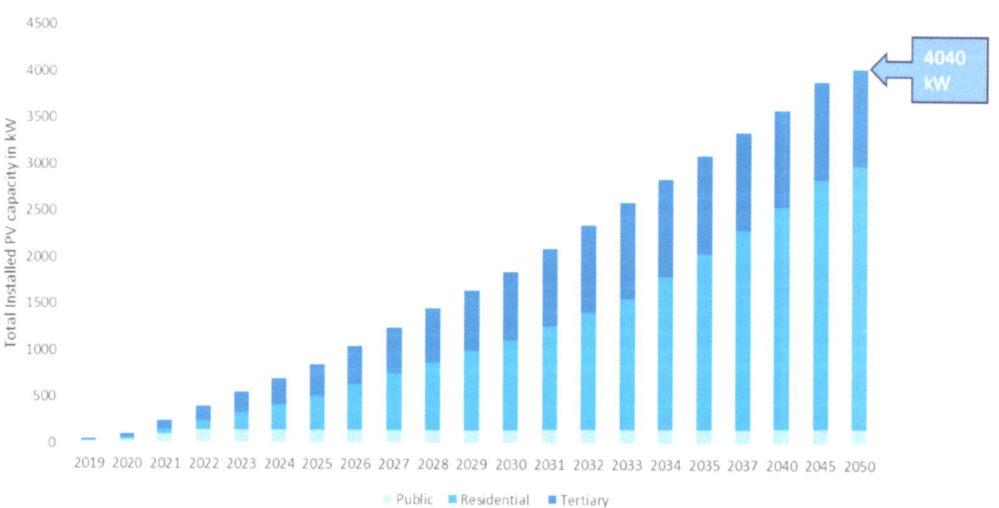

Figure 6. Total installed PV evolution for the HIGH and HIGH_STG scenarios.

The impact of the integration of self-consumption on the island's demand for imports (ELC Grid) is shown in Figure 7. With the HIGH scenario, the self-sufficiency of the island was enhanced, with reduced imports, especially during hours of PV generation. The benefit of coupling storage to PV was noticeable during peak hours, namely, in the afternoon and the evening.

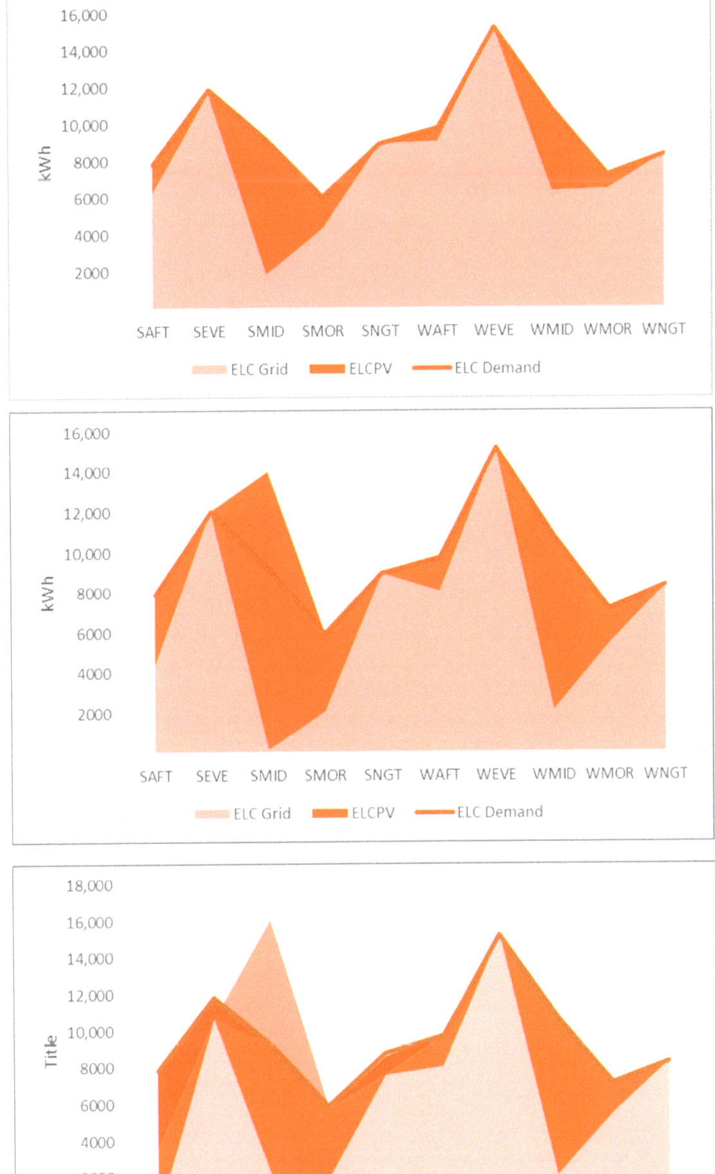

Figure 7. Daily load supply mix per time slice in 2050 for LOW, HIGH, and HIGH_STG scenarios and two seasons (S = Summer, W = Winter, I = Intermediate, NGT = Night, MOR = Morning, MID = Midday, AFT = Afternoon, and EVE = Evening).

It was also observed that the introduction of significant PV capacity, even when coupled with storage technologies, failed to mitigate imports during winter evenings. Nevertheless, as per local authorities, the highest demand was projected for the summer, a trend not mirrored in the existing data. This is mainly due to tourism activities, which cause congestion problems on the grid. Measures that can decrease imports during summer peak times are useful in this case. The increase in the shares of PVs or the decrease in imports directly relate to the decarbonization of the island, as the only two possible supply technologies are PV and imports. In the case of the HIGH scenario, during the summer midday time slice, it can be seen from Figure 7 that PV electricity covered all consumption, even in sectors where no PV was installed, such as agriculture and industry. When storage investment was allowed in the model, the electricity mix changed: PV electricity was stored for later use in the evening and afternoon when demand was high, involving long-term storage investment.

Next, the decision was made to incorporate dynamic electricity pricing, considering Italy's pioneering implementation of time-of-use (TOU) tariffs in 2010. This pricing reform became mandatory in mid-2010 for residential customers subscribing to basic contracts and touched 42% of the population [24]. It was chosen to start with a difference of 10% (set to be increased [68]) of the price between peak and off-peak (peak price being the highest) based on [24], since this is the only difference able to incentivize around 60% of Italian households to shift their power demand. According to [69], the Autorità di Regolazione per Energia Reti e Ambiente (ARERA) sets the different time slots for the pricing structure. There are three time slots, F1, F2, and F3, associated with a specific period of the day or an entire day. In view of this tariff scheme, and according to the model structure built in five time slices, a TOU can be implemented where the prices of imported electricity in the NGT and MOR are lower than in "peak hour" time, thus approaching the F2 and F3 slots for domestic users [69]. Note that in the current modeling structure, storage did not charge from the grid. This decision was mostly made to reduce reliance on the grid, as there are congestion problems involved in acquiring electricity through the submarine cable.

In this configuration, the renewable electricity price remained competitive on average. Thus, the peak price was further increased to a 20% difference between peak and off-peak, and the scenario including storage technologies was used. This scenario would allow the assessment of price-based mechanisms and a comparison between measures for integrating high shares of renewable energy in the grid. Figure 8 shows the impact of TOU on technology investments on the island: the model invested globally in higher capacities of Li-ion batteries compared to a scenario with a constant electricity price. This amounted to almost 1.5 times more total installed capacity when reaching the end of the horizon, compared to HIGH_STG, due to the appearance of investments in tertiary-dedicated batteries. Moreover, this led to increased investments in batteries for the residential sector earlier in the time horizon. In Figure 9, a decreasing pattern is seen in the share of electricity imports from the mainland with respect to the total supply, for the scenarios with constant and changing electricity prices. The difference between them can be observed until 2040. The share of electricity imports in the electricity supply reached 73.5% and 65.2% at the end of the horizon for HIGH_STG and HIGH_STG_TOU, respectively.

Conversely, the choice of investing in batteries changed the energy supply mix for the residential sector, since it is the most energy-intensive sector on the island. Figure 10 shows the charge and discharge of residential batteries, corresponding to a total installed capacity ranging from 320 kWh to 510 kWh (Figure 8).

In comparison with Figure 7 for the HIGH_STG scenario, the TOU pricing structure provides an alternative option to directly using PV. The strategy aims to meet demand with grid electricity when prices are low on winter mornings and store the electricity from PV to release later when both prices and electricity demand are higher, during the afternoon and evening. Summer midday remained interesting in all cases due to the availability of solar power.

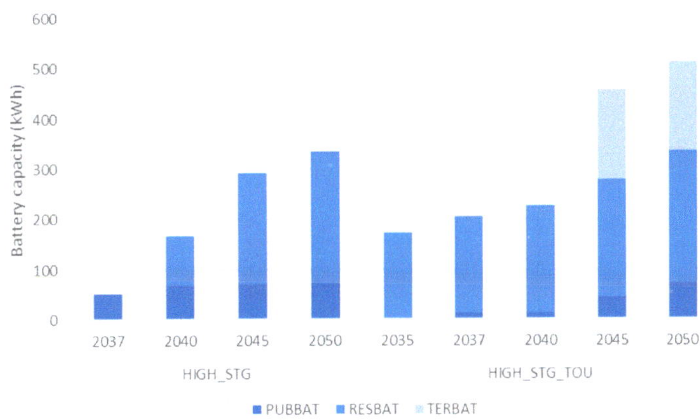

Figure 8. Comparison of the evolution of battery installed capacity for the two scenarios: HIGH_STG and HIGH_STG_TOU.

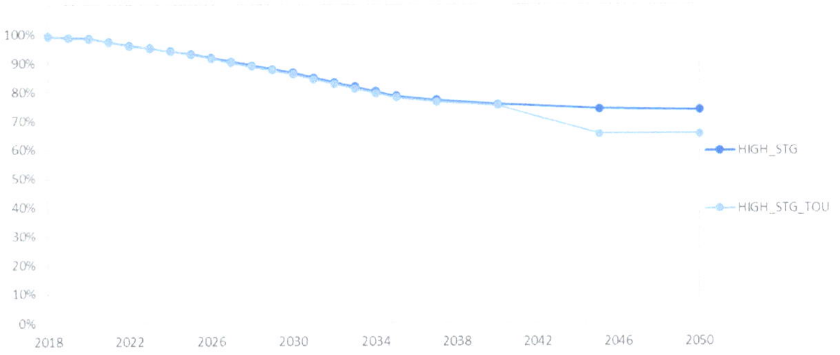

Figure 9. Absolute share of imports to the island (values compare the evolution of the shares of imports as a percentage between HIGH_STG and HIGH_STG_TOU scenarios).

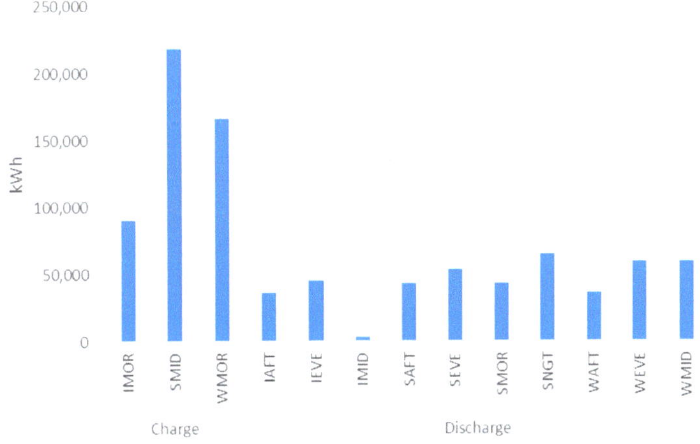

Figure 10. Storage action in the HIGH_STG_TOU scenario for the residential sector during 2050.

Due to the investments in storage, the electricity supply mix differed, especially for electricity provided by PV and batteries. In fact, for the residential and tertiary sectors, the annual production of PV directly injected into the respective sectors decreased, and more storage was used to inject electricity during high demand periods. At the system level, the TOU decreased imports at the peak hours (i.e., afternoon and evening time slices, Figure 11) in 2050 (−4% in the evening and −9.78% in the afternoon), while shifting the demand to the morning period.

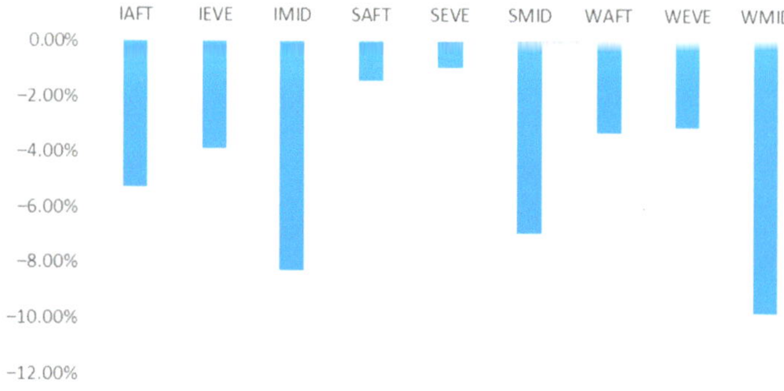

Figure 11. Electricity imports' reductions in 2050 per time slice.

A disaggregation of the time slices on a finer level would enhance this representation but is still not possible with the available data for this island. To sum up, this demonstrates one of the ways electricity storage can be used to provide flexibility and help relieve the grid during congestion periods.

4.2. Hinnøya Island

In terms of decarbonization, transport evolves toward electrification, with an increased share observed starting from 2025 (Figure 12). The country of Norway adopts taxation on fuel used by cars, which was implemented in the model by assigning a cost per kilo ton of energy used for CO_2 and nitrogen oxide (NOx) emissions related to the fuels used. In addition, diesel and gasoline are blended with a concentration of biofuel, denoted in Figure 12 as DSL blend and GSL blend, respectively. However, EVs remained the winners since no emissions are associated with this technology, considering the hydro-dominated electricity supply, with a total electrification at the end of the horizon.

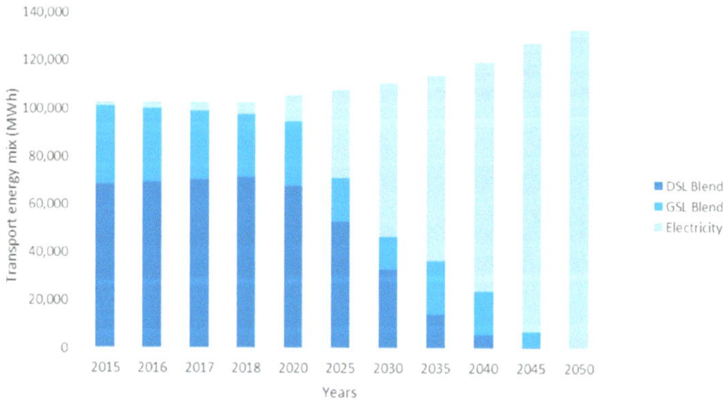

Figure 12. Transport passenger car mix throughout the horizon.

With the increase in electrification of transport passenger cars, opportunities for flexibility offered by this sector would present benefits for the grid. This can be carried out through demand-side management strategies applied to the charging of EVs. This translates into an economic approach used by TIMES, which chooses to charge the lowest electricity prices while meeting demand.

In addition, the optimization of TIMES performed on the RTP pricing structure identified the periods of charging (during low-electricity tariffs) while responding to mobility demand. The period and electricity costs corresponding to this structure are shown in Figure 13.

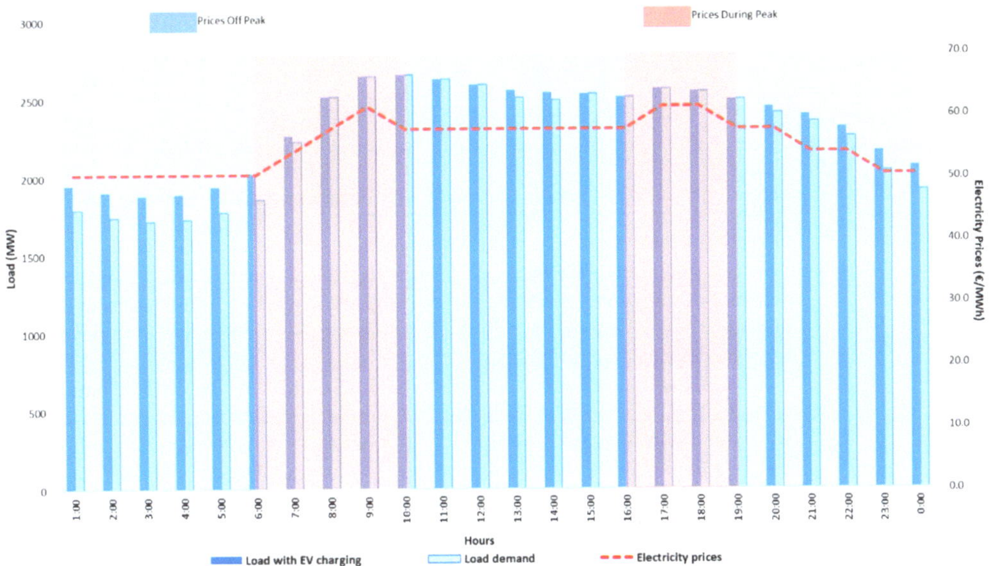

Figure 13. Peak and off-peak load variation with controlled EV charging (V1G)—winter working day in 2040.

The variation in the load curve (Figure 14) was analyzed if the prices were fixed throughout the day by taking the average prices for a day. It represents a fixed or "static" tariff structure, which lacks economic incentives for consumers to charge during off-peak hours (Figure 14). Note that the EV charging in that case will be carried out throughout the whole day, as shown by the "red" curve, responding to mobility demand but adding loads to peaks, which proved the projected increase in load due to the electrification of the transport sector without appropriate load management techniques. Nevertheless, the DSM mechanisms would benefit the grid since, for the "gray" curve, during the off-peak periods (1:00 am to 6:00 am and 7:00 pm to 12:00 am), the DSM strategy of EV charging consisted of the valley-filling phenomena. The shaded "gray" area shows that the valley-filling occurs upward in the off-peak and downward in the peak periods. At the country level, the increased electrification of the transport sector in Norway would be challenging to the grid [70]. Moreover, for the island of Hinnøya, new connection possibilities are limited (Table 1). The charging strategies, if they are controlled, provide flexibility so that electrification would be successful and lead to decarbonization of the highly emissive transport sector. No additional demand is created during peak hours, as seen in Figure 13.

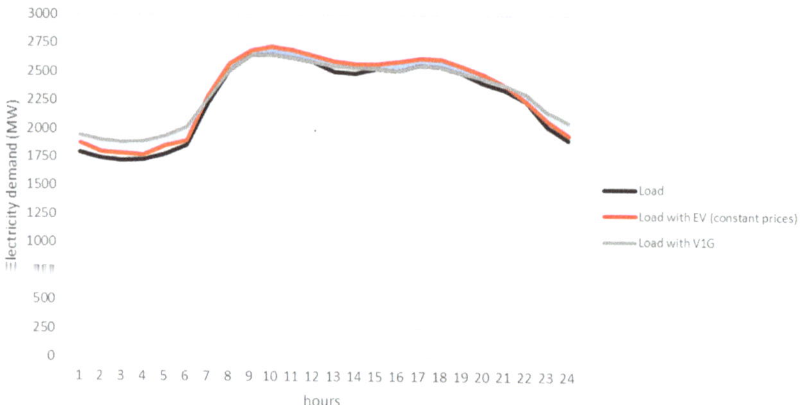

Figure 14. Load curves with and without EV charging for a typical working day in 2040.

5. Results Discussion

The possibilities for demand-side strategies in a context of decarbonizing the power systems of the two territories were investigated according to the issues and application environments where they are applied, which remains a major aspect in developing long-term power system analysis.

First, the results for Procida demonstrated that PV is cost-optimal, and that storage would alleviate imports of electricity at hours of high demand. The only way to harness PV potential in this case is rooftop PV, supporting "self-consumption". This is important because the island is "relatively isolated" in terms of its power system, as it relies only on submarine cables for its electricity supply. From the scenarios with price variation according to the time-of-use structure, the model changed the choice of investment, in terms of capacities of batteries. These investments are impacted by the different load curves of each sector where PV and storage systems would be installed. The tertiary sector sees its peak load ranging from midday to evening, where the price of electricity imports is still high, making it economically convenient to rely on batteries. However, for the public sector, peak demand starts in the morning in this scenario, and the MOR time slice (low tariff) coincides partly (for two hours) with high demand; hence, fewer batteries would be needed. The option of self-consumption (rooftop PV) decreases dependency on the grid, particularly for relatively isolated islands. Furthermore, this combination enhances the integration of solar-based electricity in the grid, thus decarbonizing the island's energy system. This structure is commonly used for managing the integration of high shares of RE, especially with high storage costs. However, it incurs costs for system operators that are not captured in this long-term planning. Short-term planning could harness the benefits of the demand variation due to varying electricity prices. Here, long-term planning was used to compare investments in new technologies that could reduce the reliance on the marine cables for the Italian case. The results demonstrate its contribution to increasing self-consumption as an active participation of the demand side.

One major challenge in decarbonizing the energy system of Hinnøya island relates to the transportation sector. It is necessary to increase the share of renewable energy for the island by harnessing emission-free electricity. However, this would increase the pressure on the grid. Then, the demand response would facilitate decarbonization. RTP, which is already installed, is one way to respond to the load. This pricing system makes it easier to manage additional demand created by the massive electrification of the transport sector. Also, it would provide a way to avoid the creation of peaks. The Norwegian case demonstrated an opportunity to decarbonize the transport sector, which is an important aspect of development for the island and the country. With the policies implemented, such as taxation on fuel, the optimal solution found relies on electric vehicles, although other options are available with biofuel. Modeling the whole energy system provides a global and precise indication of the transition of the transport sector by including the supply side with fuel imports, the blending processes with biofuel, and the distinction between traffic volumes and new market shares of petroleum, diesel, and electric vehicles.

The analysis of the decarbonization pathways with the active participation of the demand side showed that planning for the electrification of the emissive sector with RE is a common success factor. In fact, the same goal was studied for the two EU islands, although different approaches were required. To achieve this goal, one common practice is to establish the energy transition through new investments in technologies for electrification (solar PV, storage, and EV) and demand-side mechanisms. This is also supported by some policy instruments, such as fuel taxation and tax exemptions. The outcomes of the study demonstrated how each solution influences the other, for example, in the case of the TOU pricing structure and storage investments. The main characteristics outlining the outcome of the study are shown in Table 6. Therefore, the success of this transition is dependent on the coordination between solutions; here, renewable energy development and large-scale electrification. Further, other solutions can be adopted, such as energy efficiency, which might impact demand-side strategy outcomes. Moreover, the two territories present contrasting strucutures in terms of geography and challenges. For instance, reducing imports for Procida is in line with the energy autonomy of the island, while for Hinnøya, relying on imports has an important role in the decarbonization pathways of the island.

Table 6. Comparative summary of the outcomes.

	Procida	Hinnøya
Strategies analyzed	• Self-consumption • Electricity price TOU	• Eletric vehicle charging • Electricty price RTP
Implementation challenges	Renewable energy is limited • The small size of the population-dense island • Protected areas surrounding the territory	• Limited possibilities of new grid connections • Diesel generators to mitigate this issue
Success factors	• Local production of electricity with PV • Storage solutions applicable with rooftop PV	• Cross-sectoral flexibility solutions with the electrification of the transport sector with policy support • Load control
Policy implications and cross-regional lessons	• Interconnected systems that will see further integration of renewable energy • Socioeconomic impacts	

This study also examined building-interconnected energy systems and better integrated grid support in the framework of the EU's clean energy transition [71]. Indeed, microgrids or weakly connected grids often encounter similar technical challenges to those found on islands [72]. This underlines the cross-regional lessons learnt from the two case studies. Another lesson pointed out when demand-side strategies were applied was the impact of these solutions on sectors such as agriculture, commerce, etc., where participation is sometimes difficult to establish, as it affects the quality of the products.

One part of this double-territory analysis was to determine recommendations for the replicability of the analysis in other territories. Replication would require producing an energy profile of the islands, the assessment of the main energy and climate challenges alongside their long-term objectives, for instance, in decarbonization, the territory's implementation plans, and the potential roles and stakeholders of the participating sectors. It was also noted that some factors determine the employment of solutions, such as storage. This depends on the type of interconnection between the island and the mainland system, the amount of RES installed or foreseen, and the size of the island in terms of population. For the case of available interconnection, the choice of storage solution relies on information about the energy prices and the grid's resilience, and the need to reduce the use of and dependency on conventional fuels. Another important aspect for relatively isolated islands is imports of electricity, which, in terms of decarbonization, remain the responsibility of the main country to reduce their emissions.

The demand-side strategies evoked in this paper are dependent on consumers' engagement in three dimensions: investing in new and enabling technologies, enrolling in programs, and actively reducing/shifting loads [73]. As part of the climate mitigation solutions highlighted by the IPCC [74], demand in energy plays a key role in achieving the Paris Agreement. Demand-side strategies would incentivize consumers but require their participation for their success. For example, instead of participating by charging their vehicles, consumers might opt for public transportation. This remains in line with decarbonization strategies if proper policies are implemented. Nevertheless, behavior change studies are particularly beneficial for the transportation sector [75]. These studies would induce "behavioral realism in energy models" and capture the adoption of "novel technologies" through two main methods. The first involves endogenously representing modal choices and infrastructure availability, driving patterns or new mobility trends in the energy model. The second involves linking/coupling between models, which requires data exchanges. Representing consumers' behavior would account for the preferences and incomes of consumers and actors and can be provided by general equilibrium models, such as GEM-E3 [76]. The monetary value of the strategies would also play an important role in their adoption. Models such as TIMES can be linked to lifestyle and macroeconomic models [77], which capture the overall energy mix and emissions output. Such linkings highlight the interdependencies between economy, technology, and lifestyle, but do not capture the impact of strategies (carbon tax, electricity pricing, etc.) on consumers' reactions. Nevertheless, dynamic models, such as system dynamics (SD) and agent-based models (ABMs), can include feedback loops that cause endogenous changes in the system based on the decisions of the actors.

Finally, there were certain limitations evident in the results, stemming from data availability, which in turn impacted the selection of time slices in the TIMES models. Disaggregated data on the level of hours are helpful to follow the change in the load curve due to the applied strategies, but they increase the computation time. Moreover, real data that reflect the situation of the island enable a more specific judgement of the solutions (for instance, sector-specific data and seasonal demand variations). In addition, energy models often overlook non-economic dimensions, such as indirect emissions, critical raw materials, land availability issues, and impacts on land, water, and biodiversity, limiting their effectiveness in informing energy policy [78]. Implementing such factors affords more credibility to the results obtained from energy models. In fact, investment in energy projects in the EU sometimes receives social resistance, which delays their implementation. It is suggested to depict and endogenize factors such as regional density and the sizing of existing windfarms and make them drivers of the analysis. Concerning environmental factors, it is possible to manipulate life-cycle data in Python environments and thus obtain life-cycle assessment (LCA) calculations, as seen in the PREMISE model, which adapts to various future scenarios of renewable energy use and technological advancements. Other ways to improve policy evaluation include the agent-based technology adoption model (ATOM), which revealed the shortcomings of Greece's net-metering policy, demonstrating its inability to meet future photovoltaic targets, a finding that would have remained undetected with traditional system optimization models. In [79], the design and implementation of Floating Photovoltaic (FPV) systems with storage was evaluated for three representative islands in Indonesia considering techno-economic–socio-environmental impacts.

The research addressed the complex balance of benefits and challenges associated with FPV installations, highlighting the potential impacts on water conservation, algae control, and marine ecosystems. It also focused on understanding the social influence of floating photovoltaic systems, highlighting the economic activities, roles of men and women, and challenges faced due to the lack of electricity infrastructure.

6. Conclusions

Throughout this study, three types of enablers were the focal point for studying the integration of renewable energy and decarbonization: demand response, self-consumption, and control of electric vehicle charging. This strategy presents benefits for the flexibility of the energy system, an important factor for its operators when balancing supply and demand.

To address these issues and to handle the evolution of the two relatively isolated power systems studied, while minimizing the costs of investment and implementing political and environmental constraints, the bottom-up TIMES model generator approach was applied. This involved TIMES-Procida for the Italian island of Procida and TIMES-Hinnøya for the Norwegian island of Hinnøya. The prospective analysis highlighted some key points for consideration in planning the evolution of the energy systems of these two islands, in terms of investments and policies that align with their local plans. The findings can be used on a broader level, which identifies possible opportunities for replicability. For instance, the decarbonization of energy systems is facilitated by flexibility solutions and mechanisms in different end-use sectors, hence creating synergies between the sectors and finally allowing increased renewable energy integration. Electricity storage is effective in managing electricity supply, particularly with intermittent renewables, and can be applied across various power system sectors. Electric vehicles also offer advantages for renewable integration on islands, and to reduce pollution, as they have low emissions and enable flexibility from prosumers, though demand-side participation is currently limited but promising for reducing grid congestion and costly investments.

This study, based on long-term energy planning, also shed light on the local energy needs of islands, which require particular attention when it comes to strategically planning for their future development and monitoring their energy transition. The topics of energy and climate are usually handled at the national level but need to be addressed differently when it comes to islands. From the analysis, the characteristics requiring attention for such territories are the spatial and grid constraints, demographics, and policies.

Nevertheless, public willingness to participate in flexibility measures is a determining aspect for obtaining effective results. The presented conclusions would impact choices and encourage the active participation of the demand side and are relevant for all relatively isolated islands; however, practical on-the-ground methods are needed. For instance, in [80], it was proven that decision-making by end-users would rely on incentives and coordination with other stakeholders. In [81], solutions such as Virtual Power Plants would allow such coordination and valorization of the flexibility provided by the participants.

Author Contributions: Conceptualization, S.C., S.S. and N.M.; methodology, S.C. and S.S.; formal analysis, S.C.; investigation, S.C.; resources, S.C., S.S. and N.M.; data curation, S.C. and S.S.; writing—original draft preparation, S.C.; writing—review and editing, S.C. and S.S.; visualization, S.C.; supervision, S.S. and N.M.; project administration, S.S. and N.M.; funding acquisition, N.M. All authors have read and agreed to the published version of the manuscript.

Funding: This research was supported in the framework of the GIFT project (Geographical Islands FlexibiliTy), an innovative project, as part of the H2020 research program of the European Union, under grant agreement No. 824410. This research was also supported by the Modeling for Sustainable Development Chair. It is driven by Mines Paris—PSL and École des Ponts ParisTech, with support from ADEME, EDF, GRTgaz, RTE, SCHNEIDER ELECTRIC, TotalEnergies, and the French Ministry of Ecological and Solidarity Transition. The views expressed in this paper or any public documents linked to the research program are attributable only to the authors in their personal capacity and not to the funders.

Data Availability Statement: The original contributions presented in the study are included in the article, further inquiries can be directed to the corresponding authors.

Acknowledgments: The authors want to thank the partners in the GIFT project for their collaboration in terms of input data and insights.

Conflicts of Interest: The authors declare no conflicts of interest.

References

1. European Commission. Renewable Energy Targets. Available online: https://energy.ec.europa.eu/topics/renewable-energy/renewable-energy-directive-targets-and-rules/renewable-energy-targets_en (accessed on 26 March 2024).
2. European Commission; Joint Research Centre. *The POTEnCIA Central Scenario: An EU Energy Outlook to 2050*; Publications Office: Luxembourg, 2019.
3. Chlela, S.; Grazioli, G.; Marchal, F.; Selosse, S.; Maizi, N. Technological Scenarios and Recommendations (GIFT Deliverable 2.4). 2021. Available online: https://www.gift-h2020.eu/wp-content/uploads/2021/07/GIFT_Deliverable-2_4_v5.pdf (accessed on 26 August 2021).
4. Guerassimoff, G.; Maizi, N.; Mastère, O.S.E. *Iles et Energies: Un Paysage de Contrastes*; Presses des MINES: Paris, France, 2008.
5. European Union. Clean Energy for All Europeans. 2019. Available online: https://energy.ec.europa.eu/topics/energy-strategy/clean-energy-all-europeans-package_en (accessed on 1 January 2024).
6. European Commission. Clean Energy for EU Islands. Energy—European Commission, 2017. Available online: https://ec.europa.eu/energy/topics/markets-and-consumers/clean-energy-eu-islands_en (accessed on 13 March 2021).
7. Selosse, S.; Ricci, O.; Garabedian, S.; Maïzi, N. Exploring sustainable energy future in Reunion Island. *Util. Policy* **2018**, *55*, 158–166. [CrossRef]
8. Marczinkowski, H.M.; Alberg Østergaard, P.; Roth Djørup, S. Transitioning Island Energy Systems—Local Conditions, Development Phases, and Renewable Energy Integration. *Energies* **2019**, *12*, 3484. [CrossRef]
9. IRENA. *Transforming Small-Island Power Systems*; IRENA: Masdar City, United Arab Emirates, 2018; 150p.
10. IRENA. Antigua and Barbuda: Renewable Energy Roadmap. 2021. Available online: https://www.irena.org/Publications/2021/March/Antigua-and-Barbuda-Renewable-Energy-Roadmap (accessed on 11 March 2021).
11. Katsaprakakis, D.A.; Thomsen, B.; Dakanali, I.; Tzirakis, K. Faroe Islands: Towards 100% R.E.S. penetration. *Renew. Energy* **2019**, *135*, 473–484. [CrossRef]
12. Marczinkowski, H.M.; Østergaard, P.A.; Mauger, R. Energy transitions on European islands: Exploring technical scenarios, markets and policy proposals in Denmark, Portugal and the United Kingdom. *Energy Res. Soc. Sci.* **2022**, *93*, 102824. [CrossRef]
13. Grazioli, G.; Chlela, S.; Selosse, S.; Maïzi, N. The Multi-Facets of Increasing the Renewable Energy Integration in Power Systems. *Energies* **2022**, *15*, 6795. [CrossRef]
14. Zhou, W.; Hagos, D.A.; Stikbakke, S.; Huang, L.; Cheng, X.; Onstein, E. Assessment of the impacts of different policy instruments on achieving the deep decarbonization targets of island energy systems in Norway—The case of Hinnøya. *Energy* **2022**, *246*, 123249. [CrossRef]
15. Cochran, J.; Miller, M.; Zinaman, O.; Milligan, M.; Arent, D.; Palmintier, B.; O'Malley, M.; Mueller, S.; Lannoye, E.; Tuohy, A.; et al. *Flexibility in 21st Century Power Systems*; National Renewable Energy Lab. (NREL): Golden, CO, USA, 2014. [CrossRef]
16. Mohler, D.; Sowder, D. Chapter 23—Energy Storage and the Need for Flexibility on the Grid. In *Renewable Energy Integration*; Jones, L.E., Ed.; Academic Press: Boston, MA, USA, 2014; pp. 285–292. [CrossRef]
17. Lund, P.D.; Lindgren, J.; Mikkola, J.; Salpakari, J. Review of energy system flexibility measures to enable high levels of variable renewable electricity. *Renew. Sustain. Energy Rev.* **2015**, *45*, 785–807. [CrossRef]
18. Paterakis, N.G.; Erdinç, O.; Catalão, J.P.S. An overview of Demand Response: Key-elements and international experience. *Renew. Sustain. Energy Rev.* **2017**, *69*, 871–891. [CrossRef]
19. Iqbal, S.; Sarfraz, M.; Ayyub, M.; Tariq, M.; Chakrabortty, R.K.; Ryan, M.J.; Alamri, B. A Comprehensive Review on Residential Demand Side Management Strategies in Smart Grid Environment. *Sustainability* **2021**, *13*, 7170. [CrossRef]
20. Eurelectric. *Eurelectric Views on Demand-Side Participation*; Eurelectric: Brussels, Belgium, 2011.
21. ADEME. *Effacement de Consommation Electrique en France*. 2017. Available online: https://librairie.ademe.fr/energies-renouvelables-reseaux-et-stockage/1772-effacement-de-consommation-electrique-en-france.html (accessed on 1 March 2024).
22. IEA-International Energy Agency. Demand Response—Analysis. IEA. Available online: https://www.iea.org/reports/demand-response (accessed on 22 October 2020).
23. Xu, Z. *The Electricity Market Design for Decentralized Flexibility Sources*; Oxford Institute for Energy Studies: Oxford, UK, 2019. [CrossRef]
24. Eurelectric. Dynamic Pricing in Electricity Supply. 2018. Available online: https://www.eemg-mediators.eu/downloads/dynamic_pricing_in_electricity_supply-2017-2520-0003-01-e.pdf (accessed on 14 November 2020).
25. Interreg Europe. Renewable Energy Self-Consumption. 2020. Available online: https://www.interregeurope.eu/find-policy-solutions/policy-briefs/energy-self-consumption (accessed on 29 July 2021).
26. IRENA. *Renewable Power Generation Costs in 2018*; International Renewable Energy Agency: Abu Dhabi, United Arab Emirates, 2019. Available online: https://www.irena.org/publications/2019/May/Renewable-power-generation-costs-in-2018 (accessed on 1 March 2024).

27. Peiseler, L.; Wood, V.; Schmidt, T.S. Reducing the carbon footprint of lithium-ion batteries, what's next? *Next Energy* **2023**, *1*, 100017. [CrossRef]
28. IRENA. Power System Flexibility for the Energy Transition. 2018. Available online: https://www.irena.org/publications/2018/Nov/Power-system-flexibility-for-the-energy-transition (accessed on 21 January 2021).
29. Knezović, K. Active Integration of Electric Vehicles in the Distribution Network—Theory, Modelling and Practice. Ph.D. Thesis, Technical University of Denmark, Kongens Lyngby, Denmark, 2016; 229p.
30. Babrowski, S.; Heffels, T.; Jochem, P.; Fichtner, W. Reducing computing time of energy system models by a myopic approach. *Energy Syst.* **2014**, *5*, 65–83. [CrossRef]
31. Javed, H.; Muqeet, H.A.; Shehzad, M.; Jamil, M.; Khan, A.A.; Guerrero, J.M. Optimal Energy Management of a Campus Microgrid Considering Financial and Economic Analysis with Demand Response Strategies. *Energies* **2021**, *14*, 8501. [CrossRef]
32. Prina, M.G.; Manzolini, G.; Moser, D.; Nastasi, B.; Sparber, W. Classification and challenges of bottom-up energy system models—A review. *Renew. Sustain. Energy Rev.* **2020**, *129*, 109917. [CrossRef]
33. Sánchez Diéguez, M.; Fattahi, A.; Sijm, J.; Morales España, G.; Faaij, A. Modelling of decarbonisation transition in national integrated energy system with hourly operational resolution. *Adv. Appl. Energy* **2021**, *3*, 100043. [CrossRef]
34. Desport, L.; Selosse, S. An overview of CO_2 capture and utilization in energy models. *Resour. Conserv. Recycl.* **2022**, *180*, 106150. [CrossRef]
35. Joint Research Centre (European Commission); Després, J.; Keramidas, K.; Schmitz, A.; Kitous, A.; Schade, B.; Diaz Vazquez, A.; Mima, S.; Russ, H.; Wiesenthal, T. *OLES-JRC Model Documentation: 2018 Update*; Publications Office of the European Union: Luxembourg, 2018. Available online: https://publications.jrc.ec.europa.eu/repository/handle/JRC113757 (accessed on 1 March 2023).
36. IEA-ETSAP | Times. Available online: https://iea-etsap.org/index.php/etsap-tools/model-generators/times (accessed on 16 January 2022).
37. LEAP. Available online: https://leap.sei.org/ (accessed on 19 January 2022).
38. Stockholm Environment Institute. SEI. Available online: https://www.sei.org/ (accessed on 19 January 2022).
39. MESSAGE-IIASA. Available online: https://docs.messageix.org/en/latest/ (accessed on 3 March 2024).
40. IIASA-International Institute for Applied Systems Analysis. Available online: http://iiasa.ac.at/ (accessed on 19 January 2022).
41. Barnes, T.; Shivakumar, A.; Brinkerink, M.; Niet, T. OSeMOSYS Global, an open-source, open data global electricity system model generator. *Sci. Data* **2022**, *9*, 623. [CrossRef]
42. KTH Royal Institute of Technology in Stockholm. KTH. Available online: https://www.kth.se/en (accessed on 19 January 2022).
43. Brown, T.; Hörsch, J.; Schlachtberger, D. PyPSA: Python for Power System Analysis. *JORS* **2018**, *6*, 4. [CrossRef]
44. Brown, T.; Victoria, M.; Zeyen, E.; Hofmann, F.; Neumann, F.; Frysztacki, M.; Hampp, J.; Schlachtberger, D.; Hörsch, J. PyPSA-Eur: An Open Sector-Coupled Optimisation Model of the European Energy System. 2023. Available online: https://pypsa-eur.readthedocs.io/en/latest/ (accessed on 1 March 2024).
45. Loulou, R.; Goldstein, G.; Kanudia, A.; Lettila, A.; Remme, U. Documentation for the TIMES Model-Part I. 2016. Available online: https://iea-etsap.org/docs/Documentation_for_the_TIMES_Model-Part-I_July-2016.pdf (accessed on 25 March 2021).
46. Drouineau, M.; Assoumou, E.; Mazauric, V.; Maïzi, N. Increasing shares of intermittent sources in Reunion Island: Impacts on the future reliability of power supply. *Renew. Sustain. Energy Rev.* **2015**, *46*, 120–128. [CrossRef]
47. Remme, U. Overview of TIMES: Parameters, Primal Variables & Equations. Available online: https://iea-etsap.org/workshop/brazil_11_2007/times-remme.pdf (accessed on 1 March 2024).
48. Selosse, S.; Garabedian, S.; Ricci, O.; Maïzi, N. The renewable energy revolution of reunion island. *Renew. Sustain. Energy Rev.* **2018**, *89*, 99–105. [CrossRef]
49. Institute for Prospective Technological Studies (Joint Research Centre); Van Regemorter, D.; Perry, M.; Capros, P.; Ciscar, J.-C.; Paroussos, L.; Pycroft, J.; Karkatsoulis, P.; Abrell, J.; Saveyn, B. *GEM-E3 Model Documentation*; Publications Office of the European Union: Luxembourg, 2013. Available online: https://data.europa.eu/doi/10.2788/47872 (accessed on 1 March 2024).
50. Italian National Institute of Statistics. Istat Statistics. 2019. Available online: http://dati.istat.it/?lang=en (accessed on 24 August 2021).
51. HLK. 2020. Available online: https://www.gift-h2020.eu/wp-content/uploads/2021/02/D7.1-Report-on-Requirement-and-Prosumer-Analysis-with-installation-project-documentation.pdf (accessed on 23 August 2021).
52. 11342: Population and Area (M) 2007-2020-PX-Web SSB. SSB. Available online: https://www.ssb.no/en/befolkning/folketall/statistikk/tettsteders-befolkning-og-areal (accessed on 1 March 2024).
53. Electricity. SSB. Available online: https://www.ssb.no/en/energi-og-industri/energi/statistikk/elektrisitet (accessed on 13 October 2021).
54. Nordpool. List of Changes in Day-Ahead and Intraday Areas. 2008. Available online: https://www.nordpoolgroup.com/4a7b04/globalassets/download-center/day-ahead/elspot-area-change-log.pdf (accessed on 13 October 2021).
55. Italy 2023–Analysis. IEA, 2023. Available online: https://www.iea.org/reports/italy-2023 (accessed on 1 May 2024).
56. Auctions (FER1 Decree) | Clean Energy for EU Islands. Available online: https://clean-energy-islands.ec.europa.eu/countries/italy/legal/res-electricity/auctions-fer1-decree (accessed on 1 May 2024).
57. Bertsch, V.; Devine, M.T.; Sweeney, C.; Parnell, A.C. *Analysing Long-Term Interactions between Demand Response and Different Electricity Markets Using a Stochastic Market Equilibrium Model*; ESRI Working Paper; The Economic and Social Research Institute (ESRI): Dublin, Ireland, 2018.
58. Norway 2022—Analysis-IEA. Available online: https://www.iea.org/reports/norway-2022 (accessed on 1 May 2024).

59. Harstad Municipality. Klimabudsjett-Harstad Commune. 2021. Available online: https://www.harstad.kommune.no/klimabudsjett.557657.no.html (accessed on 24 August 2021).
60. Excise Duties. The Norwegian Tax Administration. Available online: https://www.skatteetaten.no/en/business-and-organisation/vat-and-duties/excise-duties/ (accessed on 2 May 2024).
61. Norwegian Ministry of Transport and Communications. National Transport Plan 2018–2029. 2018. Available online: https://www.regjeringen.no/contentassets/7c52fd2938ca42209e4286fe86bb28bd/en-gb/pdfs/stm201620170033000engpdfs.pdf (accessed on 3 May 2021).
62. IPCC. IPCC Guidelines for National Greenhouse Gas Inventories. 2006. Available online: https://www.ipcc-nggip.iges.or.jp/public/2006gl/ (accessed on 2 May 2024).
63. Norwegian Ministry of Climate and Environment. Norway's Fourth Biennial Report 2020, under the Framework Convention on Climate Change. 2020. Available online: https://unfccc.int/sites/default/files/resource/Norway_BR4%20(2).pdf (accessed on 24 August 2021).
64. Elestor. Homepage. Elestor. Available online: https://www.elestor.nl/ (accessed on 17 January 2022).
65. NVE. Langsiktig Kraftmarkedsanalyse 2020–2040. 2020. Available online: https://publikasjoner.nve.no/rapport/2020/rapport2020_37.pdf (accessed on 21 June 2021).
66. ARERA-Relazione Annuale. Available online: https://www.arera.it/it/relaz_ann/19/19.htm (accessed on 20 September 2023).
67. Sylfen. Available online: https://www.sylfen.com/ (accessed on 23 July 2021).
68. Evaluation of the Effects of a Tariff Change on the Italian Residential Customers Subject to a Mandatory Time-of-Use Tariff. Available online: https://www.eceee.org/library/conference_proceedings/eceee_Summer_Studies/2013/7-monitoring-and-evaluation/evaluation-of-the-effects-of-a-tariff-change-on-the-italian-residential-customers-subject-to-a-mandatory-time-of-use-tariff/ (accessed on 28 March 2024).
69. Confronto Tariffe Luce Enel | Enel Energia. Available online: https://www.enel.it/content/enel-it/en/supporto/faq/fasce-orarie-energia-elettrica-cosa-sono (accessed on 23 August 2021).
70. Sørensen, Å.L.; Jiang, S.; Torsæter, B.N.; Völler, S. *Smart EV Charging Systems for Zero Emission Neighbourhoods*; SINTEF akademisk forlag: Oslo, Norway, 2018; Available online: https://ntnuopen.ntnu.no/ntnu-xmlui/handle/11250/2504976 (accessed on 1 March 2024).
71. Energy and the Green Deal. 2022. Available online: https://commission.europa.eu/strategy-and-policy/priorities-2019-2024/european-green-deal/energy-and-green-deal_en (accessed on 29 April 2024).
72. EASE. EASE Study on Power System Challenges of Islands and Isolated Systems with High Shares of Variable Renewables | EASE. Why Energy Storage? 2020. Available online: https://ease-storage.eu/publication/ease-study-on-power-system-challenges-of-islands-and-isolated-systems-with-high-shares-of-variable-renewables/ (accessed on 21 January 2021).
73. Sloot, D.; Lehmann, N.; Ardone, A.; Fichtner, W. A Behavioral Science Perspective on Consumers' Engagement with Demand Response Programs. *Energy Res. Lett.* **2023**, *4*. [CrossRef]
74. IPCC. AR6 Climate Change 2022: Mitigation of Climate Change. 2022. Available online: https://www.ipcc.ch/report/sixth-assessment-report-working-group-3/ (accessed on 8 April 2022).
75. Luh, S.; Kannan, R.; Schmidt, T.J.; Kober, T. Behavior matters: A systematic review of representing consumer mobility choices in energy models. *Energy Res. Soc. Sci.* **2022**, *90*, 102596. [CrossRef]
76. E3 Modelling. GEM-E3–E3 Modelling. Available online: https://e3modelling.com/modelling-tools/gem-e3/ (accessed on 27 April 2024).
77. Millot, A.; Doudard, R.; Le Gallic, T.; Briens, F.; Assoumou, E.; Maïzi, N. France 2072: Lifestyles at the Core of Carbon Neutrality Challenges. In *Limiting Global Warming to Well Below 2 °C: Energy System Modelling and Policy Development*; Giannakidis, G., Karlsson, K., Labriet, M., Gallachóir, B.Ó., Eds.; Springer International Publishing: Cham, Switzerland, 2018; pp. 173–190. [CrossRef]
78. Süsser, D.; Martin, N.; Stavrakas, V.; Gaschnig, H.; Talens-Peiró, L.; Flamos, A.; Madrid-López, C.; Lilliestam, J. Why energy models should integrate social and environmental factors: Assessing user needs, omission impacts, and real-word accuracy in the European Union. *Energy Res. Soc. Sci.* **2022**, *92*, 102775. [CrossRef]
79. Esparza, I.; Olábarri Candela, Á.; Huang, L.; Yang, Y.; Budiono, C.; Riyadi, S.; Hetharia, W.; Hantoro, R.; Setyawan, D.; Utama, I.K.A.P.; et al. Floating PV Systems as an Alternative Power Source: Case Study on Three Representative Islands of Indonesia. *Sustainability* **2024**, *16*, 1345. [CrossRef]
80. Askeland, M.; Backe, S.; Bjarghov, S.; Lindberg, K.B.; Korpås, M. Activating the potential of decentralized flexibility and energy resources to increase the EV hosting capacity: A case study of a multi-stakeholder local electricity system in Norway. *Smart Energy* **2021**, *3*, 100034. [CrossRef]
81. Gift H2020 | Geographical Islands Flexibility. Available online: https://www.gift-h2020.eu/#overview (accessed on 11 May 2022).

Article

Optimization of Operation Strategy of Multi-Islanding Microgrid Based on Double-Layer Objective

Zheng Shi [1], Lu Yan [2], Yingying Hu [1], Yao Wang [1], Wenping Qin [3,*], Yan Liang [1,4], Haibo Zhao [1], Yongming Jing [1], Jiaojiao Deng [1] and Zhi Zhang [1]

[1] Economic and Technical Research Institute of State Grid Shanxi Electric Power Company, Taiyuan 030021, China; shizheng_sgcc@163.com (Z.S.); luobo_hu@126.com (Y.H.); wangyaoncepu@163.com (Y.W.); liangyan_lyly@163.com (Y.L.); zhb8711@163.com (H.Z.); wwwjym001@163.com (Y.J.); svfya@sina.com (J.D.); zapletree@163.com (Z.Z.)

[2] Yingda Chang'an Insurance Brokers Co., Ltd., Taiyuan 030021, China; yanlu1997@163.com

[3] Electrical and Power Engineering, Taiyuan University of Technology, Taiyuan 030024, China

[4] Electrical and Electronic Engineering, North China Electric Power University, Beijing 102206, China

* Correspondence: qinwenping@tyut.edu.cn; Tel.: +86-1333-340-7566

Abstract: The shared energy storage device acts as an energy hub between multiple microgrids to better play the complementary characteristics of the microgrid power cycle. In this paper, the cooperative operation process of shared energy storage participating in multiple island microgrid systems is researched, and the two-stage research on multi-microgrid operation mode and shared energy storage optimization service cost is focused on. In the first stage, the output of each subject is determined with the goal of profit optimization and optimal energy storage capacity, and the modified grey wolf algorithm is used to solve the problem. In the second stage, the income distribution problem is transformed into a negotiation bargaining process. The island microgrid and the shared energy storage are the two sides of the game. Combined with the non-cooperative game theory, the alternating direction multiplier method is used to reduce the shared energy storage service cost. The simulation results show that shared energy storage can optimize the allocation of multi-party resources by flexibly adjusting the control mode, improving the efficiency of resource utilization while improving the consumption of renewable energy, meeting the power demand of all parties, and realizing the sharing of energy storage resources. Simulation results show that compared with the traditional PSO algorithm, the iterative times of the GWO algorithm proposed in this paper are reduced by 35.62%, and the calculation time is shortened by 34.34%. Compared with the common GWO algorithm, the number of iterations is reduced by 18.97%, and the calculation time is shortened by 22.31%.

Keywords: microgrid; shared energy storage; optimize scheduling; negotiation price; dual objective

Citation: Shi, Z.; Yan, L.; Hu, Y.; Wang, Y.; Qin, W.; Liang, Y.; Zhao, H.; Jing, Y.; Deng, J.; Zhang, Z. Optimization of Operation Strategy of Multi-Islanding Microgrid Based on Double-Layer Objective. *Energies* **2024**, *17*, 4614. https://doi.org/10.3390/en17184614

Academic Editor: Angela Russo

Received: 19 July 2024
Revised: 4 September 2024
Accepted: 12 September 2024
Published: 14 September 2024

1. Introduction

Currently, China is building a clean and efficient new power system around the form of new energy power generation. As various distributed power generation units continue to integrate into the grid, the generation of electricity from these units exhibits inherent fluctuations and unpredictability, posing challenges to the safety and stability of the overall system operation. The generation of electric energy by distributed generation equipment has the problem of fluctuation and randomness, which will also reduce the safety and stability of system operation, and the diversification of power supply demand is difficult to meet [1]. To enhance the dependability of electricity provision and address the challenge of electricity accessibility in remote or isolated regions, the concept of a microgrid is proposed [2,3] using the Stackelberg game to solve the indeterminacy of renewable energy power generation, energy storage devices, and load. Electric vehicles are introduced into the power system as mobile energy storage to participate in regulation. Literature [4] focuses on a general two-population n-strategy evolutionary game (2PnS-EG) and applies it to investigate the evolutionary equilibrium properties of long-term strategic bidding issues in deregulated homogeneous and heterogeneous power generation-side markets (PGMs) under different market clearing mechanisms. In order to obtain an optimal joint planning and cost allocation strategy for multiple park-level integrated energy systems with shared energy storage. Literature [5] proposes a joint planning and cost allocation method for multiple park-level integrated energy systems with shared energy storage. Literature [6] focuses on the general N-population multi-strategy evolutionary games and uses them to investigate the generation-side long-term bidding issues in the electricity market. Literature [7] refers to the current situation; the indeterminacy on both sides of the signal source will cause large power imbalances and serious frequency fluctuations, threatening the stability and reliability of the operating load. Based on Nash bargaining theory and robust interval optimization, a point-to-point economic operation model is established for several microgrids. Literature [8] implemented four scheduling strategies for a microgrid system comprising solar-diesel hybrid generators, photovoltaic arrays, battery energy storage systems, and wind turbines: load tracking, generator orders, combined scheduling, and cyclic charging strategies. From the perspective of economy and reliability, a risk-based multi-grid for networked microgrids under the indeterminacy of load consumption and renewable energy generation is proposed [9].

Islanded microgrids often appear in remote areas, or when the microgrid is disconnected from the power grid due to faults, it is necessary to coordinate the output of each main body and maintain the balance of supply and demand. Its characteristic is to preferentially consume renewable energy locally and ensure equilibrium between electricity supply and demand. At present, new energy power generation still has problems, such as low utilization efficiency and weak coordination ability. The emergence of shared energy storage has solved these problems to some extent. Literature [10,11] put forward a microgrid risk-constrained scheduling strategy considering single-load demand response, introduced a bidirectional demand response mechanism and used power flexible load as a dispatchable resource to adjust the system net load curve. Literature [12] proposed a metaheuristic optimization algorithm based on fuzzy logic for the energy management of an islanded microgrid. In literature [13], a stochastic optimal scheduling strategy considering dependency relationship is proposed, and a stochastic optimal scheduling strategy is constructed by using improved scenario deep reinforcement learning. Literature [14] combined the comprehensive evaluation criteria of the economy and environmental protection to establish the optimal scheduling model of microgrid systems, including wind turbines, micro gas turbines, diesel generators, fuel cells, and batteries.

In 2018, China's Qinghai Province first proposed the concept of "shared energy storage". It is to point to being independent of the grid-centralized large independent energy storage power station. While meeting the needs of its power station, it can also provide charge and discharge services for other new energy power stations [15]. In literature [16], the Stackelberg game is employed to augment the financial returns of the shared energy storage power station. In this scenario, the user takes on the role of a follower, while the shared energy storage system acts as the leader, with the user's electricity expenditure being a key consideration in the process. The energy storage formulates the strategy according to the historical data and finally decides the transaction price. Literature [17] proposed the "shared energy storage-demand" model to track the output curve of renewable energy so that users have more choices, establish a model to optimize the net income of users and enhance the use efficiency of resources and energy storage equipment. Literature [5] specifically established a comprehensive evaluation system and gradually approached the ideal ranking by changing the weights of different indicators. An evaluation of investment costs, power distribution, and capacity sharing between centralized and distributed energy storage systems was conducted alongside the development of a multi-scenario model tailored to variations in the context of renewable energy production and user-side energy consumption [18]. The findings indicate that the centralized energy storage sharing mode is more economical.

As technology improves by leaps and bounds for the island operation scenario, considering the new scenario of shared energy storage participating in scheduling optimization, this paper establishes a typical multi-microgrid system composed of wind-solar-storage-load gas turbines. This article is mainly divided into the following contents.

(1) In terms of optimal scheduling, shared energy storage devices are introduced between islanded microgrids to integrate power resources and realize resource sharing. The improved grey wolf algorithm (GWO) is used to optimize the scheduling model. It demonstrates that incorporating shared energy storage capabilities could greatly decrease the extent of wasted wind and solar power generation and enhance the efficiency of resource utilization.

(2) From an economic point of view, a more reasonable use of shared energy storage is formulated by alternating quotations and compared with the disposition of off-grid storage. The results demonstrate a significant reduction in the capacity requirements for shared energy storage, the maximum charging and discharging capabilities of the storage system, and the overall investment costs associated with energy storage power stations.

(3) In this paper, the improved GWO is used, and the scheduling is optimized. Simulation results from typical scenarios reveal that in comparison to stand-alone off-grid energy storage configurations for individual microgrids, a shared energy storage power station enables the optimization of power flows, leading to a mutually advantageous scenario where both energy storage providers and users reap benefits.

The main structure of this paper is as follows: Section 2 builds the topology of the islanded microgrid system. Section 3 introduces the modeling process of each subject of the islanded microgrid. The optimization solution model is established in Section 4. In Section 5, it is verified that the proposed method is effective by taking the islanded multi-microgrid as the research object. Section 6 is the conclusion.

2. Modeling of Islanded Microgrid System

To enrich the application scenarios of microgrids, the microgrid model established mainly includes wind power generation equipment, PV equipment, micro-turbines, fuel cells, and load, as shown in Figure 1. Positioned as the central hub interconnecting multiple island microgrids, the shared energy storage power station effectively fulfills the specific charging and discharging requirements of each microgrid in an efficient manner. The shared energy storage can improve the user's satisfaction with electricity while reducing the rate of solar and wind abandonment and further reducing the cost of the user group [19]. Large-capacity shared energy storage could achieve coordination between regions and different periods, utilize energy storage capacity to its fullest potential, enhance the utilization of renewable energy, and alleviate grid supply pressure. The shared energy storage power station, in the case of a system emergency, can swiftly activate turbines for immediate operation, thereby sustaining the system's normal functionality [20].

The schematic diagram of the two-layer optimization decision model of an is-landed microgrid equipped with shared energy storage is shown in Figure 2. In order to describe the joint operation of new energy stations under the condition of multiple islands, a multi-island microgrid is proposed in this paper. The multi-microgrid grid-connected optimization model is mainly decomposed into the following two-stage problems for research.

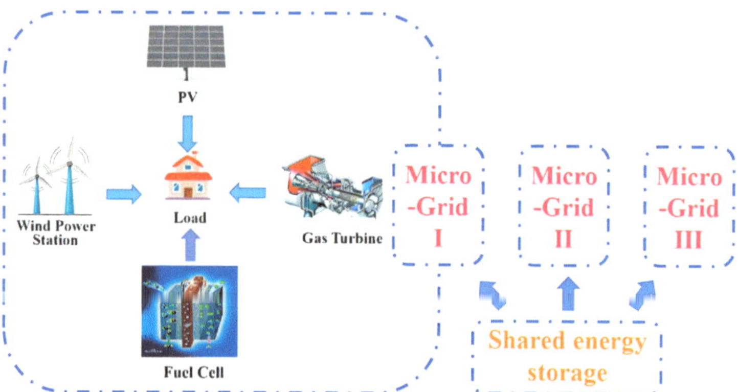

Figure 1. Multi-microgrid island operation diagram.

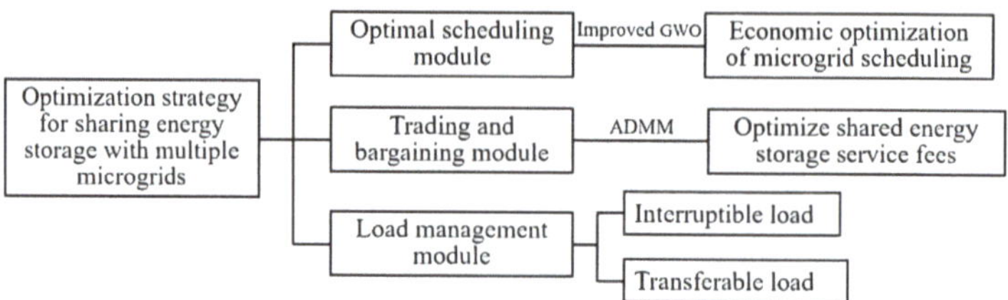

Figure 2. Schematic diagram of bi-level optimization decision-making model for islanded microgrid equipped with shared energy storage.

(1)   Optimized scheduling module

The objective of maximizing the microgrid operational economy guides the determination of each component's output literature [21]. When shared energy storage participates in the operation of multiple microgrids, the resource allocation and power flow are optimized, and the improved GWO is used to obtain the optimal scheduling results.

(2)   Transaction bargaining module

To achieve a more equitable distribution of benefits between microgrids and shared energy storage, the income distribution problem is transformed into a game problem of negotiated bargaining. In the context of this game-theoretic framework, the microgrid and shared energy storage represent the two opposing sides or players, and the alternating direction multiplier method (ADMM) is used to formulate the transaction price so as to optimize the shared energy storage service fee [22].

3. Mathematical Model of Each Subject

This section, segmented by subheadings, offers a succinct and exact portrayal of experimental outcomes, their interpretations, and the definitive conclusions that can be deduced from the experiments conducted.

3.1. Distributed Power Station Model

The radial basis neural network model [23] is used to predict the power generation of photovoltaic stations at time t is P_{pv}^t and the power generation of wind is P_{wt}^t. Satisfying the power balance constraint conditions are as follows:

$$P_{\mathrm{pv}}^t = P_{\mathrm{pv2b}}^t + P_{\mathrm{pv2L}}^t \tag{1}$$

$$P_{\mathrm{wt}}^t = P_{\mathrm{wt2b}}^t + P_{\mathrm{wt2L}}^t \tag{2}$$

$$0 \leq P^t_{\text{pv2b}}, 0 \leq P^t_{\text{pv2L}}, 0 \leq P^t_{\text{wt2b}}, 0 \leq P^t_{\text{wt2L}} \tag{3}$$

where P^t_{pv2b} and P^t_{wt2b} are the electricity sold by photovoltaic and wind power to shared energy storage, P^t_{pv2L} and P^t_{wt2L} are the electricity sold by photovoltaic and wind power to users.

The operational expenses of photovoltaic and wind power stations primarily consist of maintenance costs and the financial penalties associated with the reduction or limitation of wind and solar power generation output. The maintenance cost of the photovoltaic power station is C_{pvm}, and the maintenance cost of the wind power station is C_{wtm}:

$$C_{\text{pvm}} = \sum \tau_{\text{pv}} \sum_{t=1}^{T} \kappa_{\text{pv}} \cdot P^t_{\text{pv}} \tag{4}$$

$$C_{\text{wtm}} = \sum \tau_{wt} \sum_{t=1}^{T} \kappa_{\text{wt}} \cdot P^t_{\text{wt}} \tag{5}$$

The expression for the penalty cost incurred due to the abandonment of wind and solar energy generation is presented as follows [24]:

$$C_{\text{q}} = \begin{cases} \gamma_1 \cdot P_{\text{q}}, P_{\text{q}} \leq P_1 \\ \gamma_1 \cdot P_1 + \gamma_2 \cdot (P_{\text{q}} - P_1), P_1 < P_{\text{q}} \leq P_2 \\ \gamma_1 \cdot P_1 + \gamma_2 \cdot (P_2 - P_1) + \gamma_3 \cdot (P_{\text{q}} - P_2), P_2 < P_{\text{q}} \leq P_3 \end{cases} \tag{6}$$

where τ_{pv} and τ_{wt} are the scene probability. The scene probability τ_{pv} refers to the probability that photovoltaic power generation is P^t_{pv}. The scene probability τ_{wt} in the paper refers to the probability that wind power generation is P^t_{wt}. κ_{pv} and κ_{wt} are maintenance cost coefficients, γ_1, γ_2, γ_3 and are the penalty coefficients of each stage, P_1, P_2, and P_3 are the power standard values of the three penalty stages. P_{q} is the power of wind and light abandonment.

The overall operating expenditure of the new energy power station is recorded as follows:

$$C_{\text{new}} = C_{\text{pvm}} + C_{\text{wtm}} + C_{\text{q}} \tag{7}$$

3.2. Model of Gas Turbine

The variability and randomness of renewable energy will pose a serious hidden danger to the security and stability of microgrids. Especially in the case that energy storage technology has not been widely used, gas turbines can form complementary advantages with energy storage batteries by their good power climbing ability to achieve an 'incremental energy storage' effect. The power constraint and climbing constraints are in Equation (8).

$$\begin{cases} P_{\text{mt\_min}} \leq P^t_{\text{mt}} \leq P_{\text{mt\_max}} \\ R_{i,down}\Delta t \leq P^t_{\text{mt}} - P^{t-1}_{\text{mt}} \leq R_{i,up}\Delta t \end{cases} \tag{8}$$

where, P^t_{mt} is the output of the gas turbine at time t, $P_{\text{mt\_min}}$ and $P_{\text{mt\_max}}$ are the minimum and maximum gas turbine output, $R_{i,up}$ and $R_{i,down}$ are the maximum and minimum output of the gas turbine at time Δt.

When renewable energy power generation cannot provide users with normal operation, the gas turbine can be used as one of the reserve forces [25]. The operational cost of gas turbines encompasses three primary components: fuel expenses, maintenance costs, and environmental pollution control fees. The fuel cost is related to its output power, which is shown in Expression (9)–(11) [26].

$$C_{\text{MT}} = \sum_{t=1}^{T} \frac{C_{\text{NG}}}{\text{LHV}} \cdot \frac{P^t_{\text{MT}}}{\eta_{\text{MT}}} \tag{9}$$

$$\eta_{\mathrm{MT}} = 0.0753\left(\frac{P^t_{\mathrm{MT}}}{65}\right)^3 - 0.3095\left(\frac{P^t_{\mathrm{MT}}}{65}\right)^2 + 0.4174\left(\frac{P^t_{\mathrm{MT}}}{65}\right) + 0.1068 \tag{10}$$

where C_{MT} is the fuel expenditure of gas turbine, C_{NG} is the unit price of fuel, LHV is the energy content or heating value of fuel, P^t_{MT} is the output power of a gas turbine in t period, η_{MT} is the power generation efficiency of gas turbines. The expression of gas turbine maintenance expenditure and environmental treatment expenditure is as follows:

$$C_{\mathrm{mtc}} = \sum_{i=1}^{T} K_{\mathrm{mtc}} \cdot P^t_{\mathrm{mt}} + \sum_{i=1}^{T}\left(\sum_{j=1}^{W} \mu_j \sum_{i=1}^{N} K_{ij} P^t_i\right) \tag{11}$$

where K_{mtc} is the maintenance expenditure coefficient of microgas turbine, μ_j is the unit treatment expenditure of the jth pollutant, K_{ij} is the jth pollutant emission coefficient of the ith equipment, P^t_i is the output power of the i equipment in t period.

3.3. Fuel Cell Model

In theory, the fuel cell has high power generation efficiency. At present, the energy conversion efficiency can reach 40% to 60%, and it has often been used in microgrids because of its advantages of small environmental pollution, simple structure, and low power generation cost to remedy the volatility and uncertainty of new energy power generation and ensure that the power system can operate stably. The expression of fuel cell output power is:

$$P_{\mathrm{FC}} = V_{\mathrm{FC}} \cdot \eta_{\mathrm{FC}} \cdot \mathrm{LHV}_f \tag{12}$$

where V_{FC} is the fuel consumption, η_{FC} is the fuel cell power generation efficiency, LHV_f is the low calorific value of the fuel.

3.4. Shared Energy Storage Model

Within a system comprising multiple microgrids, the shared energy storage acts as a central hub, seamlessly connecting and dynamically managing the power flow among individual microgrids. It formulates charging and discharging schedules tailored to the operational status of each microgrid, ensuring optimal energy utilization. Compared with the independent configuration energy storage station, large-scale shared energy storage exhibits characteristics of standardized energy storage protocols and a relatively smaller aggregate capacity compared to individual installations [27]. The operation of the shared energy storage power station should meet the continuous dynamic constraints. The corresponding expression is as follows:

$$C_{\mathrm{FC}} = \frac{c_{\mathrm{gas}}}{\mathrm{LHV}_f}\sum \frac{P_{\mathrm{FC}}}{\eta_{\mathrm{FC}}} \tag{13}$$

$$\left\{\begin{array}{l} P^{\min}_{\mathrm{b\_c}} \leq P^t_{\mathrm{b\_c}} \leq P^{\max}_{\mathrm{b\_c}} U_{\mathrm{b\_c}}, \Delta E_{\mathrm{bat}}(t) > 0 \\ P^{\min}_{\mathrm{b\_d}} \leq P^t_{\mathrm{b\_d}} \leq P^{\max}_{\mathrm{b\_d}} U_{\mathrm{b\_d}}, \Delta E_{\mathrm{bat}}(t) < 0 \\ U_{\mathrm{b\_c}} + U_{\mathrm{b\_d}} \leq 1 \\ U_{\mathrm{b\_c}} \in \{0,1\} \\ U_{\mathrm{b\_d}} \in \{0,1\} \end{array}\right. \tag{14}$$

where c_{gas} is the fuel price, E^t_{bat} is the electrical power stored in the t period, $E_{\min}$ and $E_{\max}$ are the minimum and maximum energy storage, $\zeta_{\mathrm{b\_c}}$ and $\zeta_{\mathrm{b\_d}}$ are efficiency of charging and discharging processes of the t period. $P^{\min}_{\mathrm{b\_c}}$ and $P^{\max}_{\mathrm{b\_c}}$ are the minimum and maximum power capacity for charging. $\Delta E_{\mathrm{bat}}(t) < 0$ indicates that the shared energy storage is in the discharge state, $U_{\mathrm{b\_d}}$ and $U_{\mathrm{b\_c}}$ indicates the state of discharge and charge for the shared energy storage system. $P^{\min}_{\mathrm{b\_d}}$ and $P^{\max}_{\mathrm{b\_d}}$ are the minimum and maximum discharge power.

In order to ensure the continuity and sustainability of the charging and discharging, the storage capacity of the energy storage power station should return to the initial state value after a running cycle. The constraint formula is:

$$E_{\text{bat}}^{\text{T}} = E_{\text{bat}}^{0} \tag{15}$$

where E_{bat}^{0} and $E_{\text{bat}}^{\text{T}}$ denote the starting and ending points of the energy storage cycle, respectively.

Leveraging the disparity in generation and consumption timings across diverse microgrids, the shared energy storage system harmoniously integrates and dispatches power flows within a unified regional microgrid framework. This approach enhances the satisfaction of microgrid charging and discharging requirements while also reducing overall investment costs. The constraints pertaining to the charging and discharging power of the energy storage system are linearized using the large M approach. The revenue for the shared energy storage power station arises from the 28 service charges levied on microgrids for the utilization of its charging and discharging capabilities. The service expenses incurred by the microgrid are settled on a daily basis.

$$C_{\text{b}} = \sum_{t=1}^{\text{T}} \lambda_{\text{b}} \left(P_{\text{b\_c}}^{t} + P_{\text{b\_d}}^{t} \right) \tag{16}$$

where λ_{b} is the service expenditure paid when the shared energy storage provides unit charging power or discharging power.

The average annual cost C_{bat} of energy storage batteries is mainly composed of installation expenditure and operation and maintenance expenditure, which is proportional to the capacity of the energy storage configuration [28].

$$C_{\text{bat}} = C_{\text{sys}} \frac{i(1+i)^{m}}{(1+i)^{m} - 1} + C_{\text{FOM}} \tag{17}$$

where m represents the investment return period, which is mainly inversely proportional to the installation cost C_{sys}, and the expression is as follows:

$$m = \log_{i+1} \frac{1}{1 + \dfrac{iC_{\text{sys}}}{C_{\text{FOM}}}} \tag{18}$$

where i is the discount rate and C_{FOM} is operational and maintenance expenses.

3.5. Microgrid Operation Model

The user's electricity consumption in each period is P_{L}^{t} calculated using the prediction model. For the users in the island microgrid, the energy comes from PV, wind power, gas turbines, and shared energy storage. The trading power is recorded as P_{pv2L}^{t}, P_{wt2L}^{t}, P_{mt2L}^{t}, and P_{b2L}^{t}. The power balance relationship to be satisfied is as follows:

$$P_{\text{pv2L}}^{t} + P_{\text{wt2L}}^{t} + P_{\text{mt2L}}^{t} + P_{\text{b2L}}^{t} = P_{\text{L}}^{t} \tag{19}$$

(1) Transferable load

The transferable load's primary objective is to mitigate the adverse effects of load fluctuations, peak-valley differences, voltage fluctuations, and flicker, all of which are induced by the frequent operation of power electronic devices. For the power generation side, the increase of power generation output during the trough period avoids low-load fuel operation and can reduce operating costs.

$$\begin{cases} \min\| P_L^t + P_{TL}^t - P_{L\_ave} \|_2 \\ \sum_{t=1}^{T} P_{TL}^t = 0 \\ P_{L\_ave} = \dfrac{\sum_{t=1}^{T} P_L^t}{T} \\ P_{TL\_min}^t \le P_{TL}^t \le P_{TL\_max}^t \end{cases} \tag{20}$$

where P_{TL}^t is the user transferable load, P_{L_ave} is the mean load in one day, $\sum_{t=1}^{T} P_{TL}^t$ represents the total amount of transferable load in a day, $P_{TL\_min}^t$ and $P_{TL\_max}^t$ are the minimum and maximum values of transferable load. The adjusted load is:

$$P_{L\prime}^t = P_L^t + P_{TL}^t \tag{21}$$

(2)　Interruptible load considering demand response

To enhance the operational stability of island microgrids, the flexibility offered by interruptible loads is harnessed to mitigate electricity shortages. Its adjustment ability largely depends on the proportion of interruptible load and compensation cost. The mathematical conditions of load constraints are as follows:

$$\begin{cases} \sum_{t \in H_{IL}} P_L^t \le \sum_{j \in A} \Delta P_L^t U^t + \sum_{t \in H_{IL}} P_{NCL}^t \\ P_{NCL}^t \le P_L^t, t \in T \\ P_{NCL}^t = P_L^t, t \notin H_{IL} \end{cases} \tag{22}$$

where H_{IL} is the interruptible time period, ΔP_L^t is the active power of load interruption in the t period, U^t denotes the decision variable in the t period, P_{NCL}^t is uncontrollable load power. The expression of the adjusted load $P_{L\_new}^t$ is:

$$P_{L\_new}^t = P_{L\prime}^t - \Delta P_L^t \tag{23}$$

The penalty function for the interrupted load is expressed as follows:

$$C_{pen} = K_{pen} \sum_{t=1}^{T} \Delta P_L^t \tag{24}$$

where K_{pen} is the penalty coefficient of the interruption load.

Combined with the above model, the total cost expression of microgrid operation is:

$$\min C_{mg} = C_{new} + C_{MT} + C_{mtc} + C_{FC} + C_{bat} - C_b + C_{pen} \tag{25}$$

4. An Improved GWO Optimization Model Based on BAS

The GWO has the characteristics of strong global search ability but a low convergence rate. The Beetle Antennae Search Algorithm (BAS) has the characteristics of a fast search but easily falls under the local optimal solutions. Combining the advantages and disadvantages of the two, an improved GWO based on BAS is proposed. In the first stage, the purpose of the improved GWO calculation is to minimize the overall operation expenditure of the microgrid so as to plan the output of each main body. Combined with the constraints of each subject and the shared energy storage service expenditure. The second phase involves equilibrating the interests of both the microgrid and the shared energy storage power station. The non-cooperative game theory is introduced, and the ADMM is used to optimize the service expenditure.

4.1. Optimal Scheduling Model Based on Improved GWO

The GWO has the characteristics of a simple overall structure, easy programming, fast convergence speed, and high search efficiency. However, its problem of falling into local optimal solution limits its application in engineering. This paper solves this problem by adding the beetle antennae search algorithm. The process is shown in Figure 3.

(1) Input the microgrid load prediction value P_L^t, photovoltaic output prediction value P_{pv}^t, wind power output prediction value P_{wt}^t, abandoned wind and abandoned light penalty coefficient γ_1, γ_2, γ_3, and the maintenance cost coefficient of each piece of equipment, upper and lower limits of output values, and record the number of iterations as $k = 1$;

(2) With the goal of optimizing the power generation expenditure of the microgrid, the output of each subject is calculated as the Wolf individual fitness value;

(3) The calculated fitness value of each gray Wolf is ranked. The first three are recorded as α, β, and δ respectively. The corresponding location information is X_α, X_β, and X_δ respectively;

(4) If the $k > K_{max}$ iteration ends, this value is the optimal result and outputs this value; otherwise, continue to perform step 5;

(5) Updated the location of the Gray Wolf to re-elect new wolves α, β, and δ;

(6) Update iterations, $k = k + 1$;

(7) Verify whether the termination condition is met, and if so, verify whether it falls into the local optimal solution; if it is satisfied, jump out of the loop and output the final result; if it is not satisfied, use the BAS to gain a new search direction, and Revert back to step (2) and proceed with the solution.

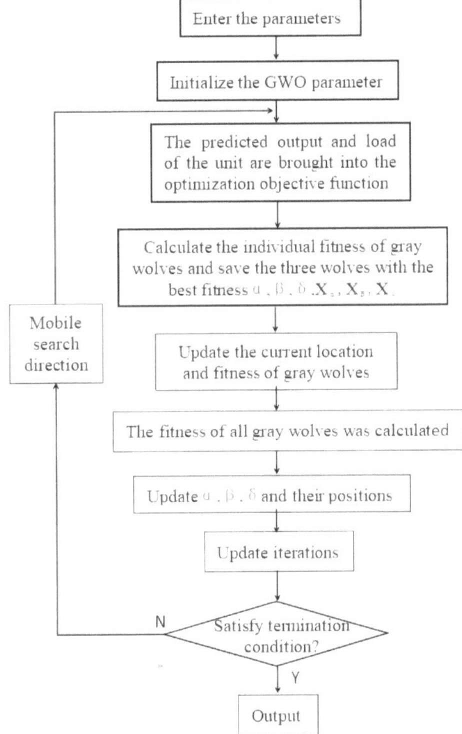

Figure 3. Improved grey wolf algorithm flow chart.

4.2. Service Fee Optimization Formulation Model Based on ADMM

During the formulation of service fees, the user charges vary depending on the micro-grid's operational state. Given the higher initial investment cost of shared energy storage, these costs are amortized on a daily basis to ensure fairness [29]. To ensure equitable benefits for all parties, the analysis initially employs Nash equilibrium to explore the influence of transaction prices on transactional modes. Subsequently, a non-cooperative game-theoretic framework is employed to determine the optimal pricing approach for shared energy storage [30].

With the intention of minimizing the costs of the shared energy storage and the operating expenditure of the microgrid [31], the optimal scheduling results obtained in the first stage are input into the service expenditure optimization model for the solution. The process is shown in Figure 4.

(1) Input the original data variables of various parameters, such as the prediction data of the PV station, wind power station, load forecast value, Lagrange penalty factor ρ, the maximum number of iterations, and convergence accuracy.

(2) For the main body of the microgrid, the mathematical model of the PV station, wind power station, gas turbine, and load is used to calculate the expected service fee r_{g2b} of the microgrid to share energy storage under the maximization of microgrid revenue.

(3) For the core entity of the microgrid, the mathematical model of the PV station, wind power station, gas turbine, and load is calculated, and the expected service fee of the microgrid to share energy storage is r_{g2b} under the maximization of microgrid revenue.

(4) Update the Lagrange multiplier and iterations.

(5) Determine whether the deviation value of r_{g2b} and $\tilde{r}_{g2b}$ is within the allowable convergence accuracy range, and if it is greater than the convergence accuracy requirement, return step 2 to continue the iteration; if it is less than the convergence accuracy requirement, the iteration ends, and the optimized electricity price is output as the result.

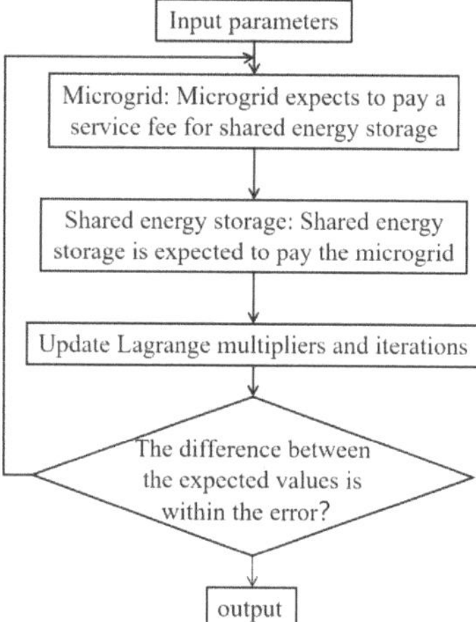

Figure 4. ADMM bargaining model flow chart.

5. Case Study

5.1. Island Multi-Microgrid Is Equipped with Independent Energy Storage

To assess the operational characteristics of diverse microgrid types and validate the model's practicality, three illustrative scenarios have been selected for analysis. Through a comparative analysis of the economic and energy flow performance between independent and shared energy storage systems across three microgrid scenarios, the superiority of multi-microgrid operation in island mode is clearly demonstrated. The three microgrids are as follows.

Microgrid I: PV + load;
Microgrid II: Wind power + load;
Microgrid III: PV + Wind power + load.

(1)   Microgrid I is configured with independent energy storage

The output of each main subject of microgrid I is shown in Figure 5. Only the participation of PV generation in microgrid operation is considered, and the PV generation is much larger than the load power consumption during the peak period.

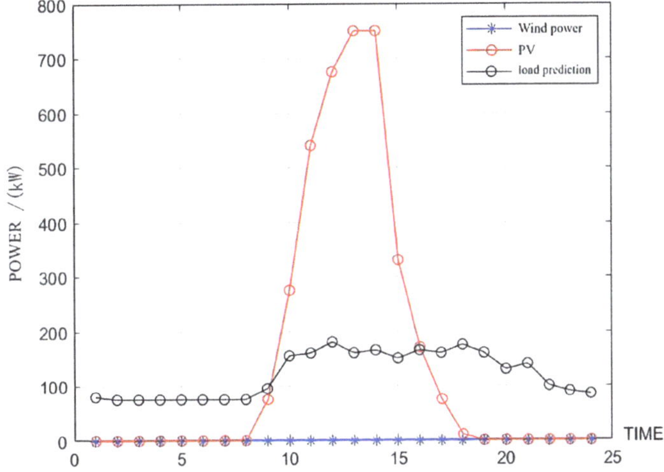

Figure 5. Output curve of each main body of microgrid I.

Figure 6 shows the curve after load optimization and adjustment. As can be seen from the figure, compared with the original load, the load curve, after taking interruptible load and transferable load into account, is more stable, which is conducive to system stability while promoting the consumption of new energy.

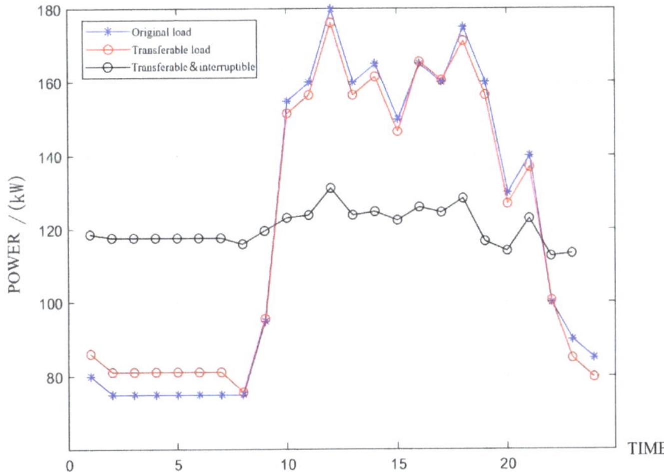

Figure 6. Microgrid I load curve after optimization and adjustment.

Figure 7 shows the energy transaction diagram of microgrid I. It can be seen that the microgrid preferentially consumes new energy power generation, and the power storage is charged when the PV generation is excess at 9:00–12:00. Due to the finite energy storage capacity, the storage system tends to reach its full capacity by 12:00. During 12:00–13:00, the energy storage is no longer charged or discharged, and the electricity for users is supplied by new energy generation and gas turbine. Resulting in the discarding and wastage of excess PV generation during the period from 13:00 to 15:00.

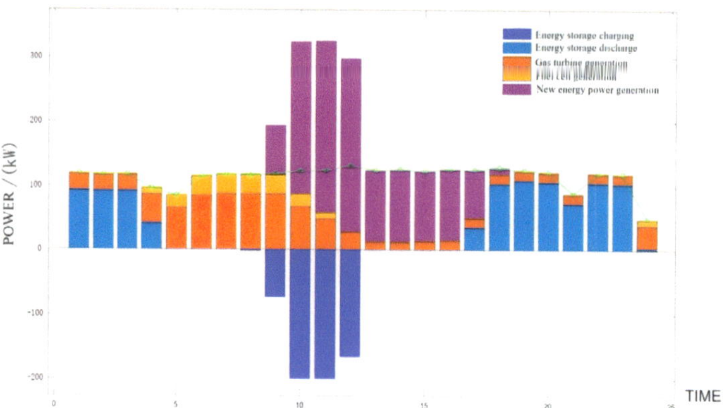

Figure 7. Microgrid I electric energy transaction diagram.

(2) Microgrid II is configured with independent energy storage

The output of each main body of microgrid II is shown in Figure 8, which only considers the participation of wind power generation in microgrid operation. Figure 9 shows the curve after load optimization and adjustment.

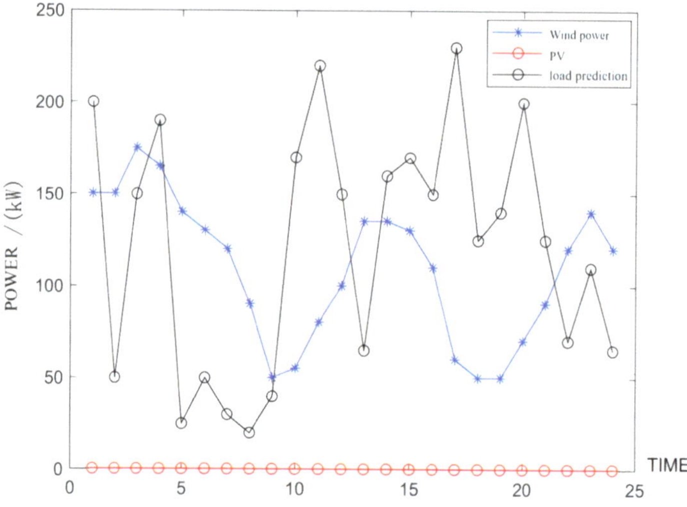

Figure 8. The output curve of each main body of microgrid II.

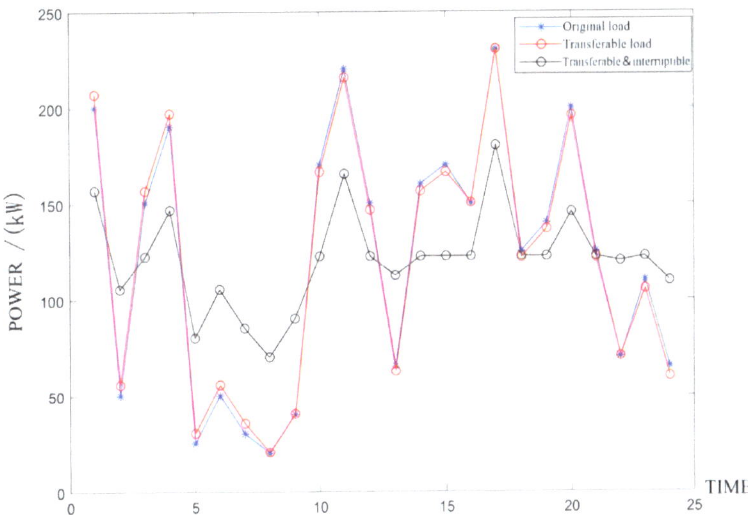

Figure 9. Microgrid II load optimization adjustment curve diagram.

Figure 10 shows the microgrid II energy transaction diagram. It shows that the new energy power generation is preferentially consumed.

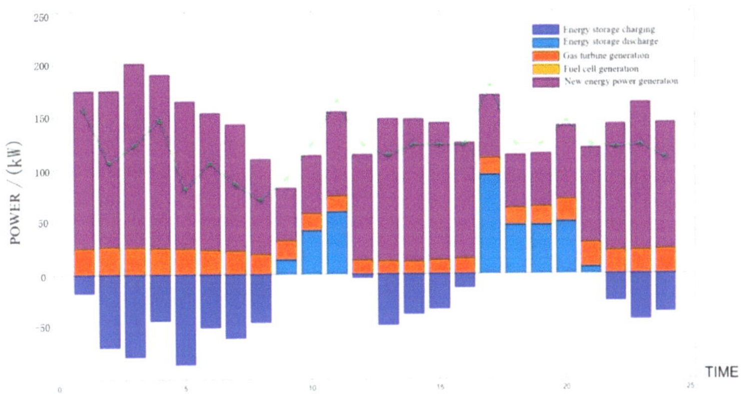

Figure 10. Microgrid II energy trading diagram.

(3) Microgrid III is configured with independent energy storage

The output of each main body of microgrid III is shown in Figure 11. Figure 12 shows the curve after load optimization and adjustment. The microgrid III energy transaction result is shown in Figure 13. The green curve represents the optimized load curve.

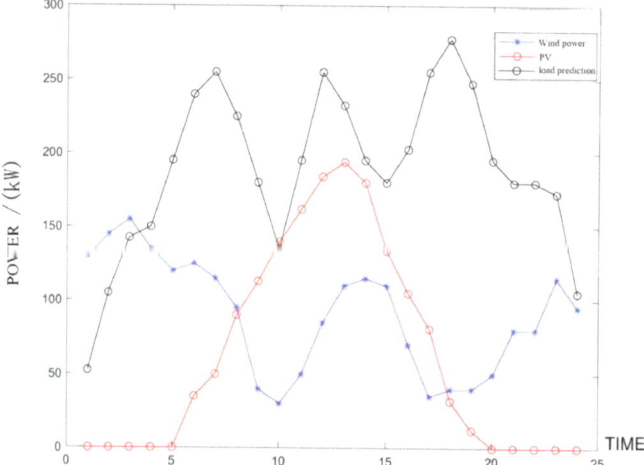

Figure 11. The output curve of each main body of the Microgrid III.

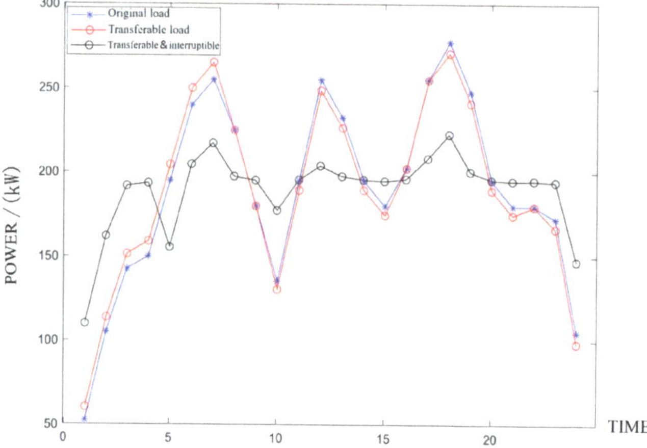

Figure 12. Microgrid III load optimization adjustment curve diagram.

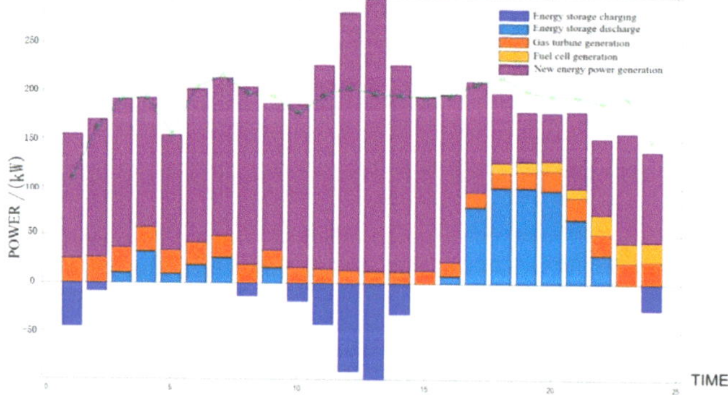

Figure 13. Microgrid III electric energy transaction diagram.

According to the specific situation of each microgrid, the three microgrids are equipped with independent energy storage of corresponding capacity, respectively. On the basis of the original operation expenditure, the average daily investment expenditure, operation expenditure, and maintenance expenditure of independent energy storage is increased [32] as follows:

$$C_{\text{inv}} = \frac{\eta_{\text{P}} \cdot P_{\max} + \eta_{\text{s}} E_{\max}}{T_{\text{s}}} + M \tag{26}$$

where η_{P} and η_{s} represent the power expenditure and capacity expenditure, respectively. $P_{\max}$ is the maximum charge and discharge power, $E_{\max}$ is the maximum capacity of energy storage, T_{s} is the expected service life of energy storage, days, m is the average daily maintenance expenditure. The improved GWO is used to calculate the optimization results when the three microgrids are configured with independent energy storage, which is shown in Table 1.

Table 1. Optimization Results for Each Microgrid.

	Energy Storage Capacity/(kWh)	Energy Storage Maximum Charge and Discharge Power/kW	Cost of Operation/Yuan
Microgrid I	2796.5625	590	1625.48
Microgrid II	739.2434	130	561.79
Microgrid III	836.8421	195.5	754.38
Total	4372.648	915.5	2941.65

5.2. Islanded Multi-Mcrogrid Is Equipped with Shared Energy Storage

A multi-microgrid system comprising three distinct microgrids in the above summary is equipped with shared energy storage as an example for analysis, and the charge and discharge service fee for the shared energy storage system has been established as 1.4 yuan/kWh. Figures 14–16 illustrate the optimal power allocation outcomes for the three microgrids. Figure 17 shows the energy state of the shared energy storage station, where the blue curve represents the energy of the energy storage station.

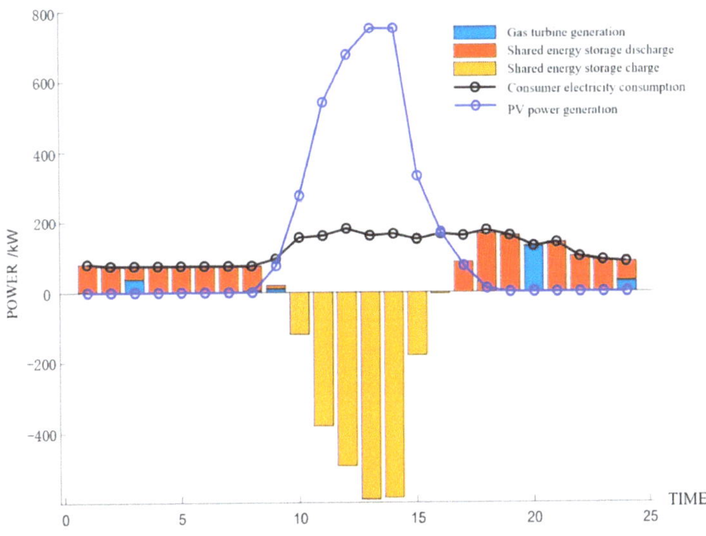

Figure 14. Microgrid I energy transaction optimization results.

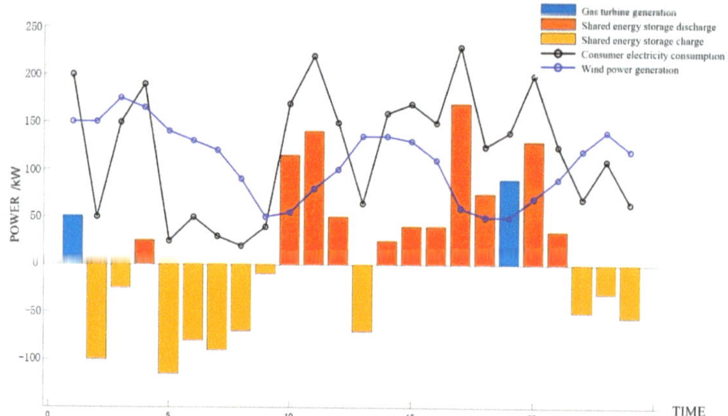

Figure 15. Microgrid II energy transaction optimization results.

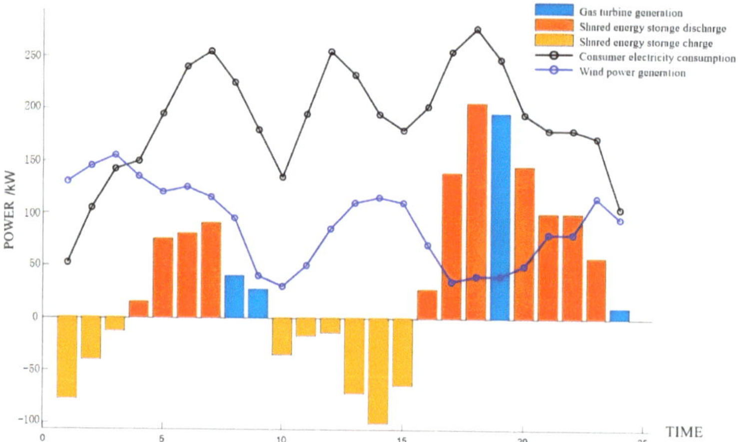

Figure 16. Microgrid III energy transaction optimization results.

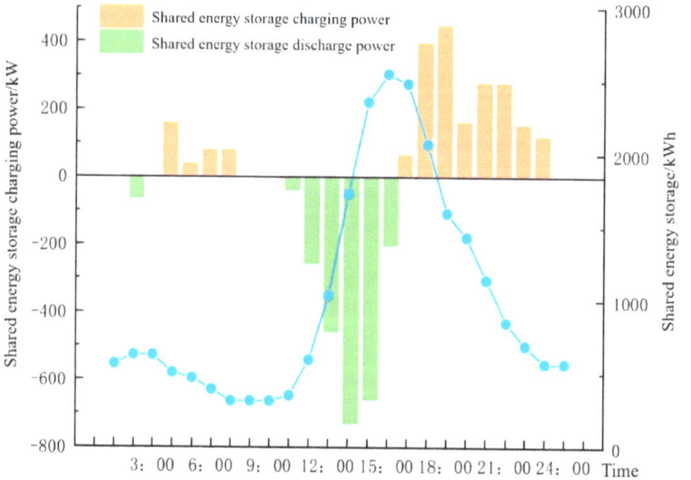

Figure 17. Charging and discharging power and power state of shared energy storage power station.

After the shared energy storage involved in the optimal scheduling among the three microgrids, the optimal capacity planning value of the shared energy storage power station is 2828.4079 kWh by using the GWO, which is about 35.316% lower than that of the independent energy storage capacity of 4372.648 kWh. The maximum charging and discharging capacity of the shared energy storage station is 732 kw, which is about 20% lower than the maximum charging and discharging power of 915.5 kW with independent energy storage. The overall payment expenditure of the three microgrids is 2487.7 yuan, which is about 15.43% lower than that of 2941.65 yuan equipped with independent energy storage. Shared energy storage greatly improves the utilization rate of new energy power generation; on the other hand, it also greatly reduces the investment expenditure of energy storage power stations and realizes multi-regional energy complementary circulation.

Based on the optimized trading electricity results obtained in the previous summary, the non-cooperative game model is used to further optimize the trading price by alternating quotations. In the initial calculation of the example, the shared energy storage is set as the price increaser, and the user is the decision maker. Taking into account the rental expenses, the ADMM is employed to address the problem, and the iterative process eventually converges to a solution. Finally, the optimal transaction price is 1.53 yuan/(kWh), which includes the average daily expenditure and maintenance expenditure of the shared energy storage investment.

In Figure 14, the two curves represent user demand and photovoltaic power generation, and the bar graph represents gas turbine power generation, energy storage discharge, and energy storage charge. The superposition of each main output is user demand. As can be seen from the figure, when the photovoltaic power generation from 1:00 to 8:00, 17:00 to 19:00, and 21:00 to 24:00 does not meet the needs of users, energy storage and discharge are shared to supply users. When there is a surplus of photovoltaic power generation from 10:00 to 15:00, the remaining power is stored by a shared energy storage power station in order to reduce the light abandonment rate and improve the level of new energy consumption. At 13:00, the photovoltaic output reaches a maximum of 753.2 kW, and the shared energy storage also reaches a maximum charging power of 590 kW. When the rest of the shared energy storage is insufficient, the gas turbine power generation supplies the load. It can be seen that more photovoltaic surplus power is stored, and the gas turbine output is significantly reduced, which greatly improves the level of new energy consumption.

As can be seen from Figure 15, when there is a surplus of wind power generation from 2:00 to 3:00, 5:00 to 9:00, and 22:00 to 24:00, shared energy storage stores the excess power to reduce the curtailment rate. When the wind power generation is insufficient from 10:00 to 12:00, 14:00 to 18:00, and 20:00 to 21:00, the shared energy storage discharge is provided to the user. At 5:00, the remaining power of wind power generation is the largest, and the wind power generation is 142.1 kW, and the shared energy storage also reaches the maximum charging power of 115.9 kW. When the rest of the shared energy storage is insufficient, the gas turbine power generation supplies the load. It can be seen that the use frequency of gas turbines decreases, and shared energy storage increases the utilization rate of new energy power generation.

As can be seen from Figure 16, when there is a surplus of new energy generation at 1:00–3:00 and 10:00–15:00, the shared energy storage starts to charge and stores the excess electricity. When the new energy generation is insufficient from 4:00 to 7:00, 16:00 to 18:00, and 20:00 to 23:00, the shared energy storage discharge supplies energy to the user. The maximum shared energy storage charging power at 14:00 is 98.76 kW. At 8:00–9:00 and 24:00, shared energy storage is not enough to support the user and the gas turbine power generation. It can be seen that shared energy storage improves the utilization rate of new energy generation, and the use frequency of gas turbines is also greatly reduced.

The improvement of the algorithm in this paper is mainly reflected in the first stage, which is to optimize the output of each body. The improved GWO algorithm combines the fast search characteristics of BAS and the strong global search ability of GWO. Compared with the traditional particle swarm optimization algorithm and GWO algorithm, the improved GWO algorithm has faster convergence speed and shorter calculation times. The comparison of algorithms is shown in Table 2.

Table 2. Comparison of calculation results of different algorithms.

Algorithm	Number of Iterations	Calculation Time/s
PSO	73	8.59
GWO	58	7.26
improved GWO	47	5.64

6. Conclusions and Prospects

6.1. Conclusions

Within the power system that incorporates renewable energy generation, energy storage is the best way to solve the time-space mismatch between new energy generation and load power consumption. In this paper, the optimal scheduling module and the transaction bargaining are studied in two stages, considering that the shared energy storage participates in the optimal scheduling. The improved GWO and ADMM are used to figure out the multi-microgrid example, which consists of three typical microgrids successively. Through case verification and result analysis, the subsequent conclusions can be deduced from the analysis:

(1) Formulate the optimal scheduling model under the multi-microgrid island operation mode. Shared energy storage gives full play to the complementarity of power consumption among microgrids to reduce wind and light abandonment rates and adopts an improved gray Wolf algorithm to optimize scheduling. According to the simulation results of the example, compared with the independent energy storage configuration of each microgrid, the shared energy storage power station can optimize the power flow and achieve a win-win situation between energy storage and users.

(2) Optimize the transaction price model under the multi-microgrid island operation mode. Using ADMM to negotiate the price, the results show that compared with the configuration of independent energy storage, the shared energy storage capacity and investment costs are greatly reduced, and the utilization rate of energy storage capacity is greatly improved.

6.2. Prospects

(1) The microgrid mathematical model established in this paper is relatively simple. In the future, it is necessary to consider the actual scenario state, consider the addition of more types of agents, and establish more abundant application scenarios.

(2) This paper only analyzes the operation status of shared energy storage participation in isolated island operation mode. In fact, shared energy storage also plays a great role in peak regulation and frequency regulation in grid-connected operation modes. In the future, it is necessary to continue to explore the study of shared energy storage participation in peak regulation and frequency regulation operation mode in grid-connected mode.

Author Contributions: Conceptualization, Y.L. and H.Z.; Data curation, Z.S.; Formal analysis, Z.S. and Y.W.; Funding acquisition, L.Y.; Investigation, J.D.; Methodology, Y.J. and Y.H.; Resources, Y.H.; Software, Z.S. and Y.L.; Supervision, Y.H., Y.W. and L.Y.; Validation, Y.W.; Visualization, Y.H.; Writing—original draft, Z.S. and Y.L.; Writing—review and editing, H.Z. and Y.W.; Innovation point, W.Q. and Z.Z. All authors have read and agreed to the published version of the manuscript.

Funding: This research received no external funding.

Data Availability Statement: The original contributions presented in the study are included in the article; further inquiries can be directed to the corresponding author.

Conflicts of Interest: Author Lu Yan was employed by the company Yingda Chang'an Insurance Brokers Co., Ltd. The remaining authors declare that the research was conducted in the absence of any commercial or financial relationships that could be construed as a potential conflict of interest.

References

1. Wei, X.; Liu, D.; Gao, F.; Liu, L.; Wu, Y.; Ye, S. Generation expansion planning of new power system considering collaborative optimal operation of source-grid-load-storage under carbon peaking and carbon neutrality. *Power Syst. Technol.* **2023**, *47*, 3648–3661.
2. Ma, X.; Pan, Y.; Zhang, M.; Ma, J.; Yang, W. Impact of carbon emission trading and renewable energy development policy on the sustainability of electricity market: A sackelberg game analysis. *Energy Econ.* **2024**, *129*, 122–135. [CrossRef]
3. Erol, Ö.; Filik, Ü.B. A Stackelberg game-based dynamic pricing and robust optimization strategy for microgrid operations. *Int. J. Electr. Power Energy Syst.* **2024**, *155*, 189–204. [CrossRef]
4. Cheng, L.; Chen, Y.; Liu, G. 2PnS-EG: A general two-population n-strategy evolutionary game for strategic long-term bidding in a deregulated market under different market clearing mechanisms. *Int. J. Electr. Power Energy Syst.* **2022**, *142*, 108182. [CrossRef]
5. Chen, C.; Liu, C.; Ma, L.; Chen, T.; Wei, Y.; Qiu, W.; Lin, Z.; Li, Z. Cooperative-game-based joint planning and cost allocation for multiple park-level integrated energy systems with shared energy storage. *J. Energy Storage* **2023**, *73*, 108861. [CrossRef]
6. Cheng, L.; Liu, G.; Huang, H.; Wang, X.; Chen, Y.; Zhang, J.; Meng, A.; Yang, R.; Yu, T. Equilibrium analysis of general N-population multi-strategy games for generation-side long-term bidding: An evolutionary game perspective. *J. Clean. Prod.* **2020**, *276*, 124123. [CrossRef]
7. Xu, J.; Yi, Y. Multi-microgrid low-carbon economy operation strategy considering both source and load uncertainty: A Nash bargaining approach. *Energy* **2023**, *263*, 125–137. [CrossRef]
8. Shezan, S.; Ishraque, F.; Muyeen, S.; Arifuzzaman, S.; Paul, L.C.; Das, S.K.; Sarker, S.K. Effective dispatch strategies assortment according to the effect of the operation for an islanded hybrid microgrid. *Energy Convers. Manag.* **2022**, *14*, 184–192. [CrossRef]
9. Zhu, Y.; Li, G.; Guo, Y.; Li, D.; Bohlooli, N. Modeling optimal energy exchange operation of microgrids considering renewable energy resources, risk-based strategies, and reliability aspect using multi-objective adolescent identity search algorithm. *Sustain. Cities Soc.* **2023**, *91*, 104–116. [CrossRef]
10. Zhang, Z.; Li, H. Hierarchical energy management strategy for islanded microgrid cluster considering flexibility. *Power Syst. Prot. Control.* **2020**, *48*, 97–105.
11. Nie, Y.; Qiu, Y.; Yang, A.; Zhao, Y. Risk-limiting dispatching strategy considering demand response in multi-energy microgrids. *Appl. Energy* **2024**, *353*, 122–131. [CrossRef]
12. Rodriguez, M.; Arcos–Aviles, D.; Martinez, W. Fuzzy logic-based energy management for isolated microgrid using meta-heuristic optimization algorithms. *Appl. Energy* **2023**, *335*, 120–132. [CrossRef]
13. Dong, W.; Sun, H.; Mei, C.; Li, Z.; Zhang, J.; Yang, H.; Ding, Y. Stochastic optimal scheduling strategy for a campus-isolated microgrid energy management system considering dependencies. *Energy Convers. Manag.* **2023**, *292*, 117–124. [CrossRef]
14. Zhang, H.; Li, G.; Wang, S. Optimization dispatching of isolated island microgrid based on improved particle swarm optimization algorithm. *Energy Rep.* **2022**, *8*, 420–428. [CrossRef]
15. Cheng, L.; Qi, N.; Zhang, F.; Kong, H.; Huang, X. Energy internet: Concept and practice exploration. In Proceedings of the IEEE Conference on Energy Internet and Energy System Integration (EI2), Beijing, China, 26–28 November 2017; pp. 1–5.
16. Chen, W.; Xiang, Y.; Liu, J. Optimal operation of virtual power plants with shared energy storage. *IET Smart Grid* **2023**, *6*, 147–157. [CrossRef]
17. Asri, R.; Aki, H.; Kodaira, D. Optimal operation of shared energy storage on islanded microgrid for remote communities. *Sustain. Energy Grids Netw.* **2023**, *35*, 101–114. [CrossRef]
18. Chen, C.; Zhu, Y.; Zhang, T.; Li, Q.; Li, Z.; Liang, H.; Liu, C.; Ma, Y.; Lin, Z.; Yang, L. Two-stage multiple cooperative games-based joint planning for shared energy storage provider and local integrated energy systems. *Energy* **2023**, *284*, 129–139. [CrossRef]
19. Wang, Y.; Yang, J.; Jiang, W.; Sui, Z.; Chen, T. Research on optimal scheduling decision of multi-microgrids based on cloud energy storage. *IET Renew. Power Gener.* **2022**, *16*, 581–593. [CrossRef]
20. Dong, H.; Fu, Y.; Jia, Q.; Wen, X. Optimal dispatch of integrated energy microgrid considering hybrid structured electric-thermal energy storage. *Renew. Energy* **2022**, *199*, 628–639. [CrossRef]
21. Duan, J.; Xie, J.; Feng, L. Gain allocation strategy of wind-solar-water-hydrogen multi-agent energy system based on cooperative game theory. *Power Syst. Technol.* **2022**, *46*, 1703–1712.
22. Xu, Y.; Ye, S.; Qin, Z.; Lin, X.; Huangfu, J.; Zhou, W. A coordinated optimal scheduling model with Nash bargaining for shared energy storage and multi-microgrids based on Two-layer ADMM. *Sustain. Energy Technol. Assess.* **2023**, *56*, 124–138. [CrossRef]
23. Xu, D. PV output prediction method based on similar day selection combined with PSO-RBF. *Electron. Meas. Technol.* **2020**, *43*, 78–83.
24. Yang, Y.; Zheng, P.; Mao, R.; Qin, H.; Wang, Y. An economic scheduling algorithm for isolated island microgrid based on uncertainty margin segmentation quantization and wind and light abandonment segmentation punishment. *Mod. Electr.* **2023**, *40*, 73–81.
25. Li, X.; Qin, W.; Jing, X. The joint optimization strategy of virtual power plants participating in the main and auxiliary markets considering uncertain risks and multi-agent coordination. *Power Syst. Technol.* **2022**, *132*, 1–16.

26. Tao, Z.; Liu, M.; He, W. Virtual power plants taking into account the uncertainty and correlation of the landscape participate in the optimal scheduling of the electric-carbon-green certificate market. *Distrib. Energy* **2024**, *9*, 55–64.
27. Liu, D.N.; Zhang, X.T.; Li, D.X.; Dong, H.; Li, C.G.; Liu, Z.M. Day-ahead optimal economic dispatch of industrial users based on shared energy storage power station. In Proceedings of the 7th International Symposium on Advances in Electrical Electronics, and Computer Engineering, Xishuangbanna, China, 18–20 March 2022; Volume 12294, pp. 174–177.
28. Wu, N.; Xiao, J.; Feng, Y.; Bao, H.; Lin, R.; Chen, W. Economic feasibility analysis of user-side battery energy storage based on three electricity price policies. In Proceedings of the 2020 IEEE Sustainable Power and Energy Conference (iSPEC), Chengdu, China, 23–25 November 2020; pp. 2034–2039.
29. Wang, Q.-Y.; Lv, X.-L.; Zeman, A. Optimization of a multi-energy microgrid in the presence of energy storage and conversion devices by using an improved gray wolf algorithm. *Appl. Therm. Eng.* **2023**, *234*, 121–141. [CrossRef]
30. Huang, E.; Zhang, T.; Zheng, F.; Lin, J.; An, X.; Shi, H. Day-ahead and real-time energy management for active distribution network based on coordinated optimization of different stakeholders. *Power Syst. Technol.* **2021**, *45*, 2299–2308.
31. Tang, W.; Wu, B.; Zhang, L.; Zhang, X.; Li, J.; Wang, L. Wang. Multi-objective optimal dispatch for integrated energy system based on device value tag. *CSEE J. Power Energy Syst.* **2020**, *7*, 632–643.
32. Xie, X.; Zhang, Y.; Meng, K.; Dong, Z.; Liu, J. Emergency control strategy for power systems with renewables considering a utility-scale energy storage transient. *CSEE J. Power Energy Syst.* **2021**, *7*, 986–995.

MDPI AG
Grosspeteranlage 5
4052 Basel
Switzerland
Tel.: +41 61 683 77 34

Energies Editorial Office
E-mail: energies@mdpi.com
www.mdpi.com/journal/energies